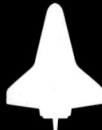

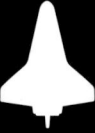

THE CREW OF CHALLENGER STS-51L

THE CREW OF COLUMBIA STS-107

FRANCIS R. SCOBEE

RICK D. HUSBAND

MICHAEL J. SMITH

WILLIAM C. McCOOL

ELLISON S. ONIZUKA

DAVID M. BROWN

JUDITH A. RESNIK

KALPANA CHAWLA

RONALD E. McNAIR

LAUREL B. CLARK

S. CHRISTA McAULIFFE

MICHAEL P. ANDERSON

GREGORY B. JARVIS

ILAN RAMON

For 30 years, so many courageous space shuttle astronauts have embarked upon missions into the hostile environment of space and returned safely. Many have taken the epic journey more than once, but some left Earth never to return home.

This book is dedicated to the brave men and women
who lost their lives while reaching for the stars.

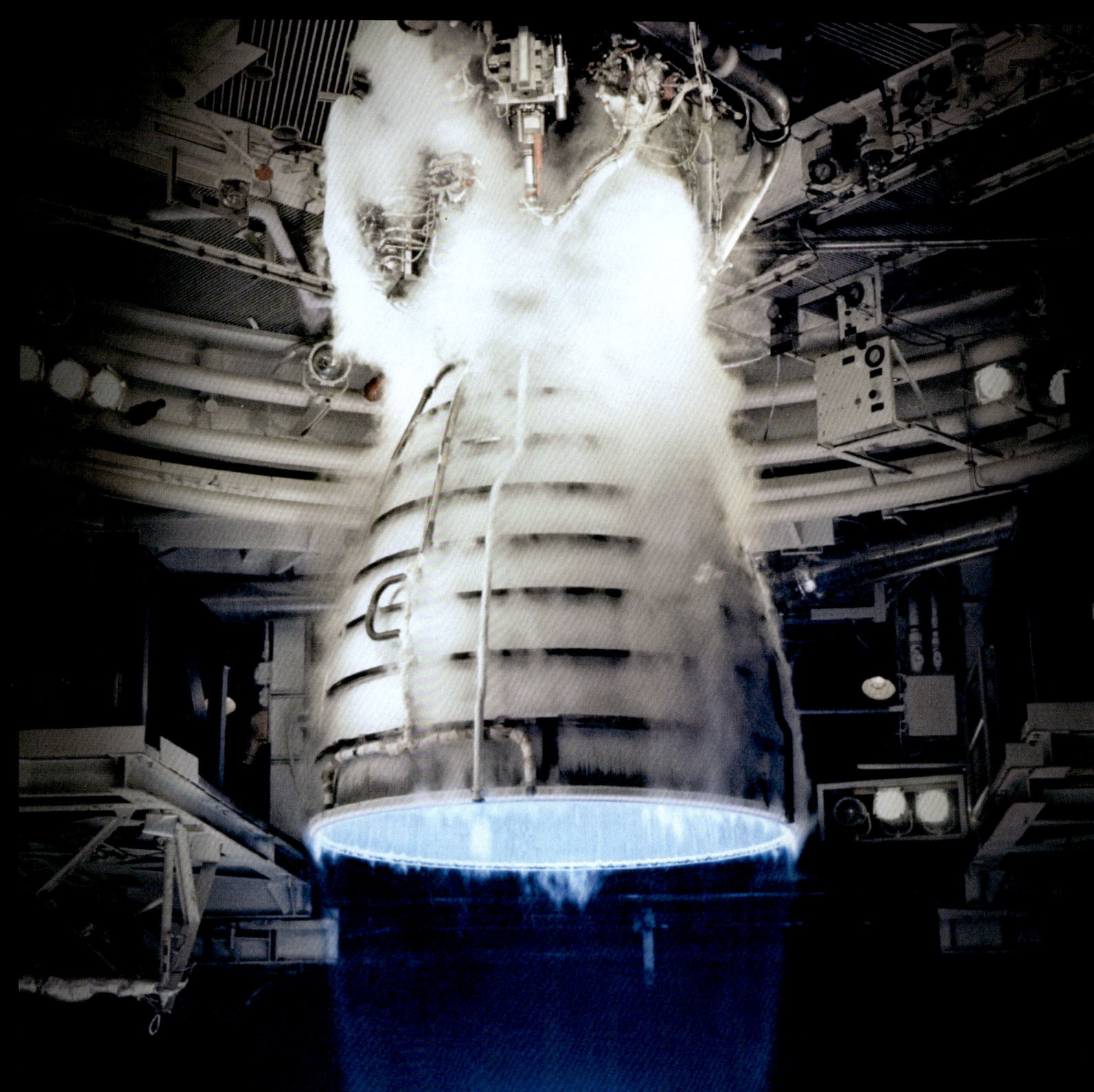

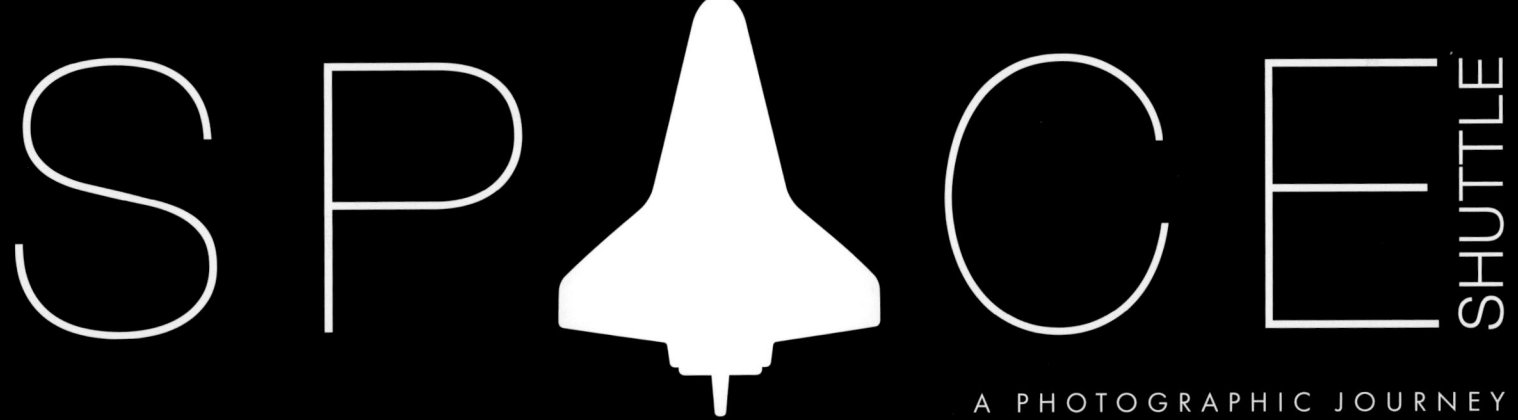

SP CE SHUTTLE

A PHOTOGRAPHIC JOURNEY

FOREWORD BY
CHRISTOPHER J. FERGUSON

LUKE WESLEY PRICE

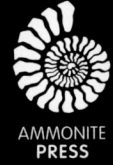

AMMONITE
PRESS

COLUMBIA
(OV-102)

1981–2003 | 300 days in space | 125 million miles | 28 missions

CHALLENGER
(OV-099)

1983–1986 | 62 days in space | 26 million miles | 10 missions

DISCOVERY
(OV-103)

1984–2011 | 365 days in space | 148 million miles | 39 missions

ATLANTIS
(OV-104)

1985–2011 | 307 days in space | 126 million miles | 33 missions

ENDEAVOUR
(OV-105)

1992–2011 | 299 days in space | 123 million miles | 25 missions

CONTENTS

FOREWORD

It is August 2018, and the sun is rising over the Space Coast of Florida. Having witnessed the early morning launch of a Delta IV (heavy) from Launch Complex 37, I have just made the short trip past Complex 39A – where we launched on STS-135 – and on to the massive Vehicle Assembly Building. While the landscape at Cape Canaveral has changed significantly since the space shuttle last left Earth, the original mission to safely launch payloads for private and government customers to Low Earth Orbit and beyond remains the same. Pad 39A, replete with modifications made to accommodate SpaceX's new rockets, looks dramatically different to when we used it to climb into Atlantis in July 2011. Other modifications throughout the expansive launch complex indicate that, from the ashes, a new phoenix of exploration is rising.

But we will never forget the space shuttle. Emerging from the burst of national adrenaline that took us to the Moon on Apollo in a period of less than 10 years, the shuttle aimed to normalize space for living and working. By every measure, it was successful. Routinely introducing up to seven astronauts to space at a time, the shuttle became an icon in itself. Personally, I will never forget my first up-close view of an orbiter as a newly minted astronaut. The sheer scale alone was enough to astound. At 122 feet long, it was in the same class as a Boeing 737... only perched on its tail and strapped to 4 million pounds of rocket fuel.

To elicit the trust of the nation to allow me and my crew to escape Earth for a few weeks on Endeavour and then Atlantis was the ultimate honour. There have been several amazing 'moments in space' on all of my three missions to the International Space Station – too many to recount here – but some were especially poignant. On my very first launch, shortly after the solid rockets separated and the ride became significantly smoother, I took a first glimpse out of the pilot's right-hand window at the US East Coast. The immediate sensation was that of incredible altitude and speed, but it also felt distinctly like flying... in almost unbelievable style.

The first view of Earth from orbit is incomparable. No amount of mental rehearsal could prepare me adequately for the sensory overload. It was almost an out-of-body disbelief of my fortune to be experiencing what so few have. But there was always much to do, and the view would have to wait. In addition to the great opportunity to command a space shuttle mission, a fringe benefit was that the commander always had a front seat (except, of course, for space walks, which I still consider on my bucket list). Compared to the next generation of human-rated spacecraft coming online, the shuttle was very crew-intensive and required almost constant 'watching' for the next scheduled activity. But this also made it wonderful for an astronaut because we knew that the success of the mission had a lot to do with how we performed.

On approach to the International Space Station the true magnitude of this orbital outpost becomes apparent. Poised a few hundred feet below ISS and in formation flights at 8 km/s, the enormity of what mankind has managed to craft comes into focus. In the few moments we have to relax, it's clear that the ISS is so massive that it cannot be seen in its entirety through any of the shuttle's windows. As the orbiter's vernier reaction jets slowly guide us into position, the reality sets in that our background, training and persistence have enabled us to be here in this moment. As the commander of the final shuttle flight and the last to experience this out-of-the-window view it's clear just how amazing humankind can be when challenged with a goal and the means to attain it. The shuttle performs flawlessly. Sure there are the occasional minor annoyances, but generally the orbiter does as it was designed to do and takes us safely to a docking with ISS. In a task that seems more like bringing a massive supply ship into port, the two vehicles become one, the hatches are opened, and the formalities are attended to.

Even while docked to the much larger ISS, the shuttle is still home. My Russian counterparts, none of whom had ever seen the inside of a shuttle on Earth let alone in space, are enthralled to look around. For test pilots, admiration of a beautiful craft knows no international border. The shuttle was our place to eat, sleep, exercise and work. The quiet din of fans and pumps lulled us to sleep. It was... home.

Following an early morning landing at Kennedy Space Center and some well-deserved recognition of the team of custodians who made space shuttle flights look easy, Atlantis is powered down... for the final time. But the evidence of the overwhelming success of the shuttle lives on in the form of the International Space Station, Hubble Space Telescope and Chandra X-ray Observatory, and in the findings from Galileo, Compton Gamma Ray Observatory and many more. We also remember those who gave their lives in the name of space exploration. Their dream will live on. The thrill and pride of being a shuttle astronaut was a most rewarding experience; we were lucky to have played a small part.

Christopher J. Ferguson, 2019

Christopher J. Ferguson is a former NASA astronaut who was commander of the space shuttle Atlantis for flight STS-135, the final mission of the space shuttle program. Previously, on his first space mission (STS-115) in September 2006, he had been the pilot of Atlantis, and he commanded STS-126 aboard the space shuttle Endeavour in 2008. He retired from NASA in 2011 and is now director of Crew and Mission Operations for Boeing's Commercial Crew Program.

INTRODUCTION

"The dream is alive." These triumphant words were spoken by legendary astronaut John Young after successfully bringing home the first space shuttle from its maiden orbital flight in April 1981. It was a landmark in piloted space travel due to the fact that NASA had succeeded in making the world's first reusable space vehicle. I was just two years old when the renowned Gemini and Apollo veteran made this historic space flight as commander of space shuttle Columbia on STS-1, which would pave the way for a long and distinguished career for America's new state-of-the-art spacecraft.

My own earliest memory of a real space launch is of a cold day in January 1986. I was six years old and my school had arranged for the children and teachers to gather in a large assembly room to witness space shuttle Challenger launch into space for its 10th mission on STS-51L – the 25th shuttle launch in total, and now somewhat routine – but this one was different. This was a significantly unique mission as it marked the first time an ordinary civilian would fly aboard a space shuttle. There was a real sense of excitement around the school due to the fact that the civilian chosen to rocket into space would be a 37-year-old teacher named Christa McAuliffe. As a young boy, I was aware of astronauts and space travel, but I had never actually witnessed a launch, or at least not to my recollection. What we witnessed that day was a terrible and unexpected tragedy as Challenger exploded right in front of our eyes, just a few minutes after lift-off. I can remember teachers quickly directing us away from the unfolding events and swiftly transposing lessons onto a different subject. However, the Challenger incident stuck with me and although it didn't seem to bother me at the time – probably because I didn't quite understand the true nature of what had just occurred – it highlighted the start of what would eventually become an important interest in my life.

Twenty-five years later, 2011 marked the last ever space flight of the truly remarkable NASA Space Transportation System vehicle, the space shuttle. This book was compiled to commemorate 30 years of piloted space flight aboard an American treasure, a world phenomenon and arguably the most technologically advanced vehicle ever made.

The book pays tribute to the five space-worthy orbiter vehicles built by NASA – Columbia, Challenger, Discovery, Atlantis and Endeavour – and exhibits them in some of the most amazing space shuttle images you will ever see. It proceeds from the anticipation on the launch pad, through the dangerous launch sequence – ripping through Earth's atmosphere and reaching for the stars – and on to some of the most vivid mission photographs of extraordinary events such as space walks, astronaut maintenance work and docking with the International Space Station, before the signature return and landing. My intention was to show a 'best of' collection of photographs from NASA's vast archives compiled from the shuttle's complete history. I have endeavoured to make sure every image that made my final selection had a little extra something that made it stand out from the rest, something that would evoke a reaction from the viewer. Some images are naturally older than others and some are sharper, but I didn't want this to turn into a showcase for the best cameras or photographic lenses. I wanted it to reflect on 30 years of inspiring NASA images from the space shuttle's impressive history: an impressionistic photographic journey spanning the moment any of the shuttles arrives at the Florida launch complex through to the moment they touch back down on American soil.

Even though this project is primarily about the orbiters themselves, it is also important to acknowledge each and every crew member who flew aboard the space shuttle; because, as beautiful and technologically advanced as it may be, it is an experienced and valiant crew that truly brings the shuttle to life. At the back of this book you will find details of every STS mission launched, including documentation of every astronaut to serve aboard any of the five orbiters. You will also find a section archiving the space shuttle mission patches designed for every STS mission. These exquisite pieces of work are often forgotten, but each is a vibrant representation of the mission at hand and the crew that served on board. More often than not, the patches were designed or inspired by the crew themselves, just as the old mission insignias from past space flights were. I hope you have fun looking through them, I personally think they are fantastic and hold a great deal of nostalgia. My thanks go out to every person who put their time and creativity into producing these little gems.

What started out as a hobby, after a hard day's graft at my graphic design studio, slowly became my obsession. After contacting the NASA image department in 2009 and setting out my intentions, I was kindly put in contact with the sources to start making my dream a reality. I made it my mission to search through thousands upon thousands of official NASA photographs, changing my mind constantly but ultimately making sure I stayed true to my original concept. As you can imagine, there is an unbelievable wealth of subject matter held by NASA, but unfortunately there just aren't enough trees in the world to make the paper to print them all on. Okay, maybe that's not entirely true; nevertheless, I painstakingly scoured the vaults in order to eventually end up with what I consider to be my personal favourites, as well as a selection that I felt would inspire the reader, no matter what age or previous space knowledge they may have, long after the shuttle has taken its rightful place in the history books.

The main rule for inclusion in this collection was that the images had to focus primarily on the shuttle itself. I really wanted to use this opportunity to pay homage to the orbiters and entice the reader to take some time to appreciate the sheer beauty of the design, detail and elegance of the spacecraft. Many of these images have rarely been seen by the general public as we are often subjected to the same NASA favourites. Some of those classic images are included in this book as I never tire of looking at them, just as I never tire of seeing Buzz Aldrin's conquering presence on the lunar surface or the famous Earthrise image from Apollo 8's historic mission to the Moon in 1968. When deemed necessary, the images within this book went through a process of post production to enhance their visual impact. For example, some of the older images required digital retouching, colour correction or cleaning where the film had left dirty marks or scratches on the original scans. Occasionally, marks or blurs encroaching into the frame may have been digitally removed to enhance the final image, if I thought the end result warranted the retouching and only if it did not detract from the original raw scan. My intention was to showcase these wonderful photographs in the best possible way.

It seemed about time that NASA's greatest shuttle images were put together in one great collection, for people all over the world to remember and marvel at this historic accomplishment, long after the space shuttle's retirement in 2011. This book is a creative accolade to the greatest spacecraft built to date. It's more than just a photo album, it's a commendation, a memorial if you will, a labour of love. The end result is a stunning collection of classic favourites, along with some sensational and perhaps unusual forgotten treasures that take you on a journey, which I hope you'll revisit time and time again.

So here it is, my tribute to 30 glorious years of the magnificent space shuttle.

The dream is alive!

Luke Wesley Price

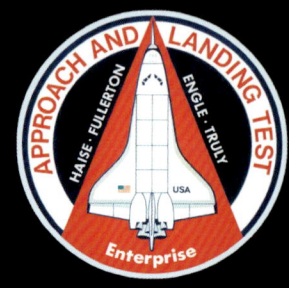

TO BOLDLY GO

In 1977, NASA conducted a series of tests for the space shuttle prototype Enterprise. The Approach and Landing Test (ALT) program was to examine the flight characteristics and aerodynamics in preparation for the very first orbital space flight of space shuttle Columbia, in April 1981.

Throughout the ALT program, Enterprise would be subjected to a series of taxi-test runs and captive flights where the prototype orbiter would be mounted on top of a Boeing 747 shuttle carrier aircraft. Following the success of these initial taxi-tests, Enterprise completed five piloted free-flights. Once a suitable altitude was reached, the shuttle, consisting of a two-man crew, would be jettisoned from the shuttle carrier aircraft and glide to a safe landing at Edwards Air Force Base in the California desert. Enterprise was never designed to be a space-worthy orbiter due to being fitted with inoperative dummy engines and a non-functional heat shield.

The space shuttle orbiter prototype was originally planned to be named Constitution, but due to popular demand it was very appropriately named after the Starship Enterprise from the ever popular TV show Star Trek.

Astronauts who flew aboard Enterprise:
FRED W. HAISE JR • C. GORDON FULLERTON • JOE H. ENGLE • RICHARD H. TRULY

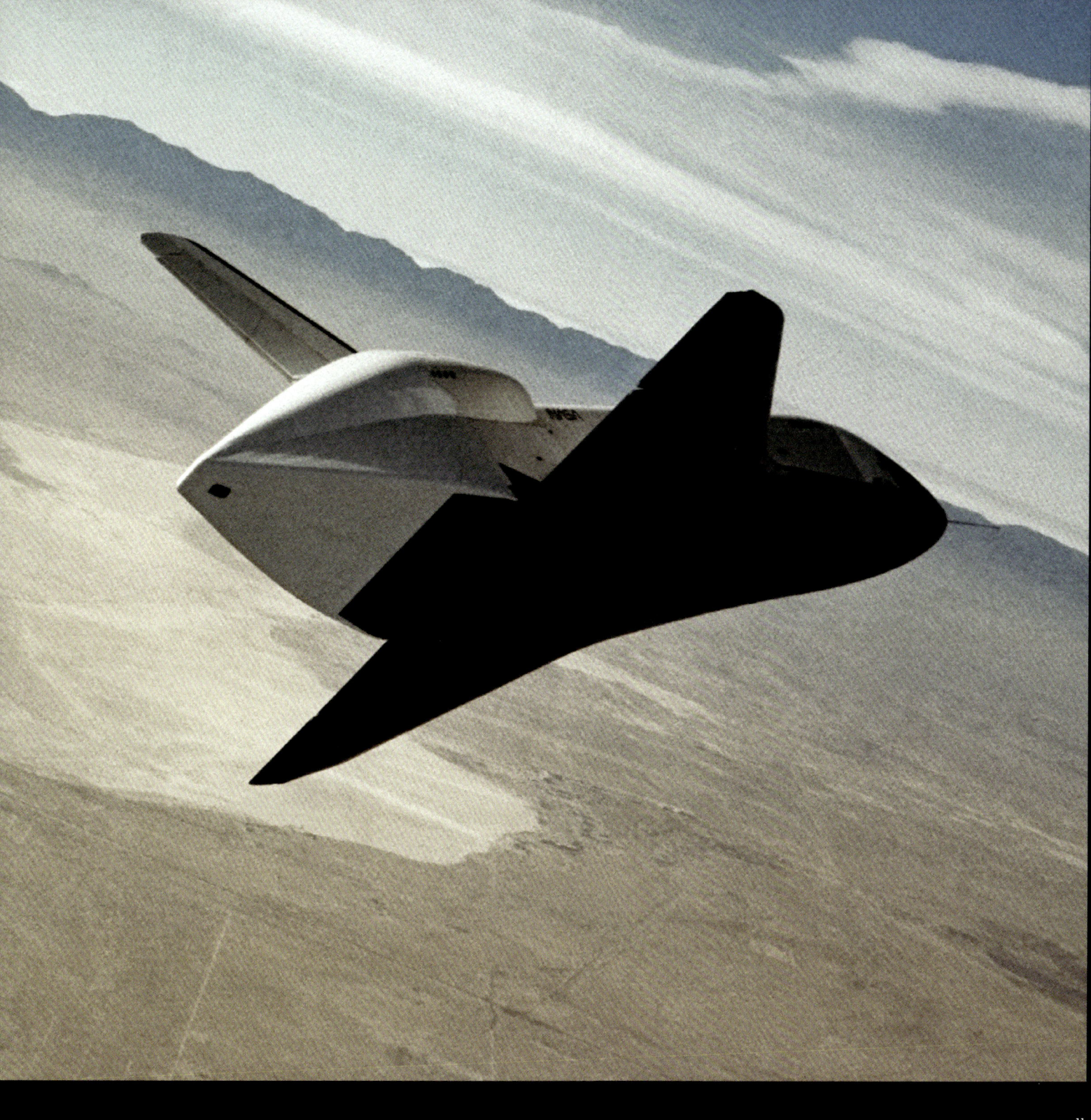

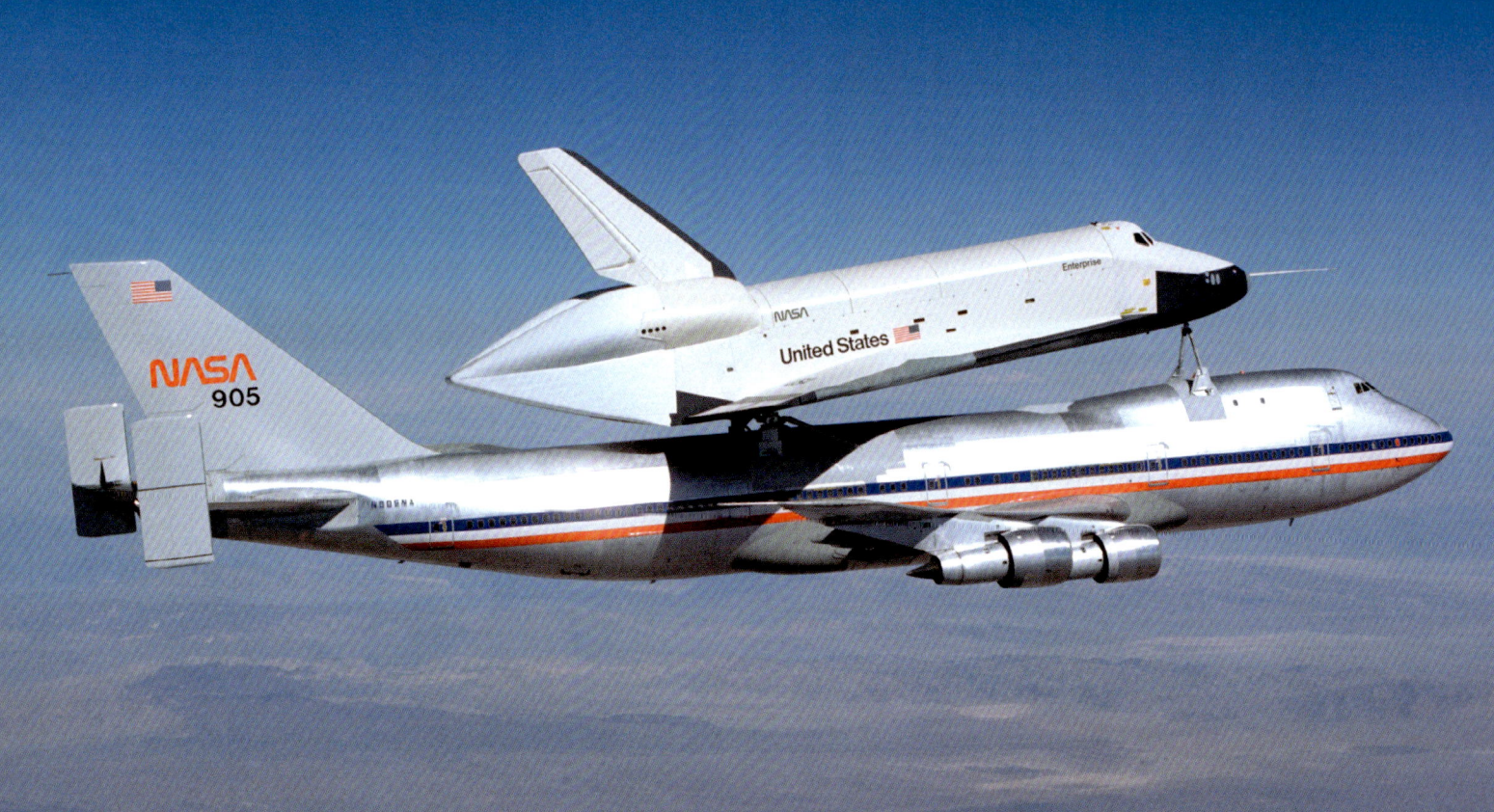

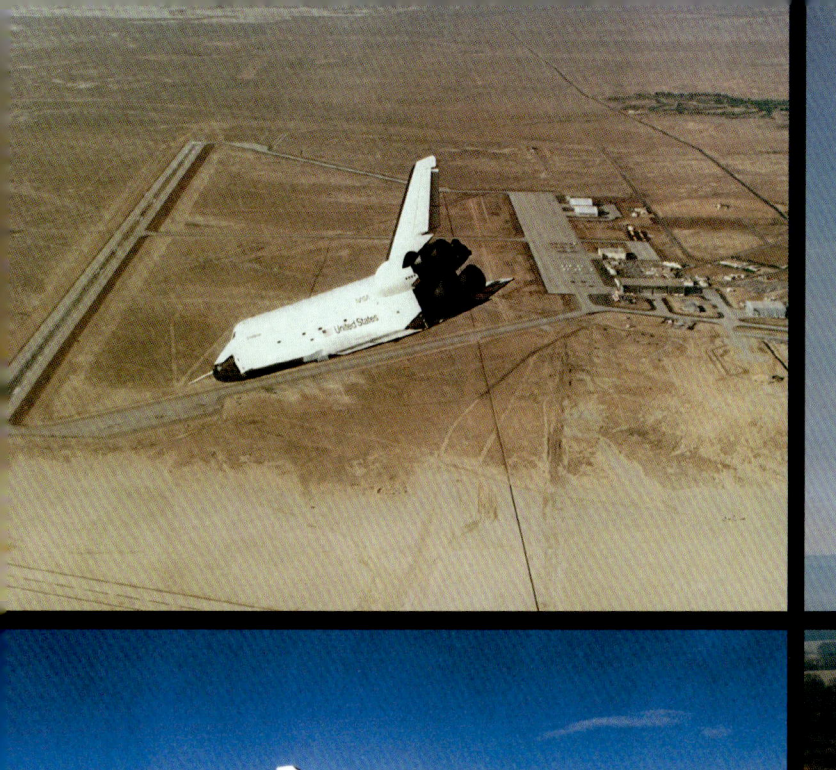
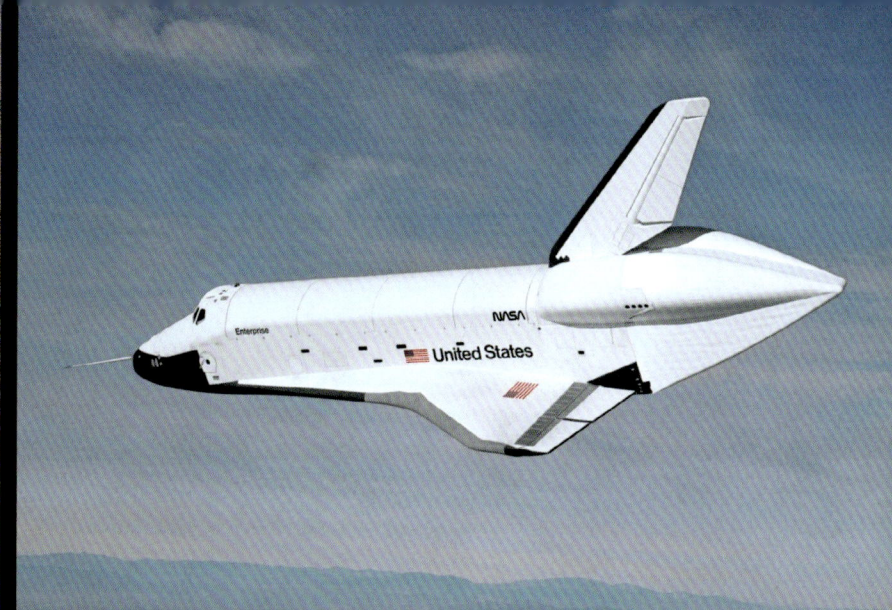
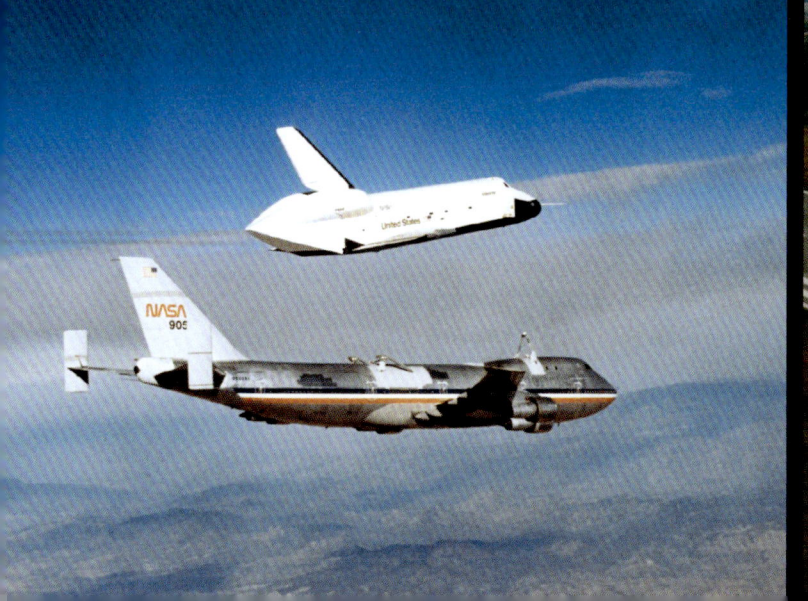
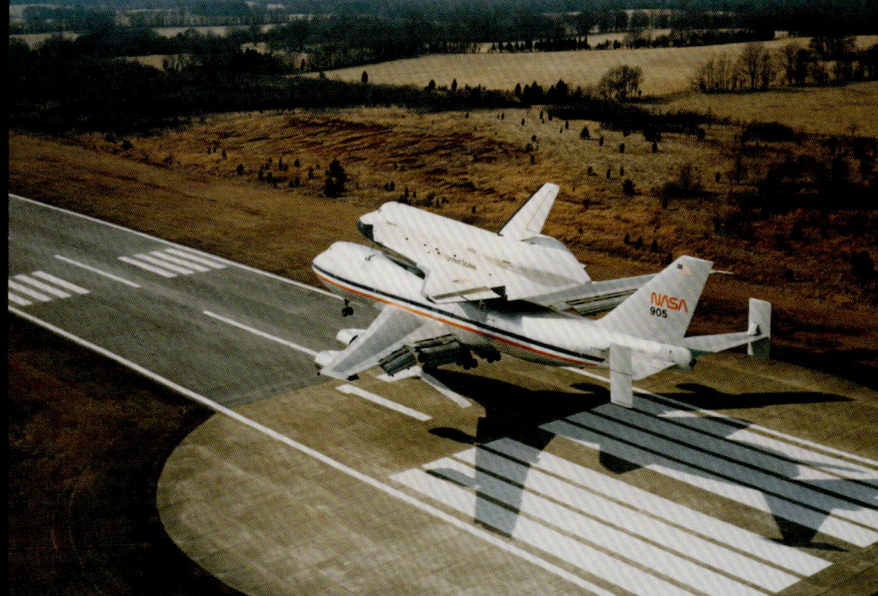

00:00:01

LAUNCH PAD

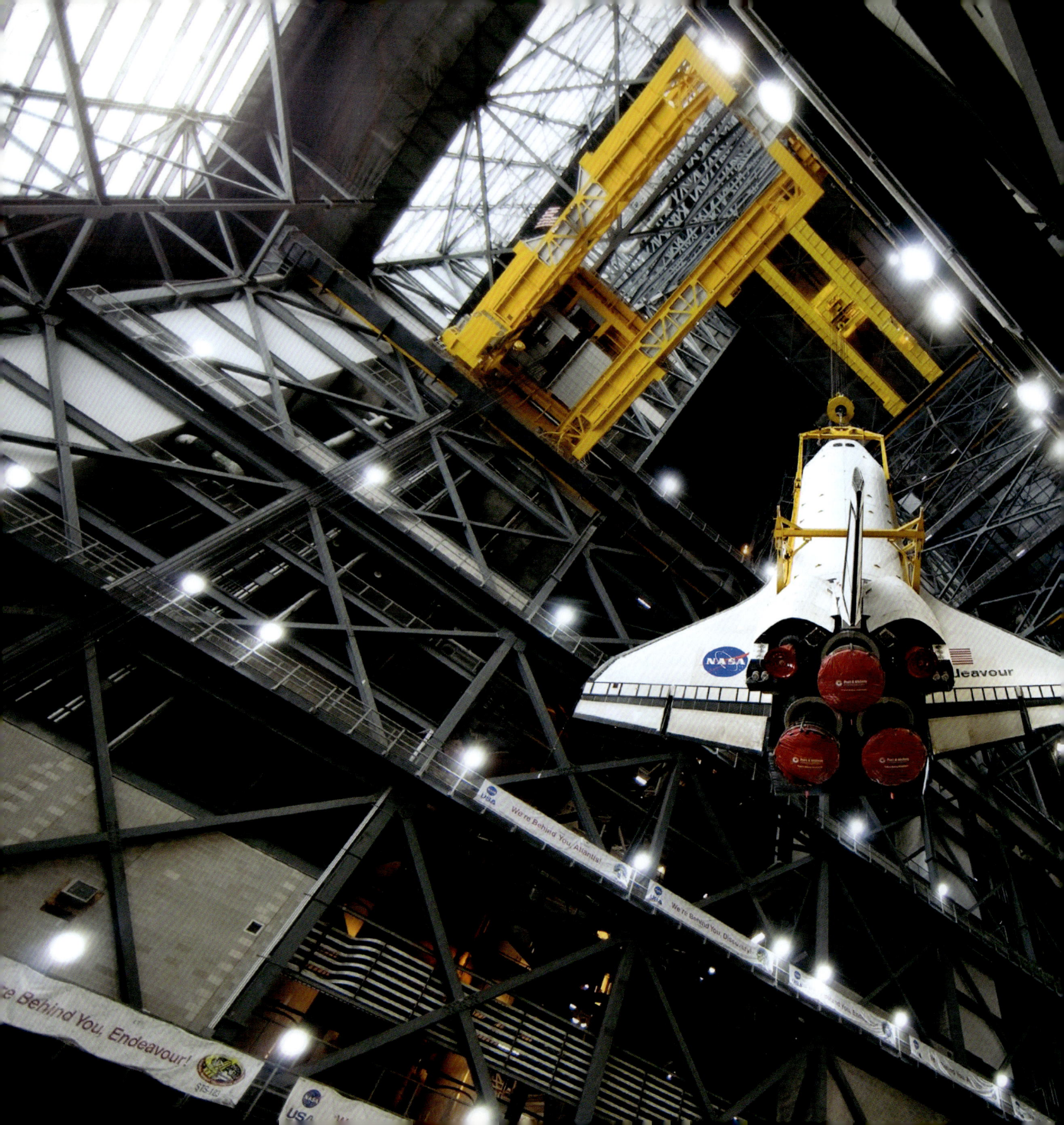

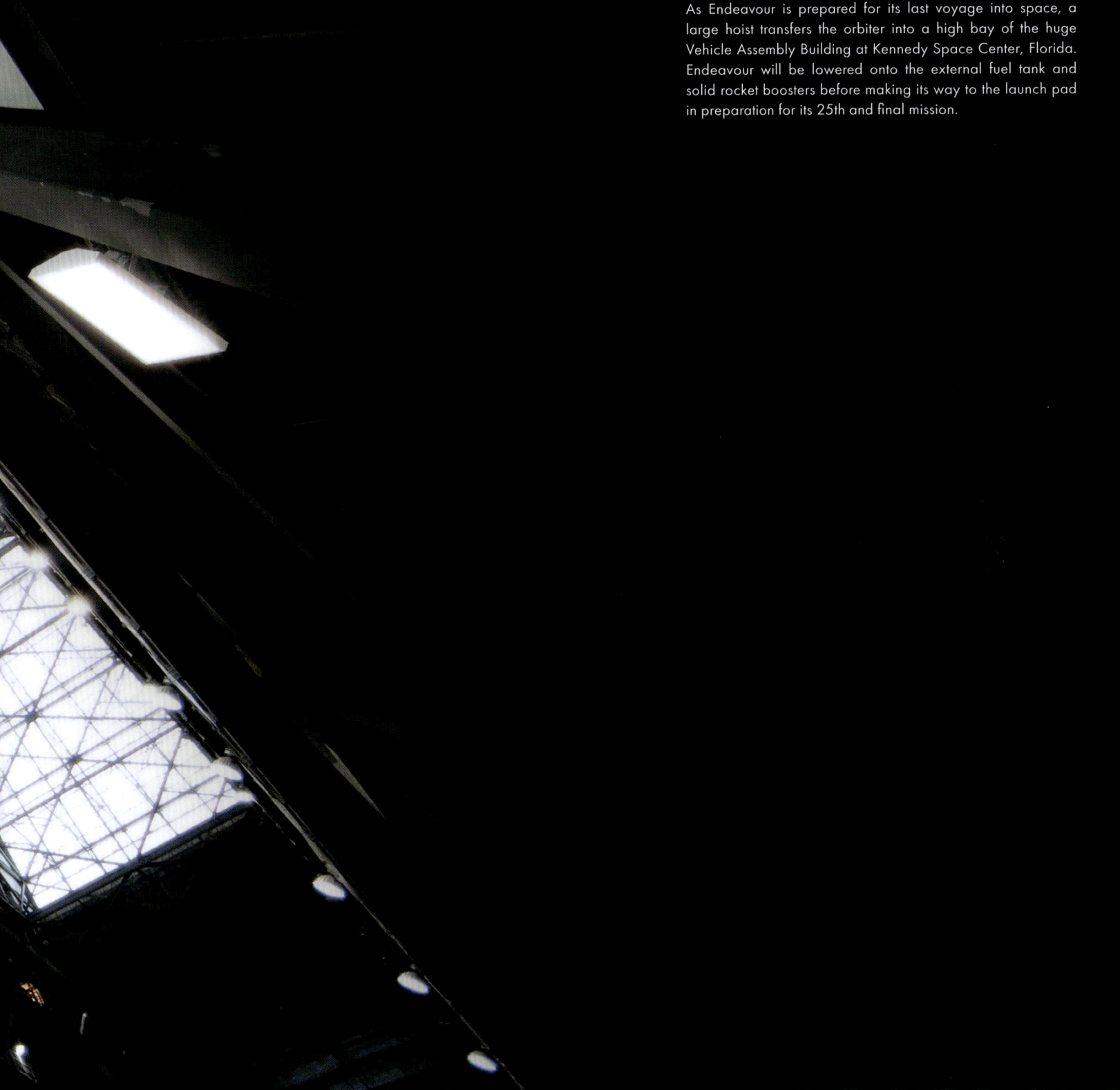

As Endeavour is prepared for its last voyage into space, a large hoist transfers the orbiter into a high bay of the huge Vehicle Assembly Building at Kennedy Space Center, Florida. Endeavour will be lowered onto the external fuel tank and solid rocket boosters before making its way to the launch pad in preparation for its 25th and final mission.

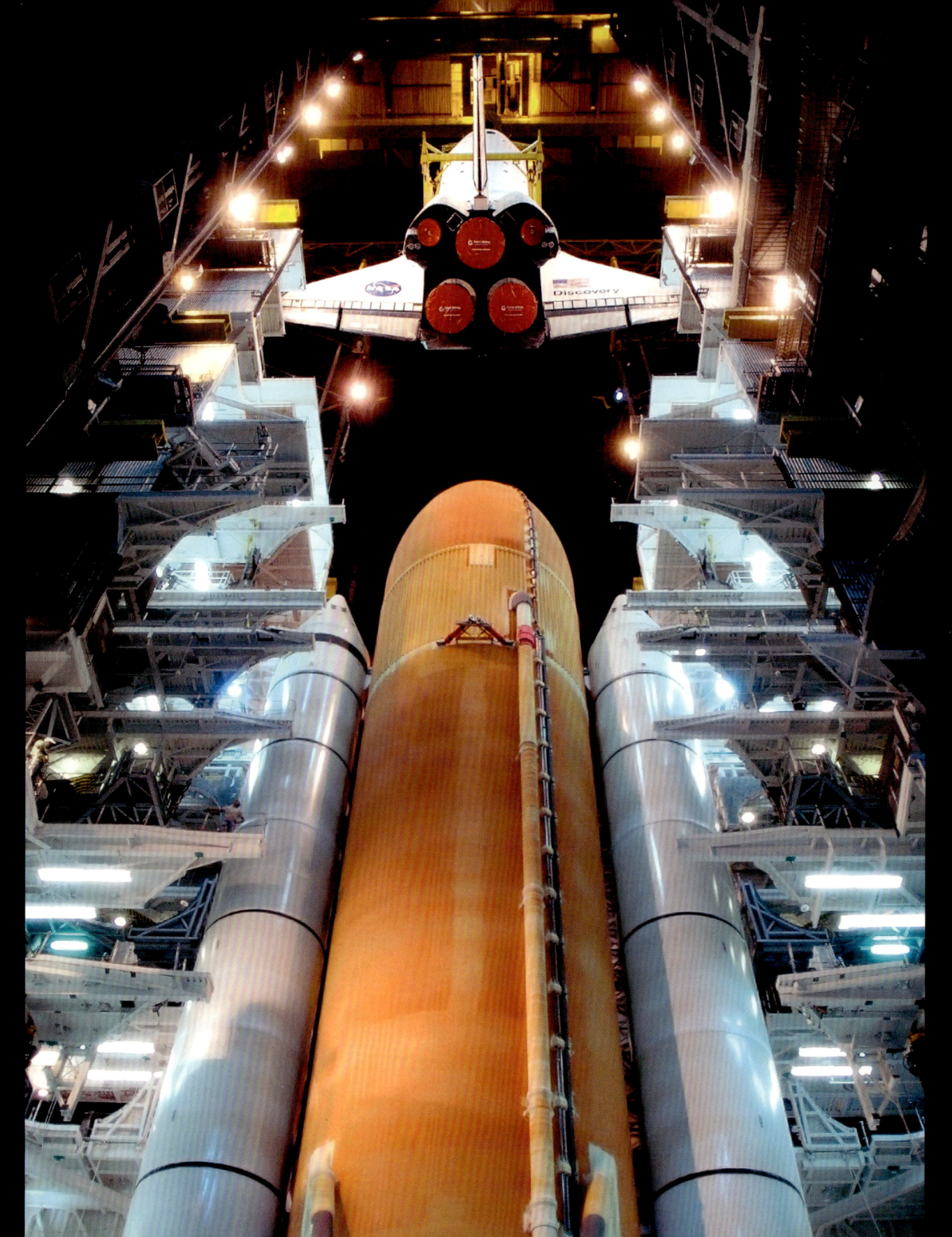

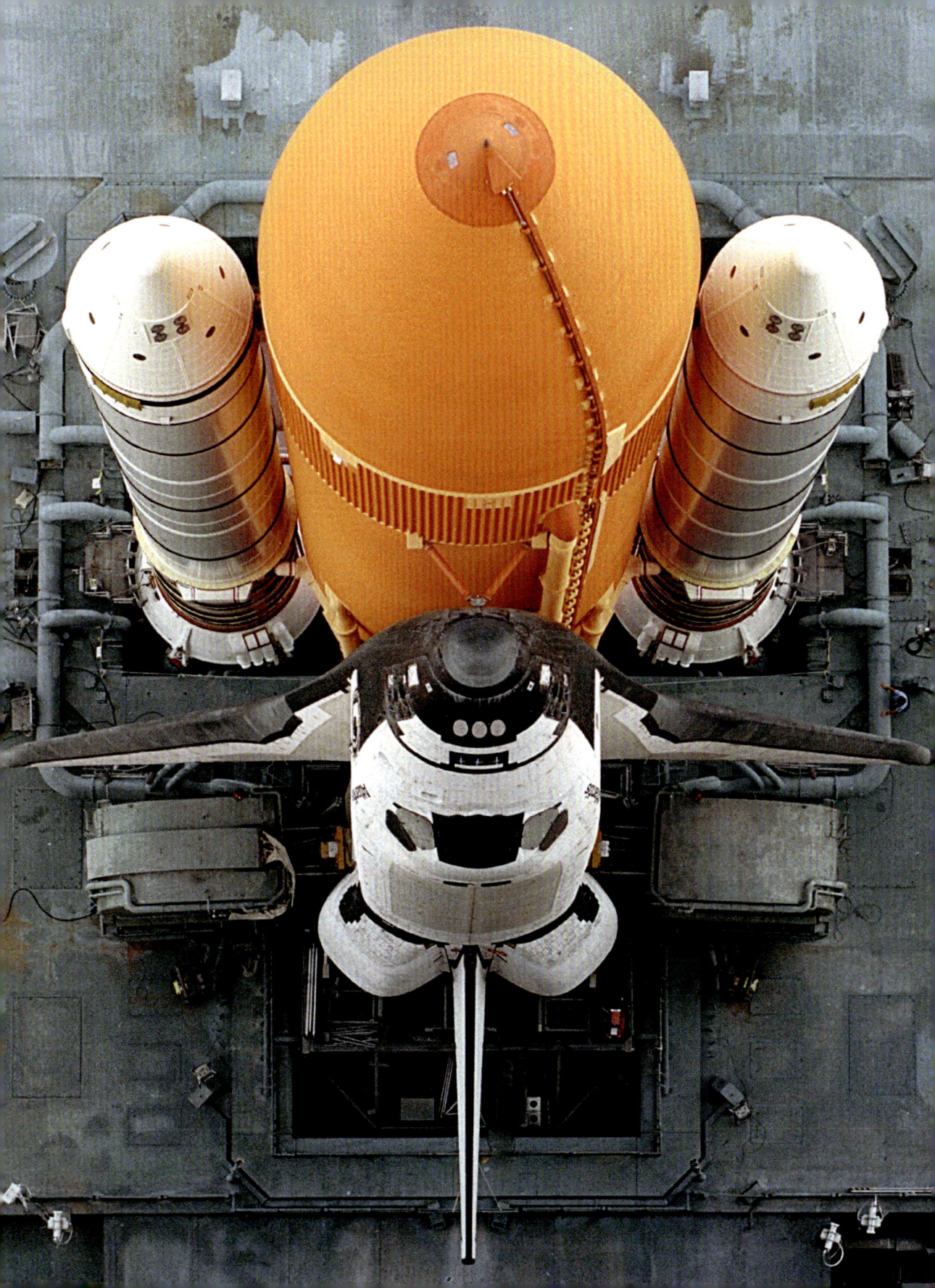

STS-124
31 May to 14 June 2008

(Previous spread: Left) Inside the Vehicle Assembly Building, Discovery is carefully attached to the orange external fuel tank and solid rocket boosters located either side. These powerful rockets provide over 80% of thrust for the first two minutes of flight, before being jettisoned and falling to the ocean by parachutes. Once recovered, the solid rocket boosters are refurbished for use on future space shuttle missions.

STS-79
16–26 September 1996

(Previous spread: Right) Space shuttle Atlantis starts a very slow journey from the Vehicle Assembly Building (VAB) to Launch Pad 39A. This dramatic image looking down on Atlantis sitting proudly atop the stack was taken on the roof of the VAB, 525 feet (160 m) above the ground.

STS-121
4–17 July 2006

(Right) Discovery emerges defiantly from NASA's Vehicle Assembly Building at Kennedy Space Center, ready to begin its long and slow journey to the launch pad. During the mission, the crew will deliver supplies and perform repairs to the International Space Station as well as test new safety and repair techniques following the shuttle disaster of 2003, in which Columbia and her brave crew perished.

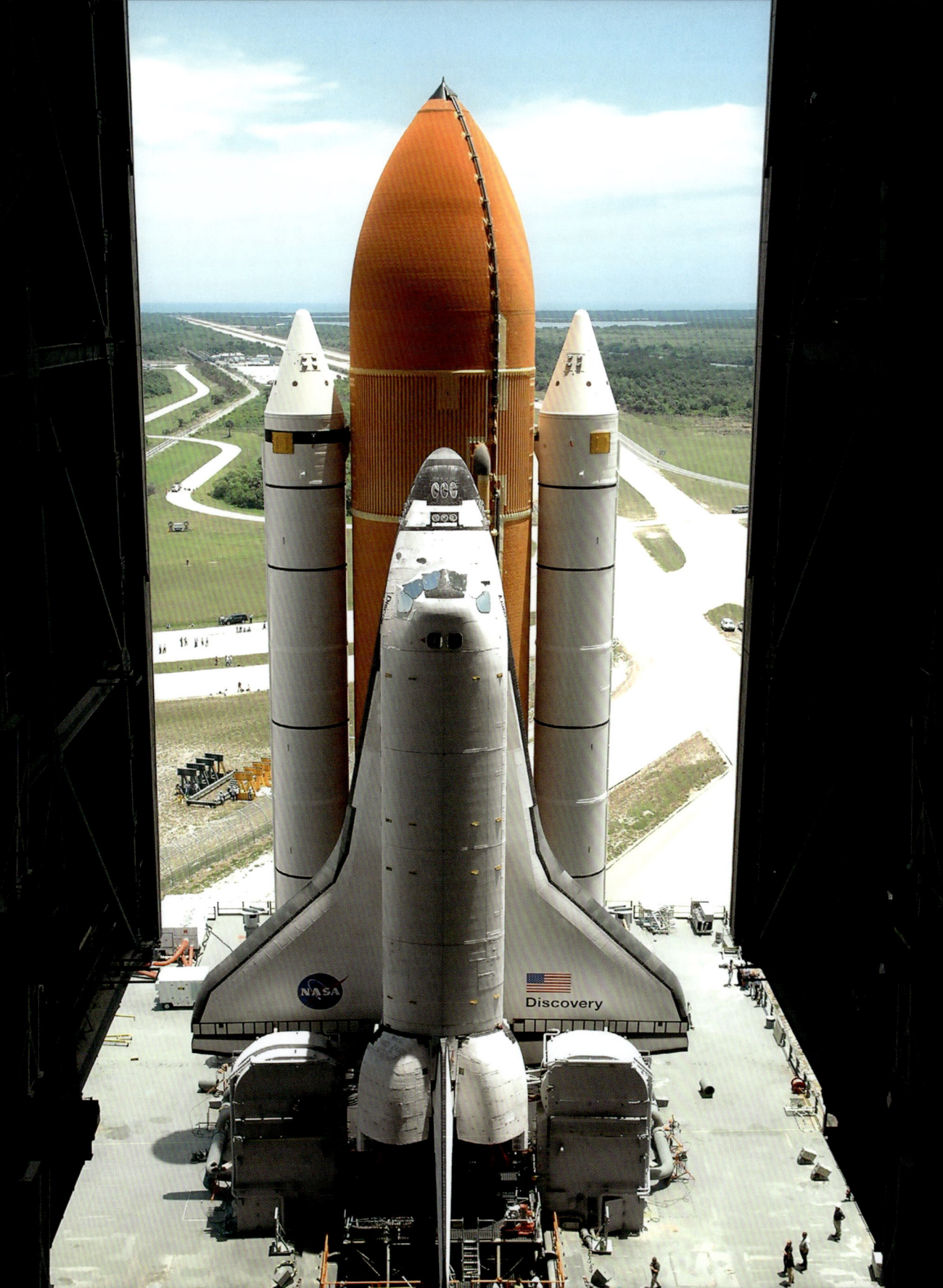

Dawn at Kennedy Space Center, January 2009. Discovery makes its long journey to the launch pad on top of the crawler transporter. The mission objectives are to install the S6 truss segment and solar arrays at the International Space Station allowing a six-person crew to live on board from May 2009.

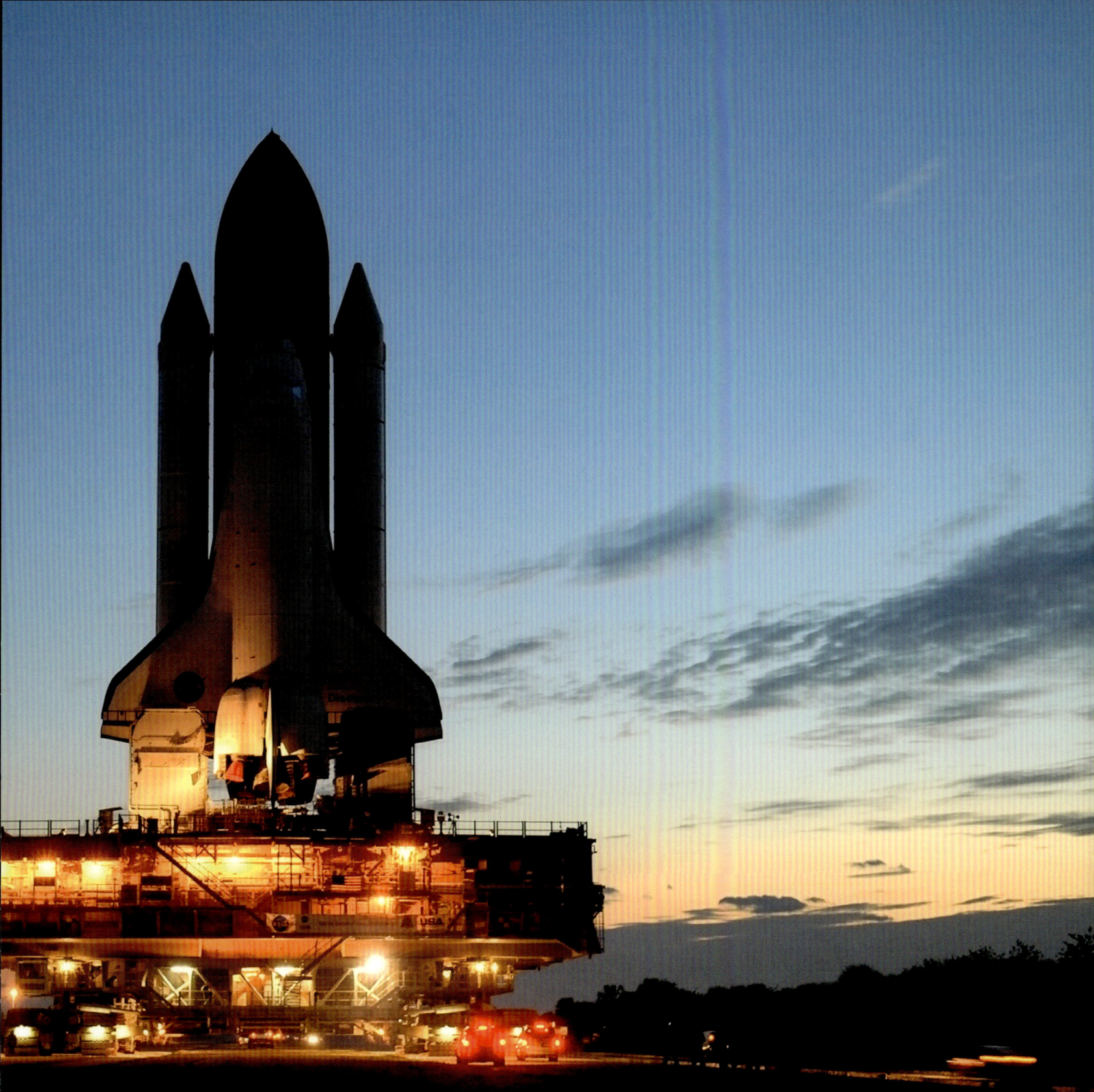

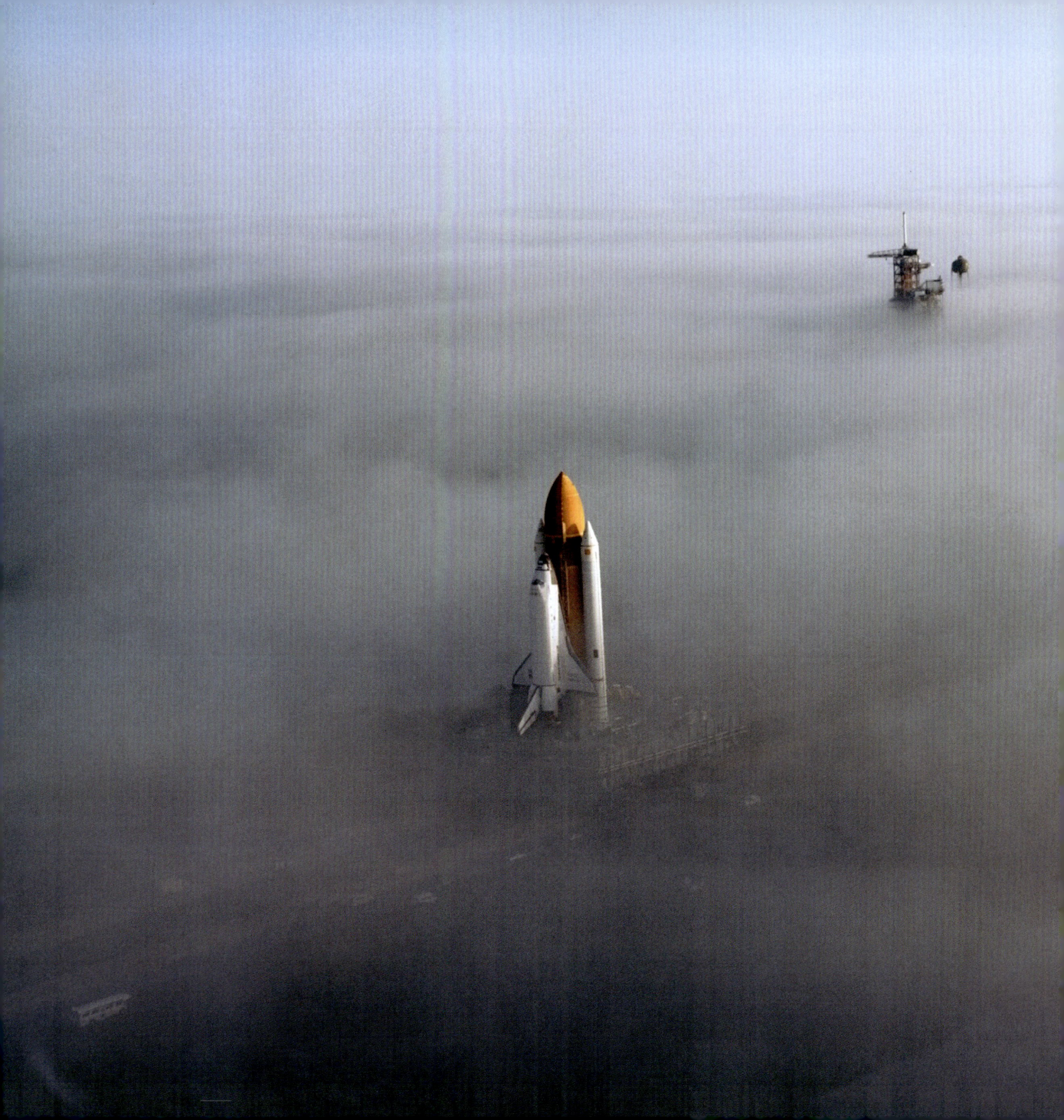

4–9 April 1983

NASA's latest orbiter slowly makes the long journey to Launch Pad 39A at Kennedy Space Center, Florida. Covered by a thick blanket of fog, Challenger can be seen towering through the mist with the launch pad visible in the distance. The launch was originally scheduled for January, but problems were discovered with Challenger's three main engines, requiring one to be replaced and the other two to be repaired. This image was taken in November 1982.

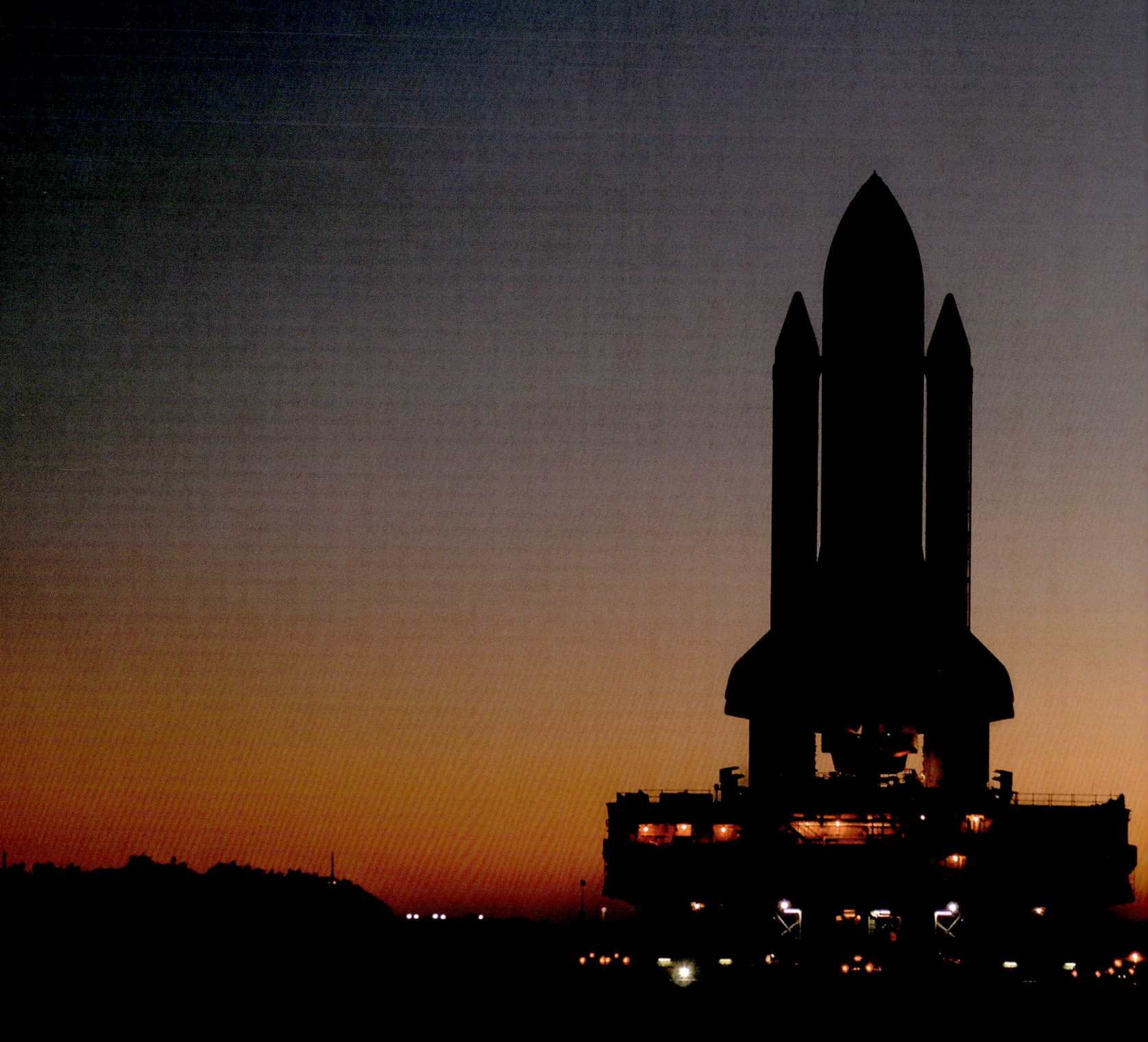

STS-122
7–20 February 2008

As the sun rises over Florida, a silhouetted Atlantis makes its way from the Vehicle Assembly Building to Launch Pad 39A. The distinctive black shape stands tall and proud as it prepares for its 13-day orbital mission where it will deliver the Columbus laboratory to the International Space Station. This image was taken in November 2007.

Sunrise at Kennedy Space Center and space shuttle Atlantis makes tracks along the 3.5-mile trek towards Launch Pad 39A for a second time. During a thunderstorm in February 2007, Atlantis suffered hail damage to the external tank as well as damage to the heat shield tiles while sitting on the launch pad. The shuttle returned to the Vehicle Assembly Building for repairs and was assigned a new launch target for June 8.

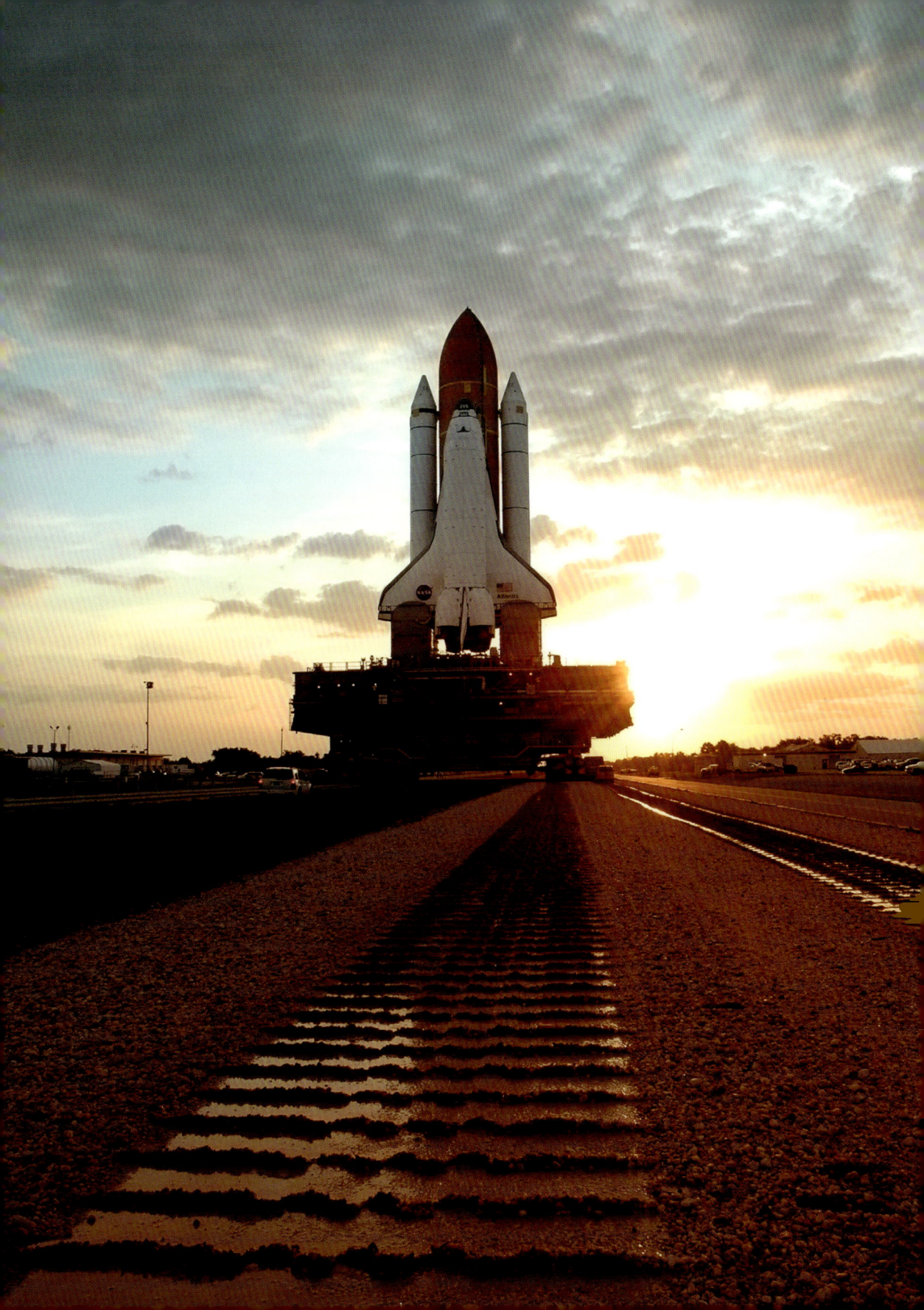

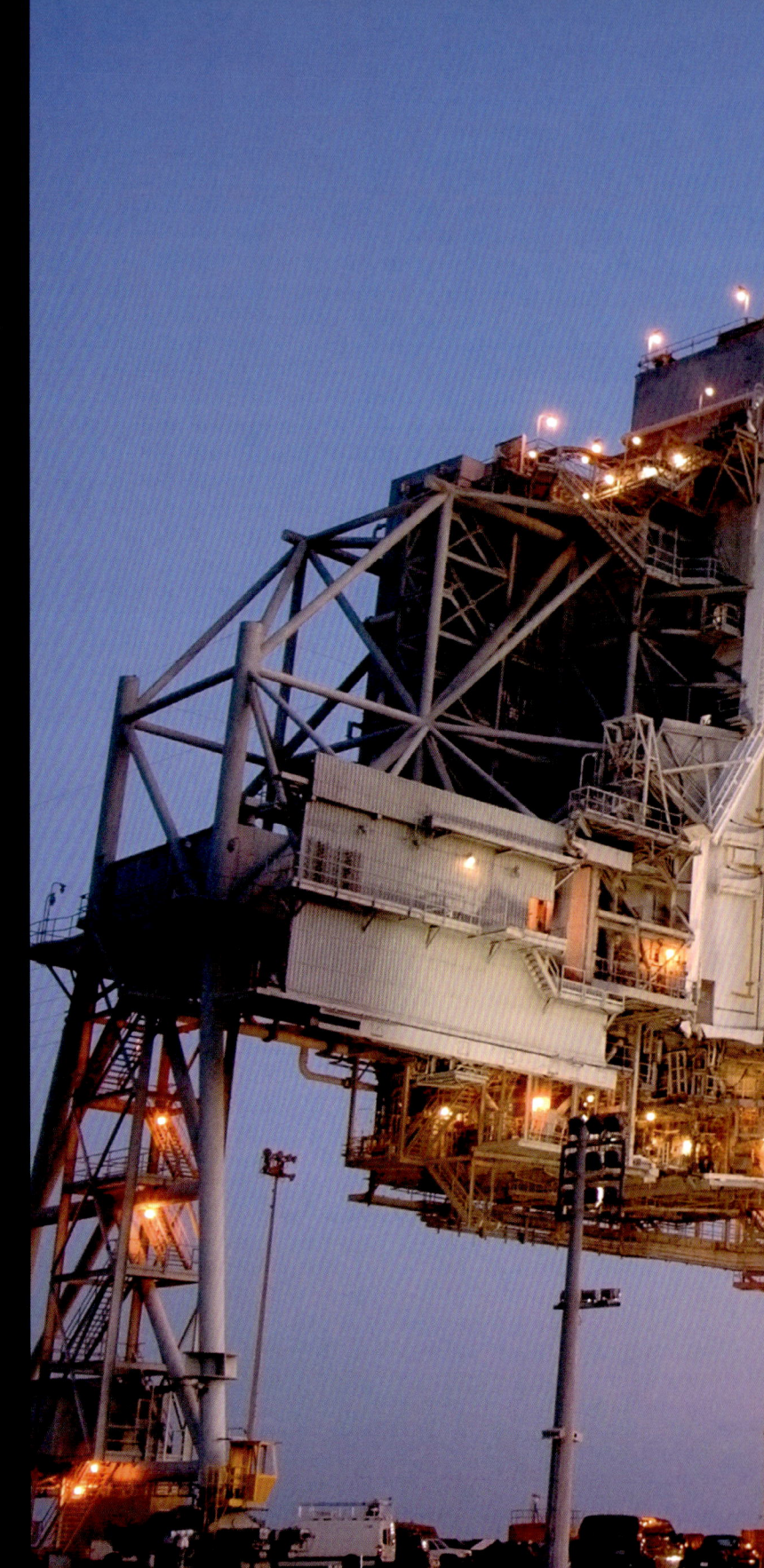

STS-126
14–30 November 2008

With Atlantis already resting on Launch Pad 39A for STS-125, Endeavour arrives at Launch Pad 39B amid a glow of lights and the beautiful dusk sky over Merritt Island in September 2008. Designated STS-400, it was originally rolled out as a Launch-On-Need rescue flight for STS-125, in the unlikely event of something happening to Atlantis during its mission to service the Hubble Space Telescope. STS-125 was postponed until 2009 due to a problem with Hubble. Endeavour was transferred to Launch Pad 39A to begin its STS-126 mission.

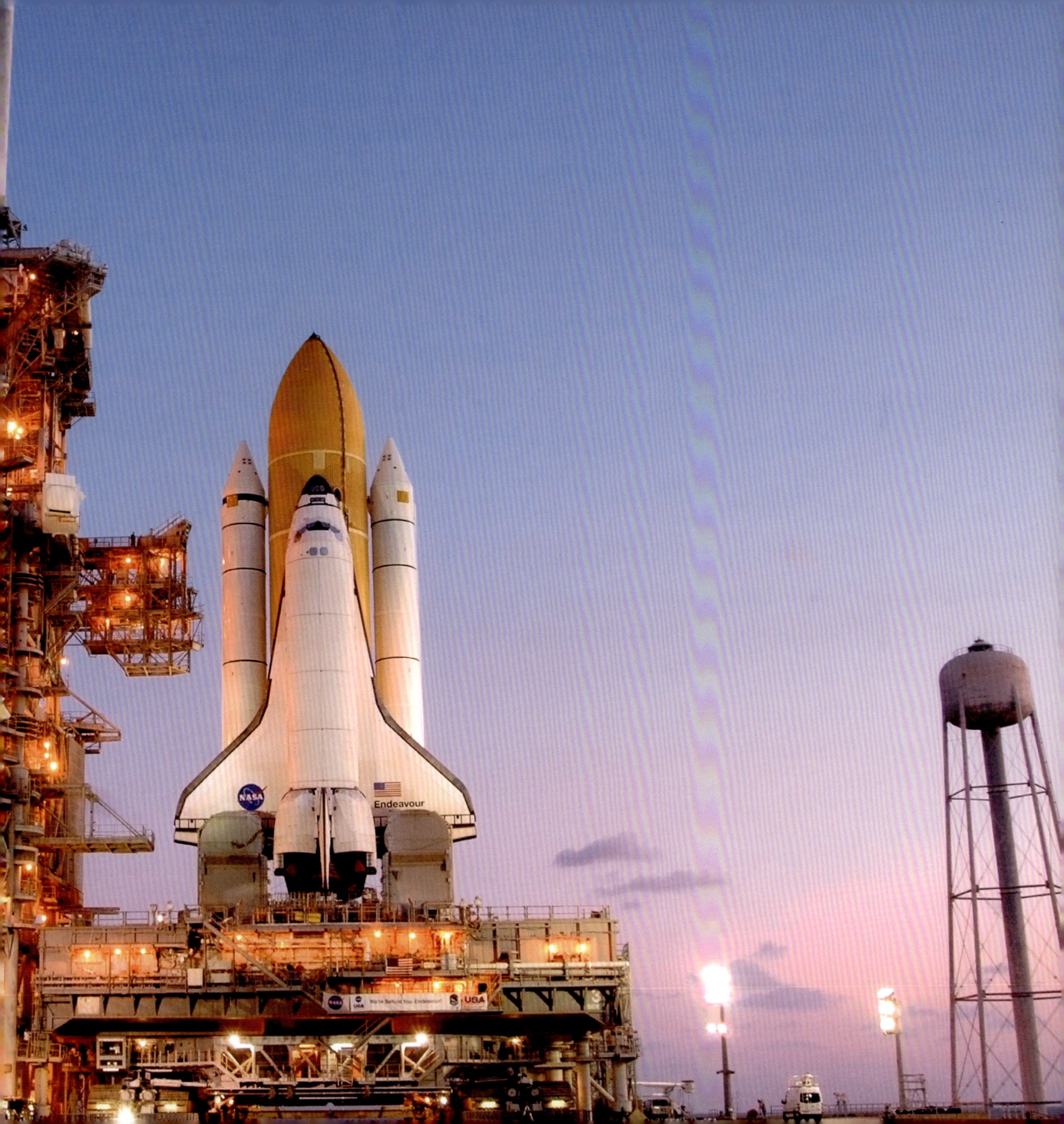

STS-1
12–14 April 1981

(Right) A timed exposure of Columbia before its first voyage into space. STS-1 was the first maiden space flight of a new NASA spacecraft to be flown with a piloted crew.

STS-1
12–14 April 1981

(Following spread) Columbia, America's first orbital space shuttle, rests patiently on Launch Pad 39A at Kennedy Space Center, ready for its first ever journey into space. Veteran Gemini and Apollo astronaut John W. Young is the Commander of this historic maiden flight alongside Pilot Robert L. Crippen. Notice the white external fuel tank also used for STS-2. In order to save weight the external tank was left unpainted from STS-3 onwards, giving the distinctive orange look that now defines the space shuttle's appearance.

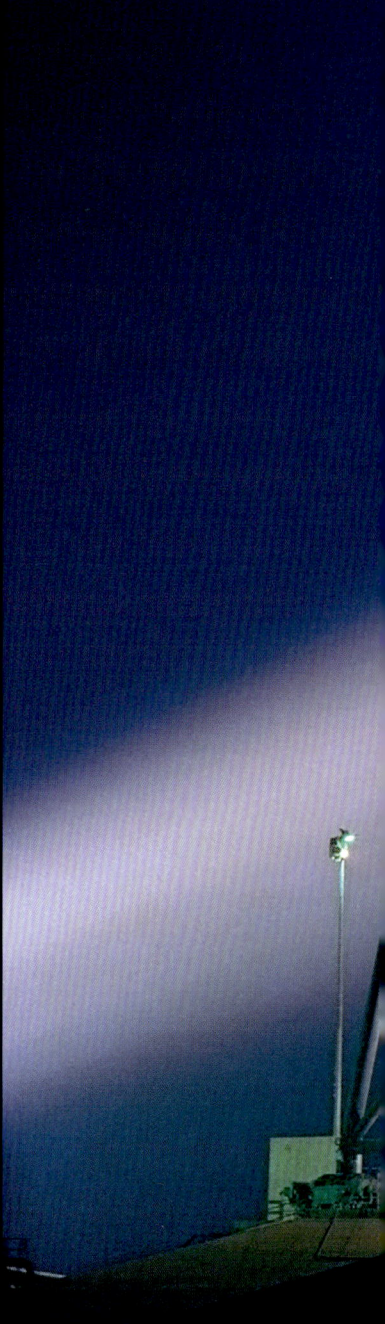

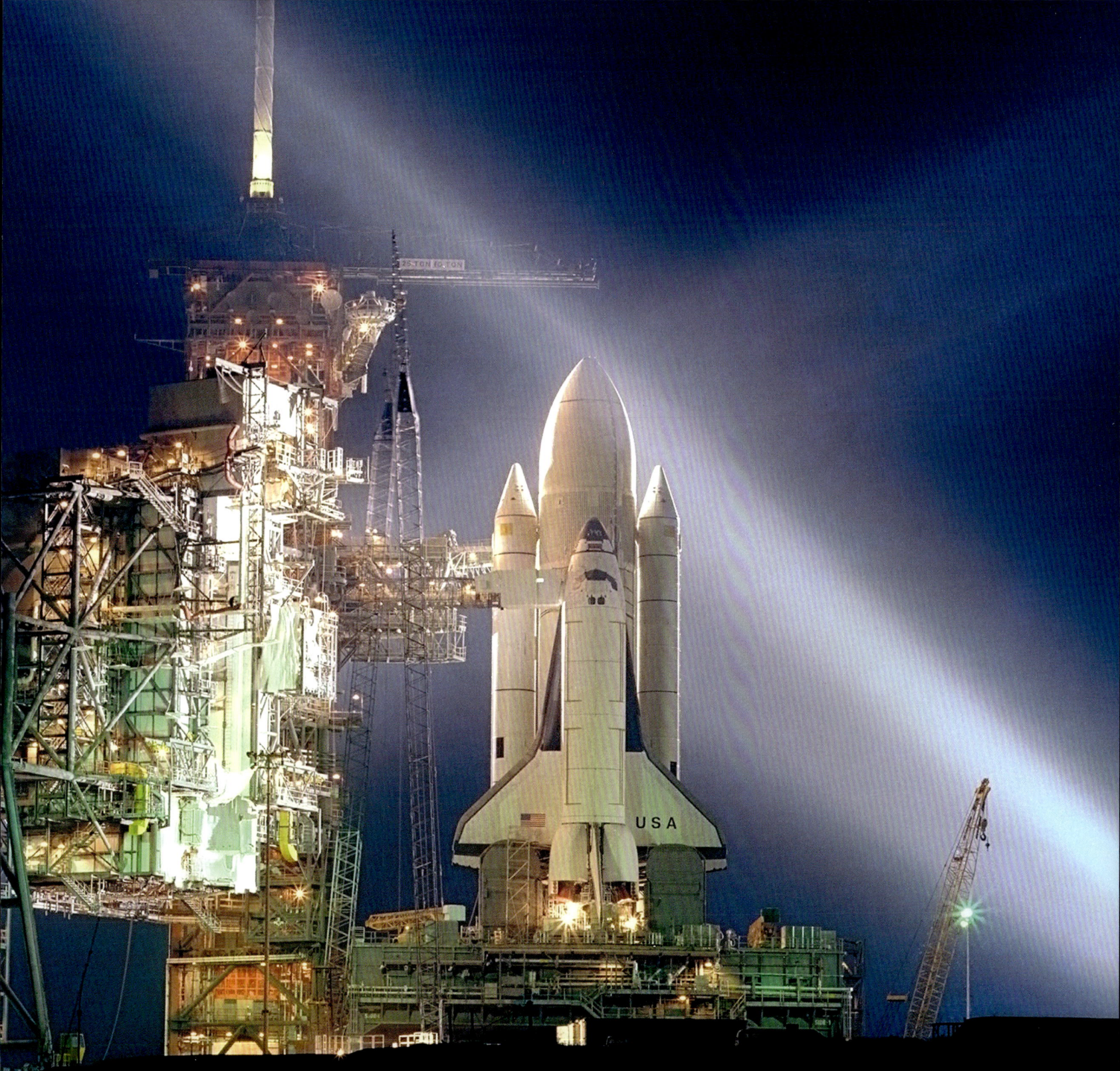

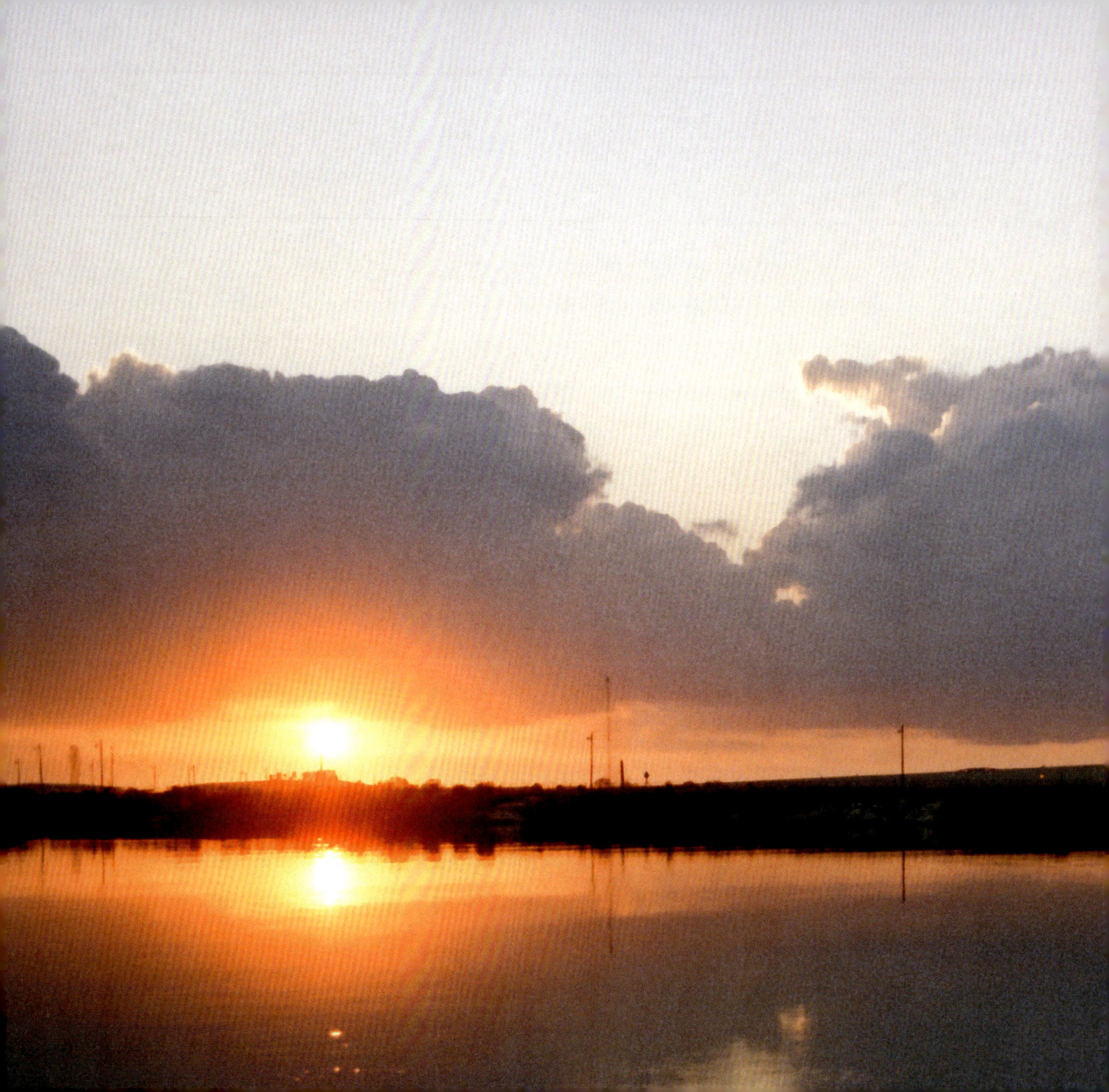

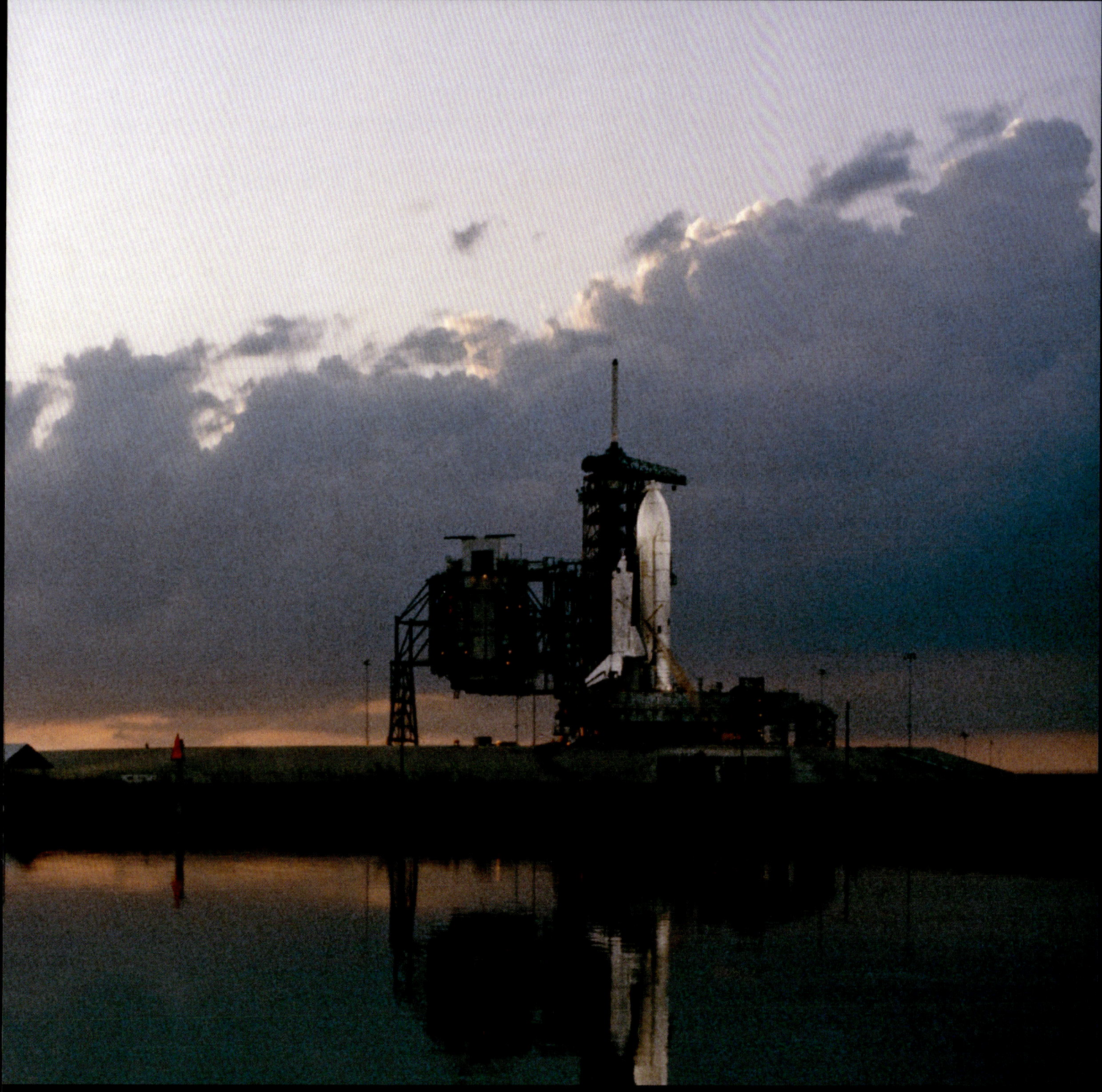

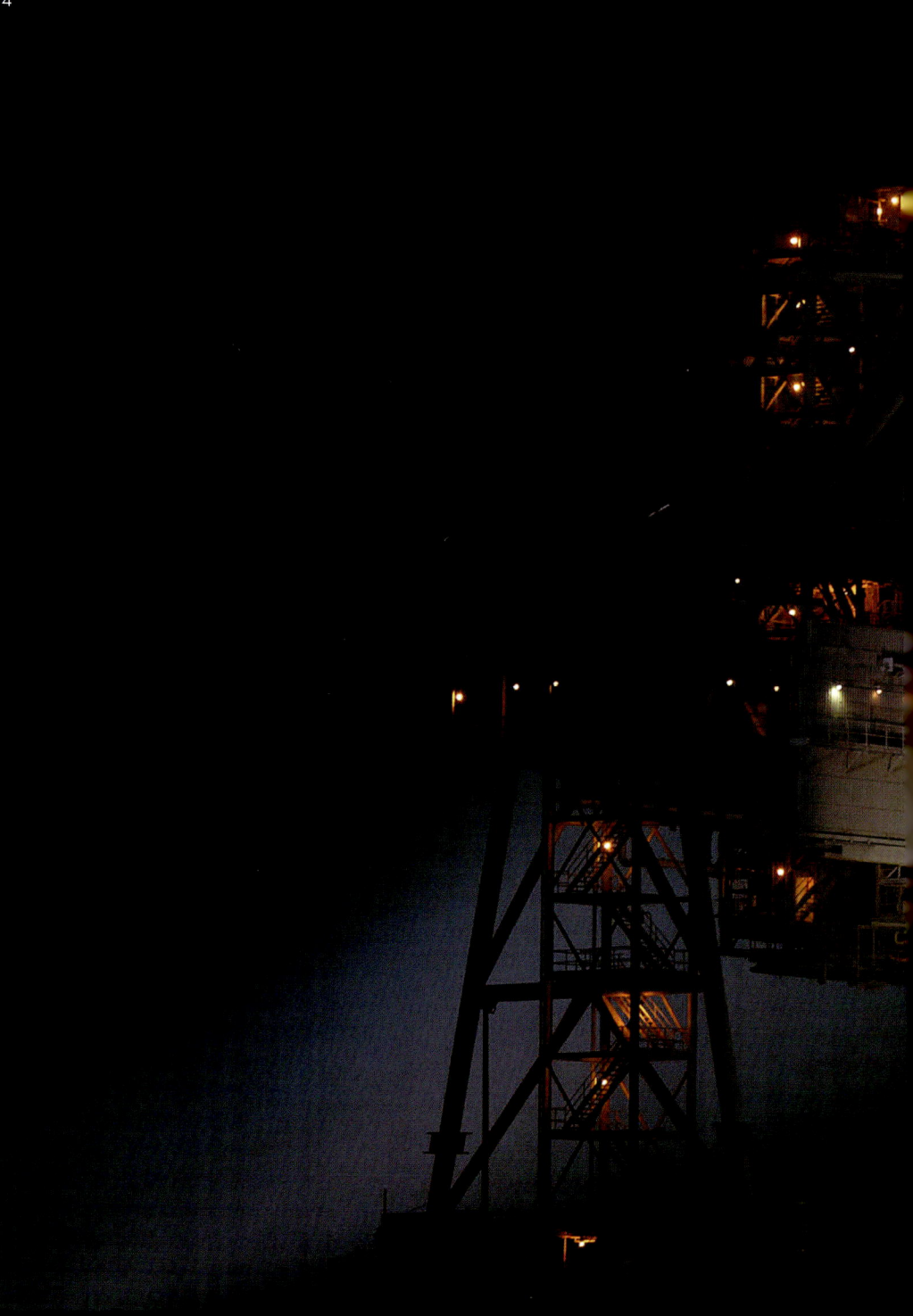

Space shuttle Discovery stands illuminated at Kennedy Space Center in preparation for its 39th and final flight. On this last trip to the International Space Station, Discovery will carry the Permanent Multipurpose Module, Express Logistics Carrier-4 and the Robonaut 2, the first humanoid robot in space.

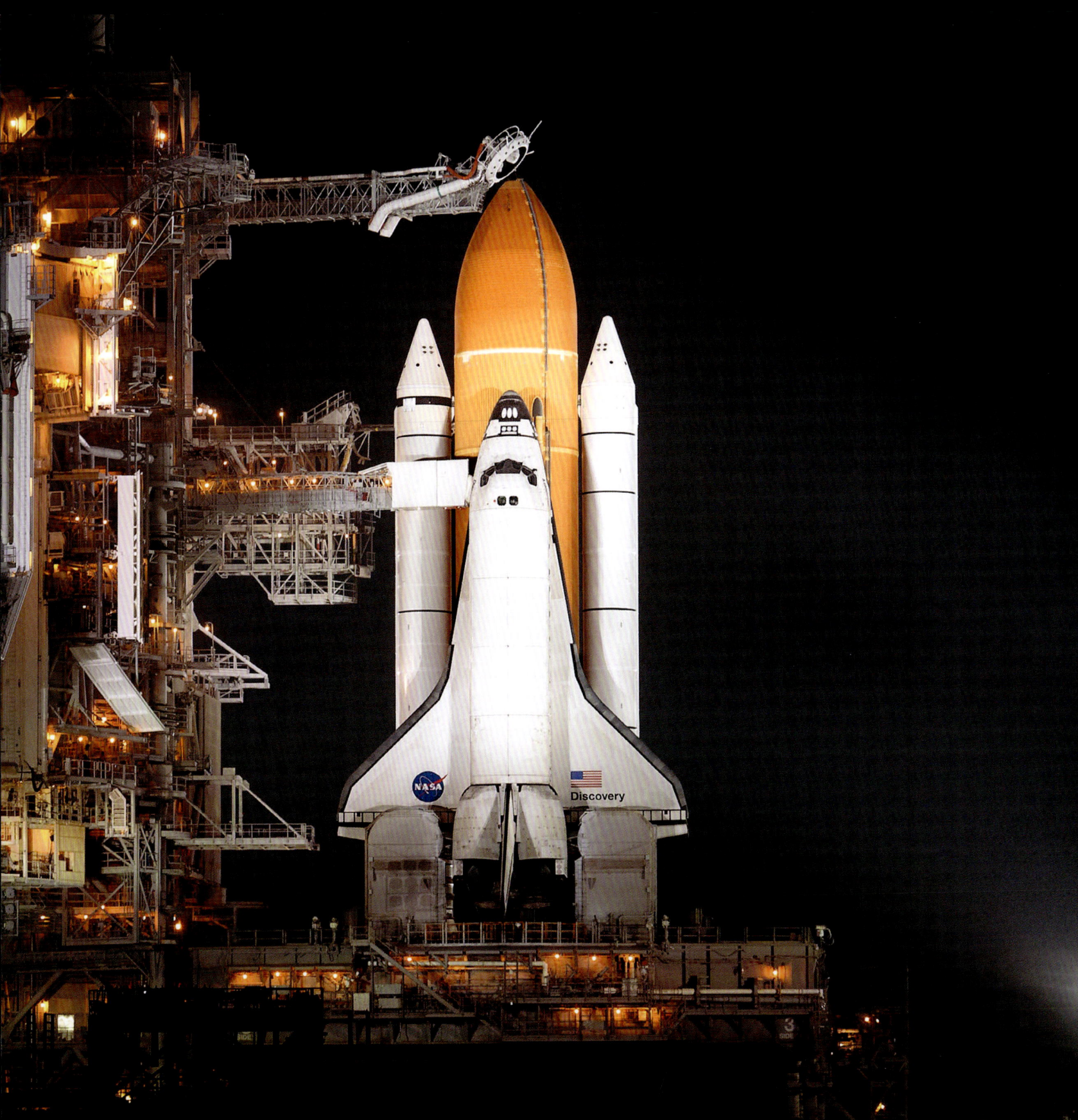

00:00:02

LIFT-OFF

30 August to 5 September 1983

The first night launch of a space shuttle, and Challenger floods the black night sky over Merritt Island with a radiant orange glow. It was the eighth shuttle launch in total and the third in a row for Challenger. A later revelation found that the launch was close to ending in tragedy due to one of the solid rocket boosters nearing catastrophic malfunction. Three years later, Challenger exploded just a few minutes after lift-off. Extreme cold temperatures had affected the function of the solid rocket booster's O-ring gaskets, which allowed leaked flames to ignite the fuel tank.

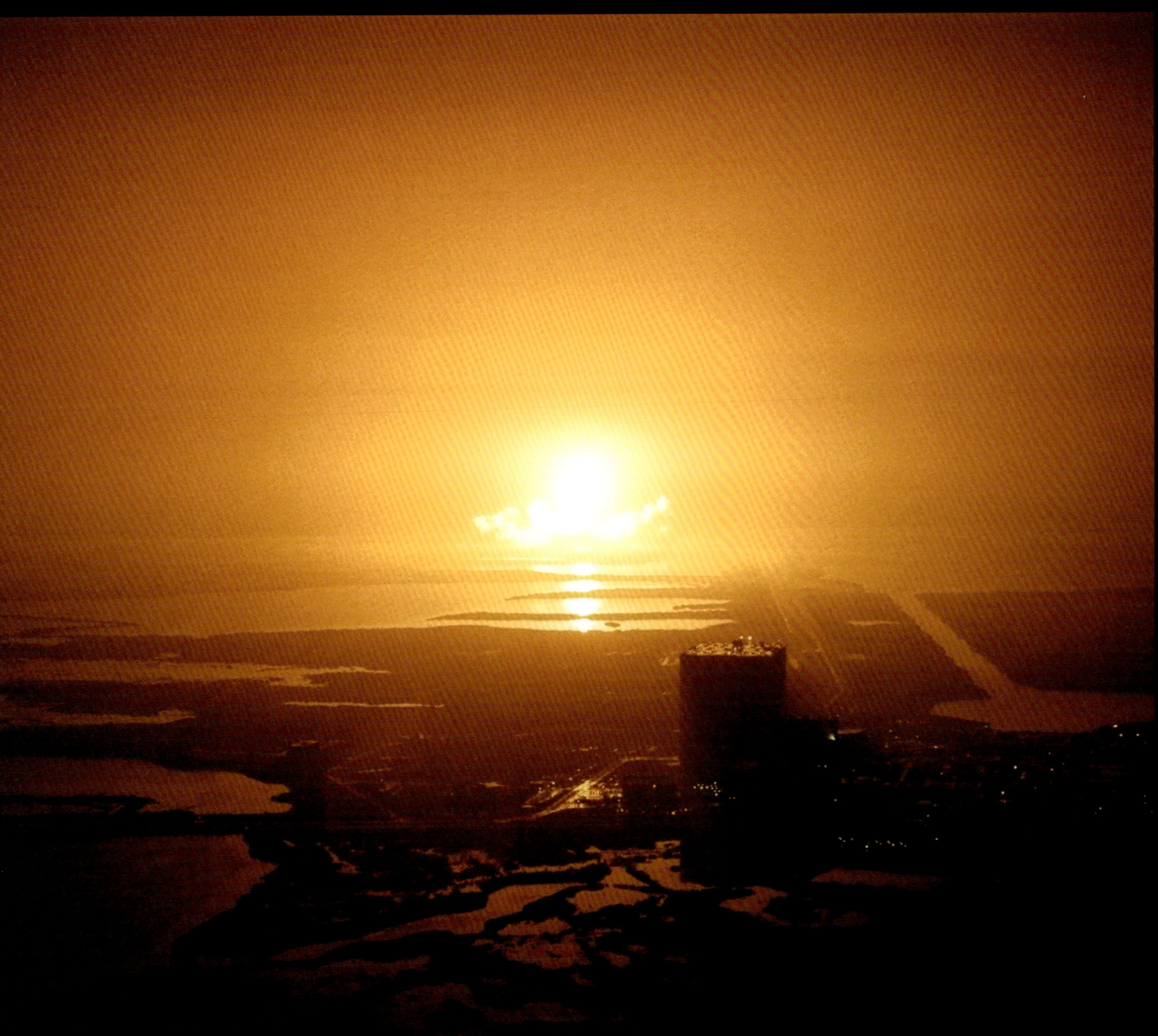

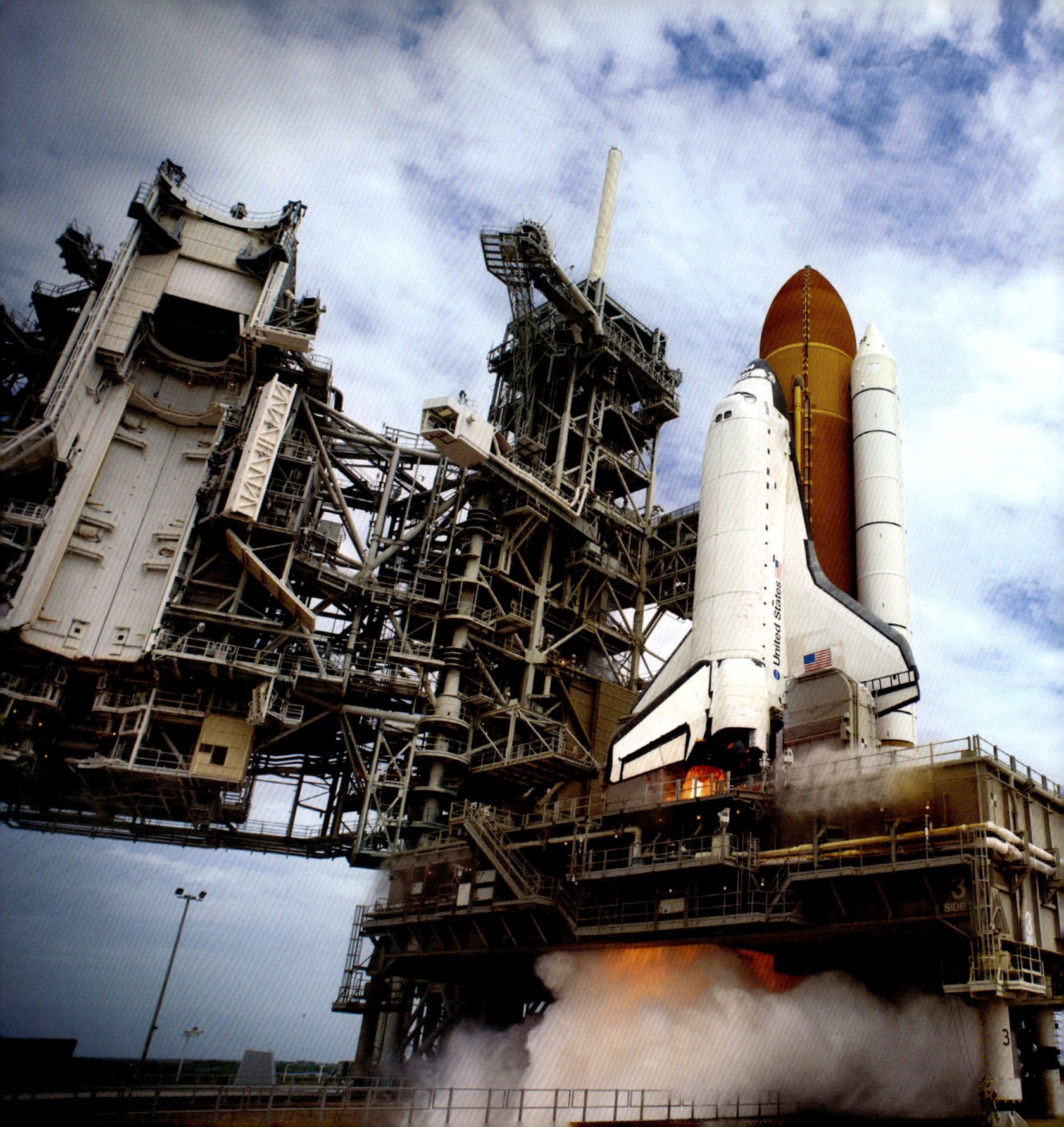

STS-135
8–21 July 2011

Atlantis's three main engines and solid rocket boosters ignite, launching the space shuttle into orbit one last time. STS-135 is the 135th and final mission of America's space shuttle program. For the first time since 1983, the shuttle carries a small, four-person crew. Their mission objectives are to deliver supplies and equipment to the International Space Station, including the Raffaello Multi-Purpose Logistics Module.

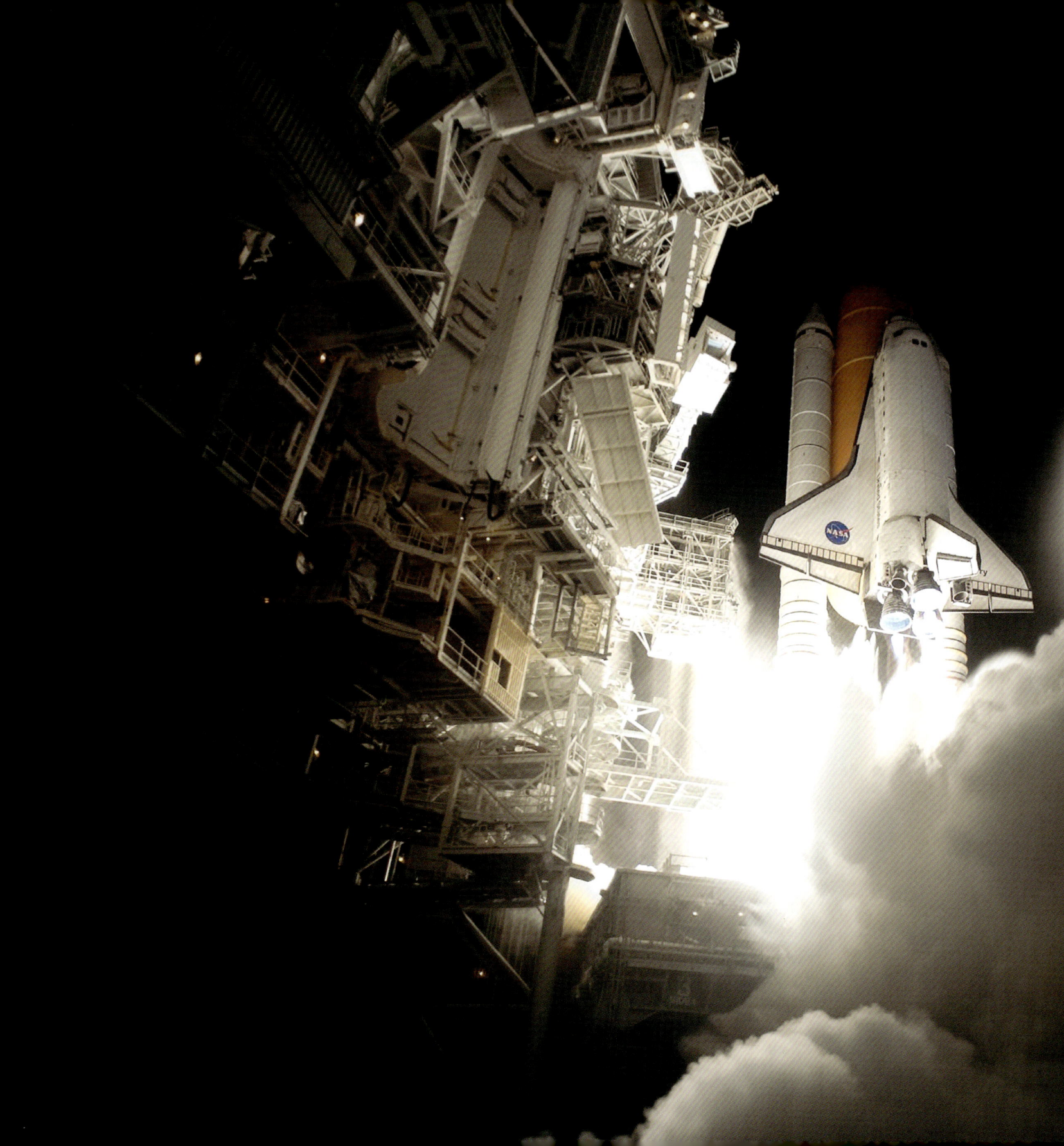

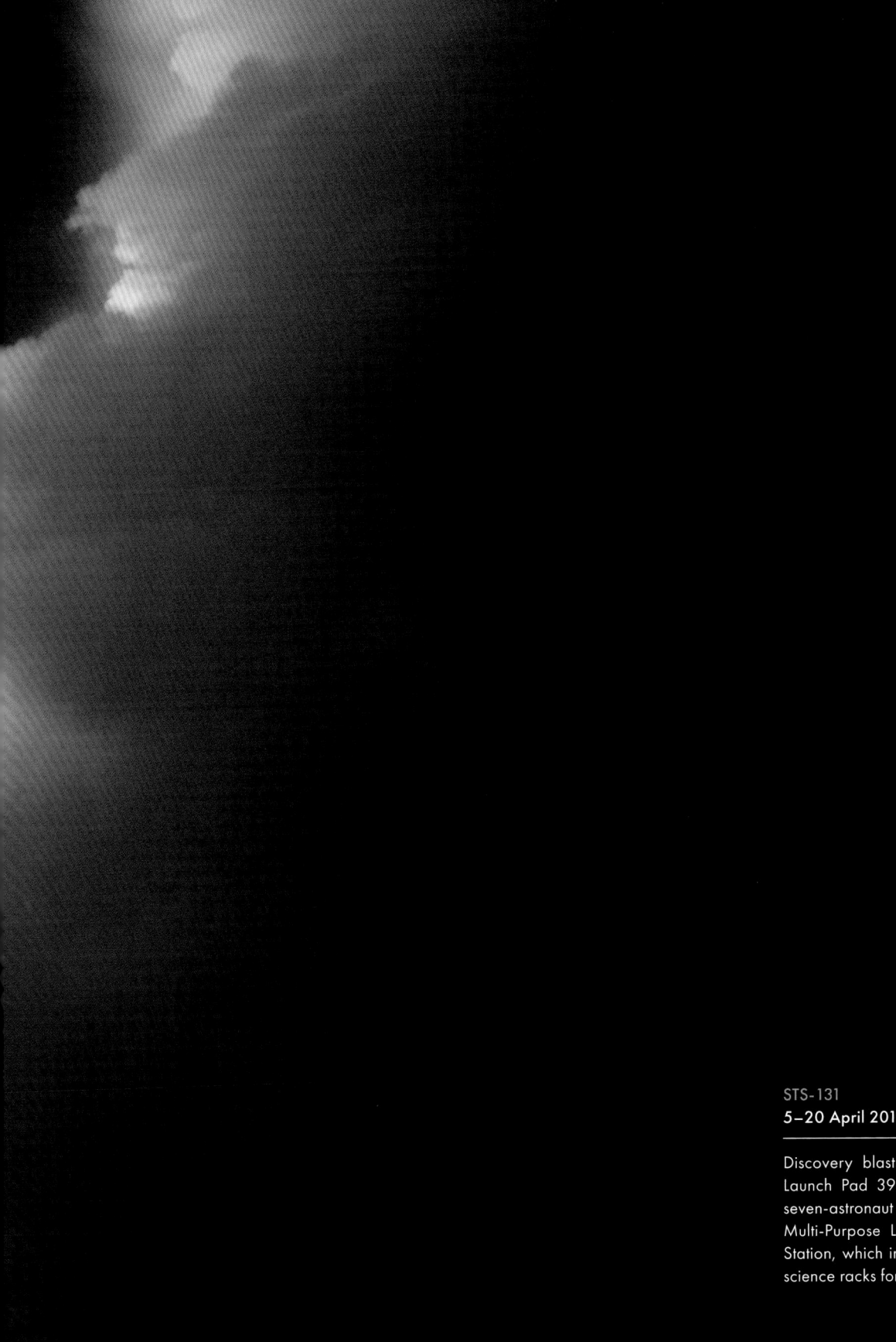

STS-131
5–20 April 2010

Discovery blasts off into the early morning sky, engulfing Launch Pad 39A at Kennedy Space Center in smoke. The seven-astronaut crew will deliver supplies and the Leonardo Multi-Purpose Logistics Module to the International Space Station, which includes new crew sleeping quarters and new science racks for the station's laboratories.

STS-1
12–14 April 1981

(Right) John W. Young and Robert L. Crippen are just seconds into the very first space shuttle launch. The mission will last over two days, complete 37 revolutions of Earth and travel a distance of over one million miles.

STS-129
16–27 November 2009

(Following spread) Atlantis emerges valiantly through a blazing mountain of smoke after a successful launch from Kennedy Space Center. The eleven-day mission featured three space walks and focused on providing new equipment to the International Space Station including a spare gyroscope.

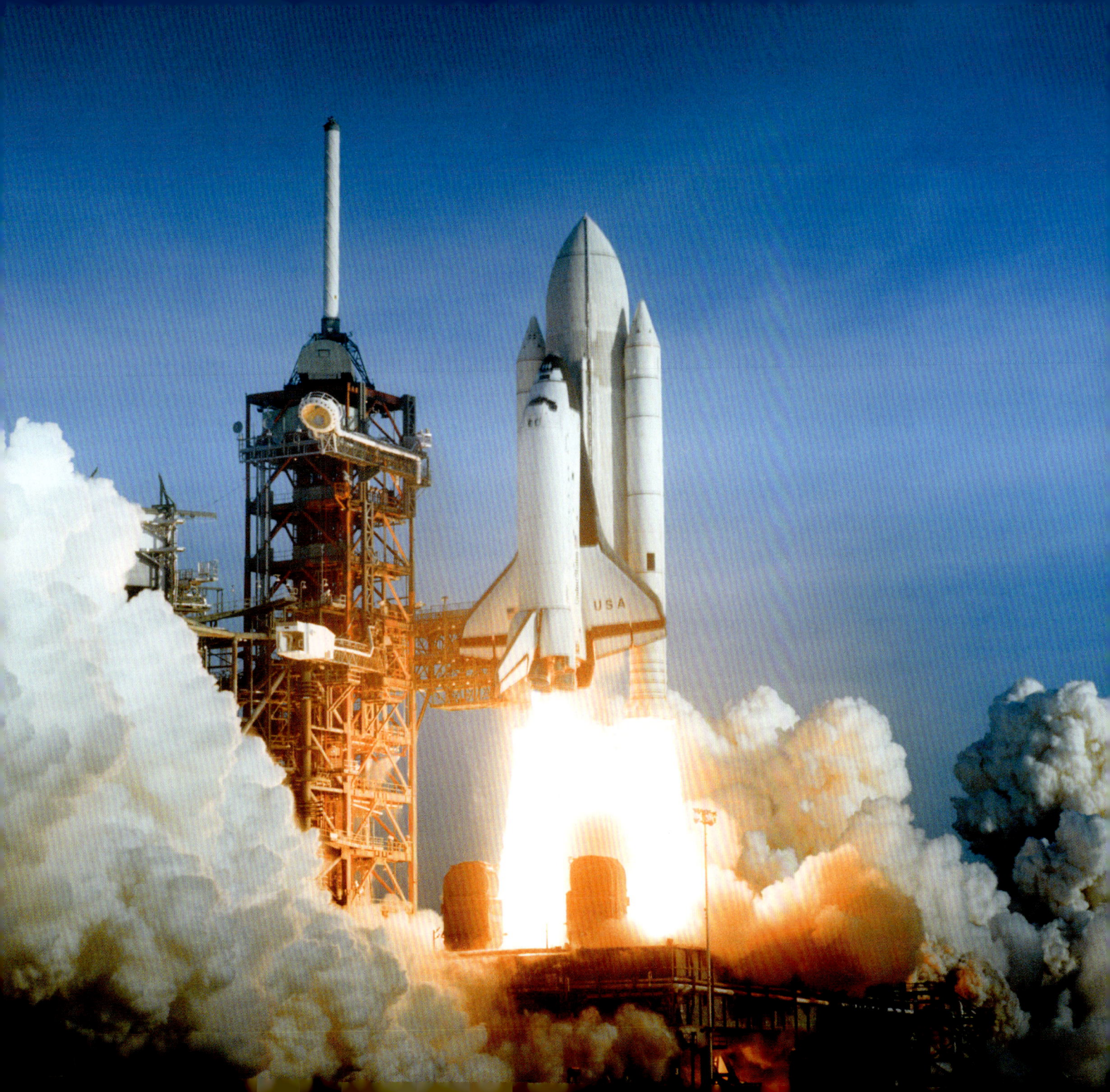

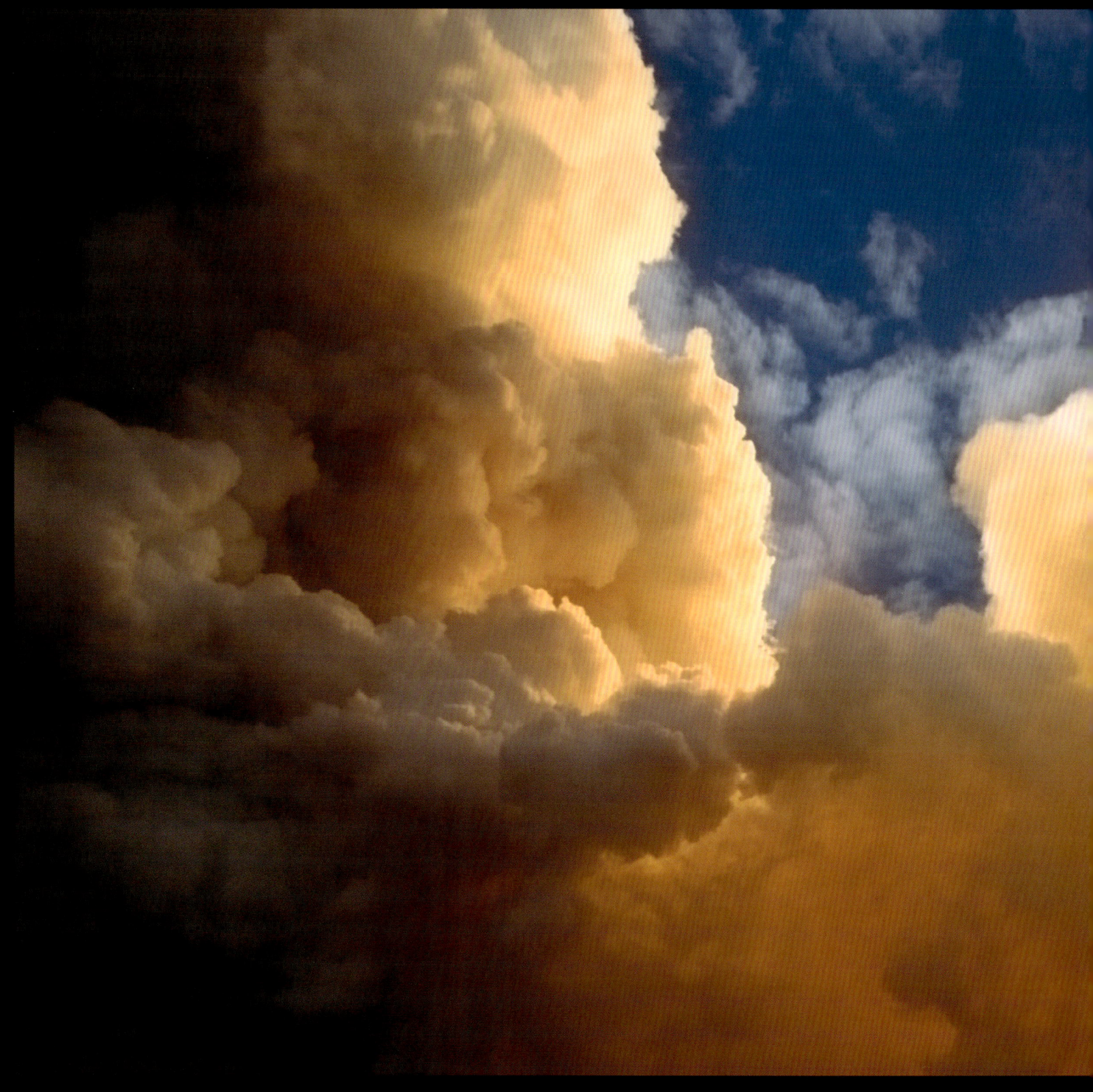

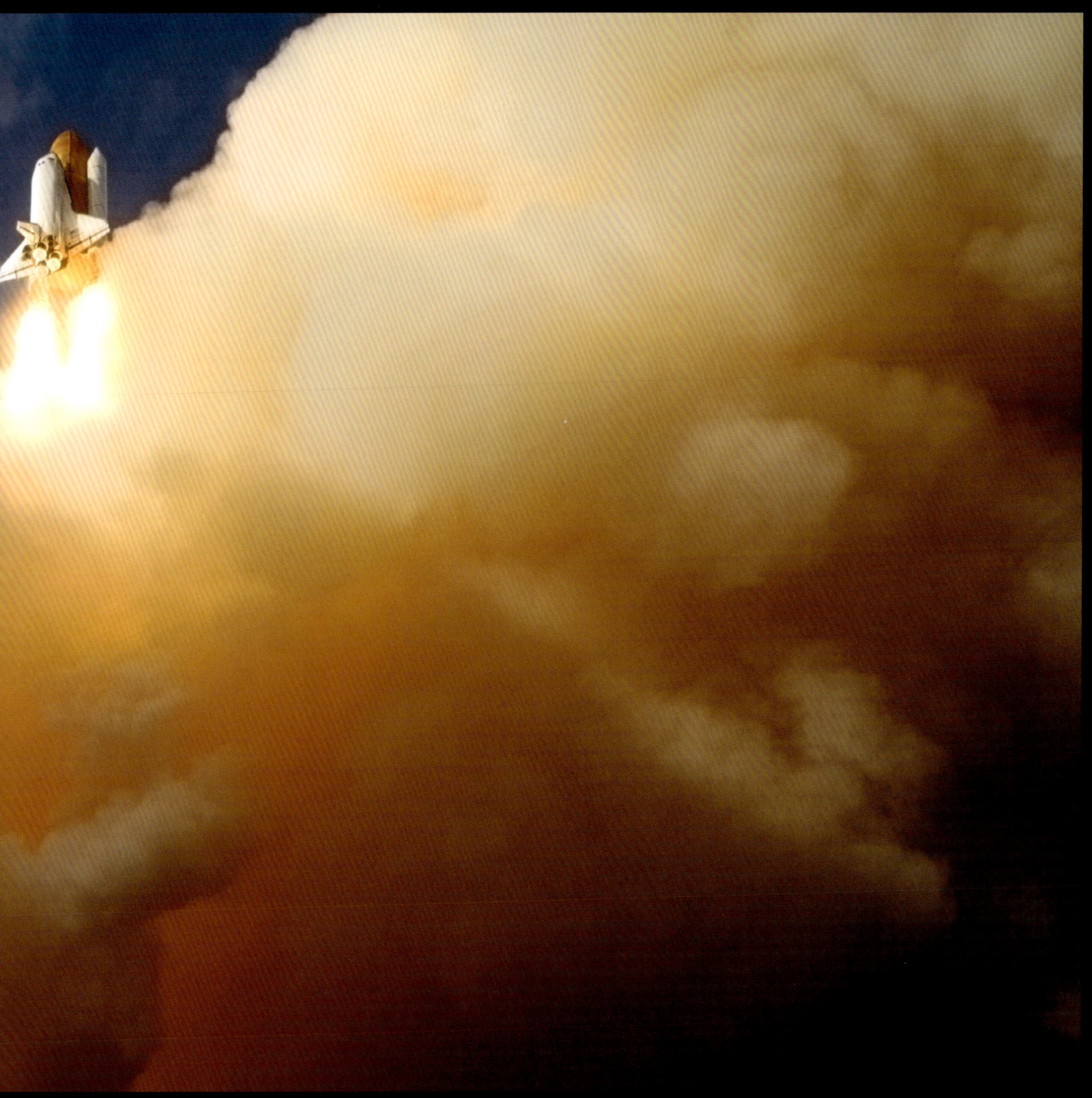

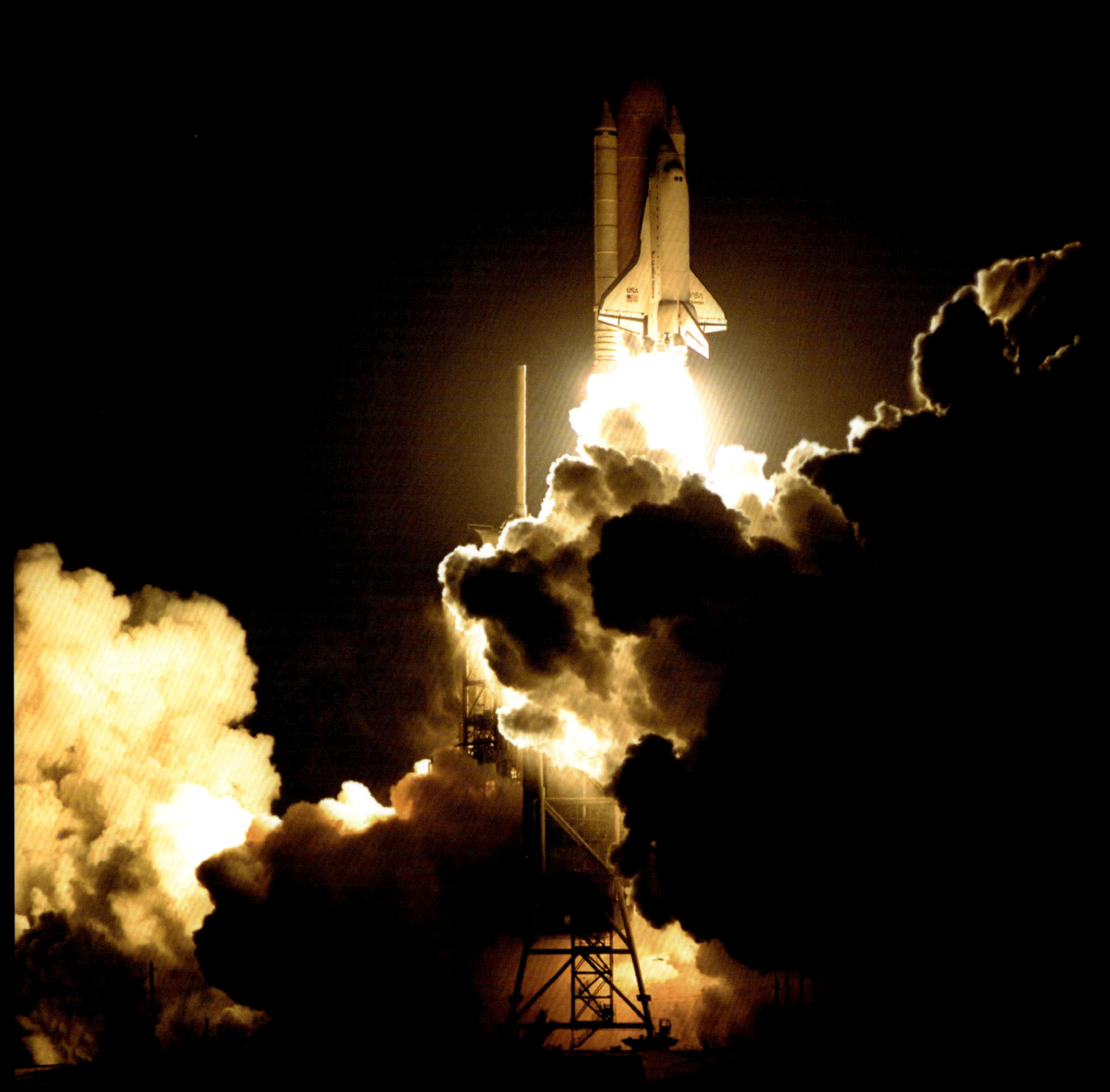

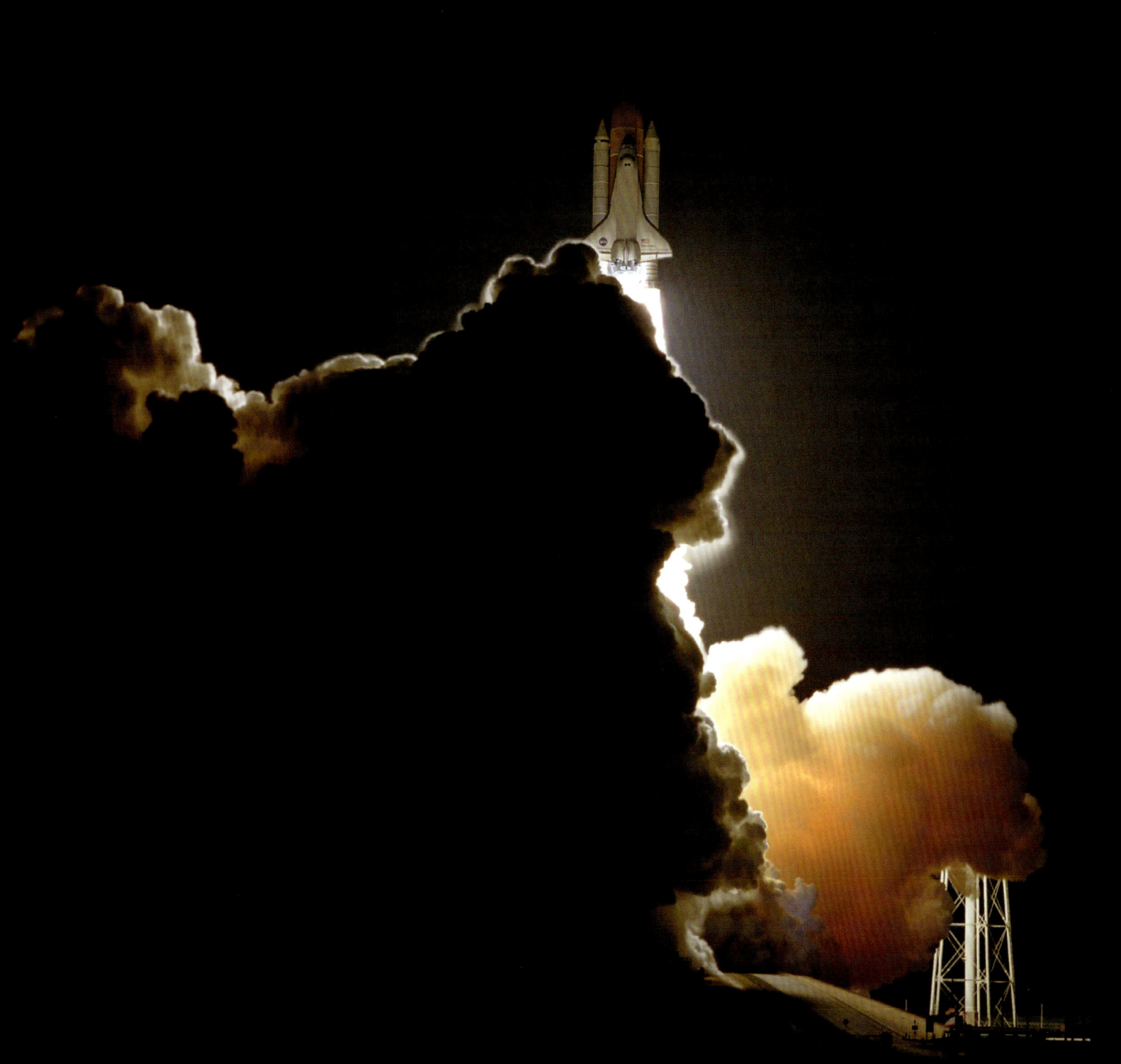

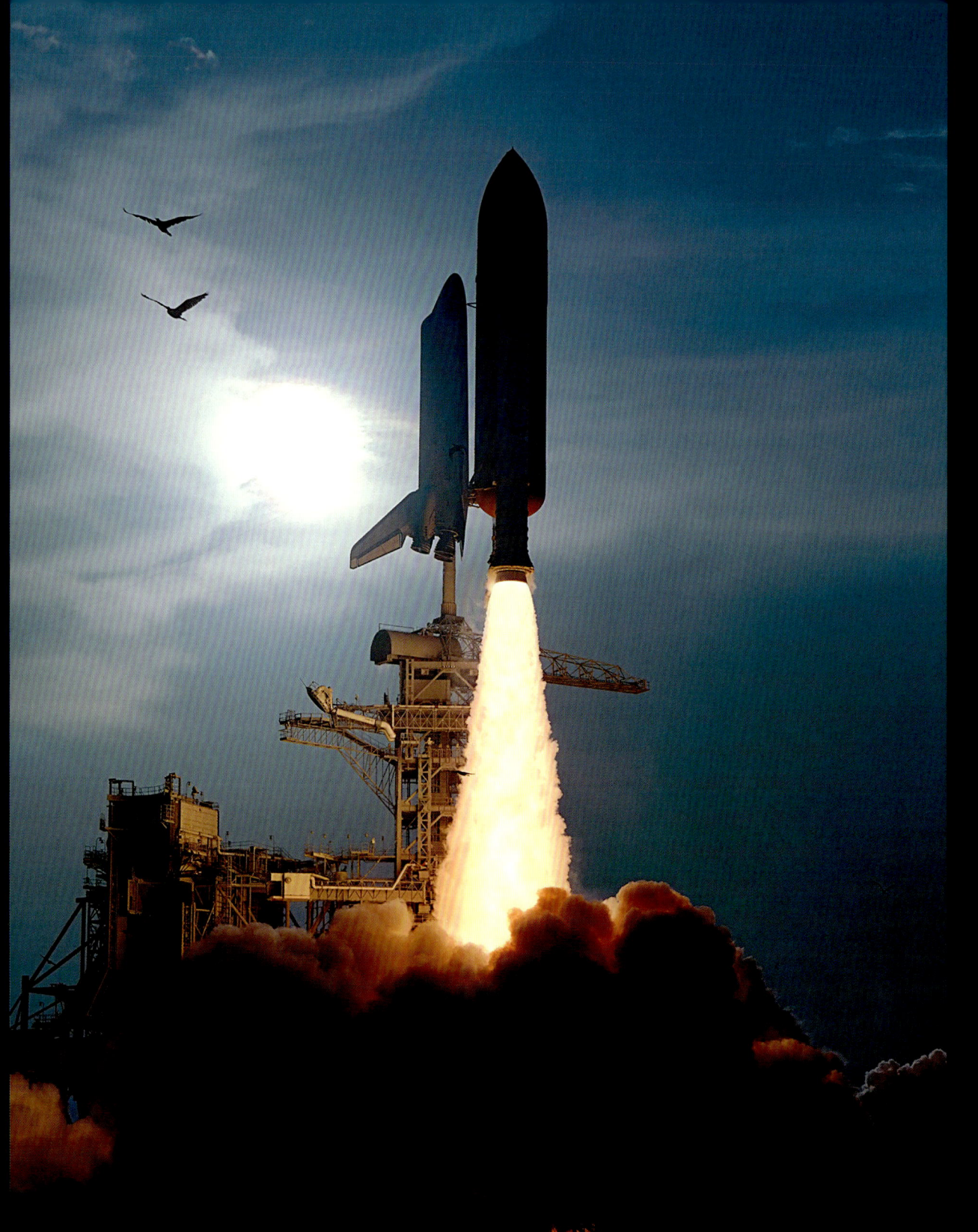

STS-89
22–31 January 1998

(Previous spread: Left) Endeavour goes for glory as it rips apart the calm night sky, en route to the Russian Mir space station. This marks the first time another space shuttle, other than Atlantis, will dock with Mir in Earth orbit.

STS-116
9–22 December 2006

(Previous spread: Right) Space shuttle Discovery, with its seven-astronaut crew, blasts off into the black night sky on its way to the International Space Station. This was the first night-time shuttle launch since 2002.

STS-64
9–20 September 1994

(Left) Seagulls receive the shock of their lives as Discovery's solid rocket boosters ignite in the early evening Florida sky, to start the 64th space shuttle mission and the 19th space flight for Discovery.

An intense fireball of light trailing from Endeavour's solid rocket boosters illuminates Merritt Island, as the eight-person crew courageously rocket into space with the Moon observing from afar. This mission will consist of four space walks and prepare the International Space Station to accommodate six crew members for longer duration stays.

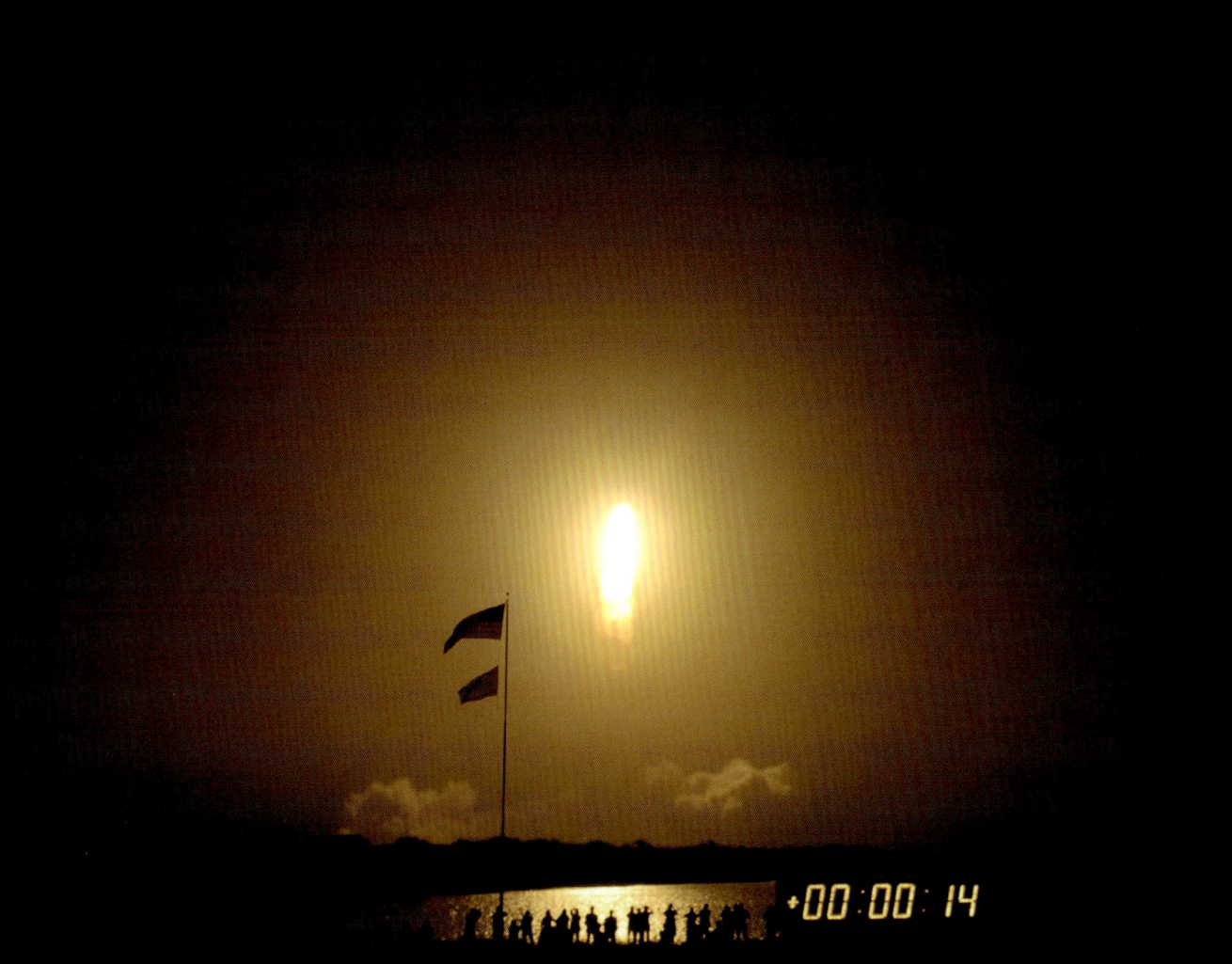

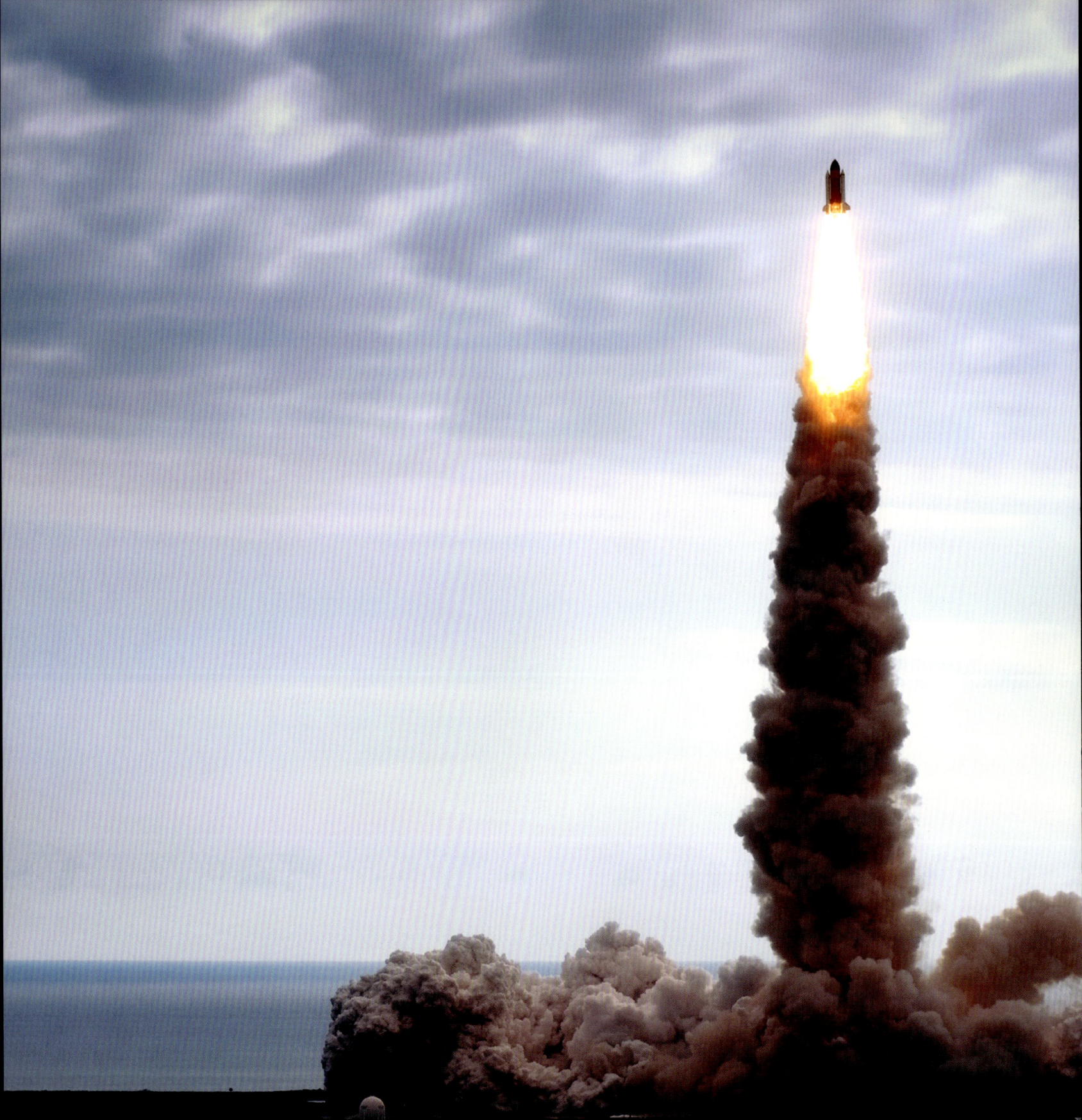

Endeavour blasts into the early morning Florida sky to begin its final mission. Its main objectives are to deliver supplies and equipment to the International Space Station before retirement, including the Alpha Magnetic Spectrometer (AMS) and the Express Logistics Carrier-3. The AMS Module is designed to measure antimatter and search for evidence of dark matter, helping scientists to gain a greater understanding of the origins of the universe.

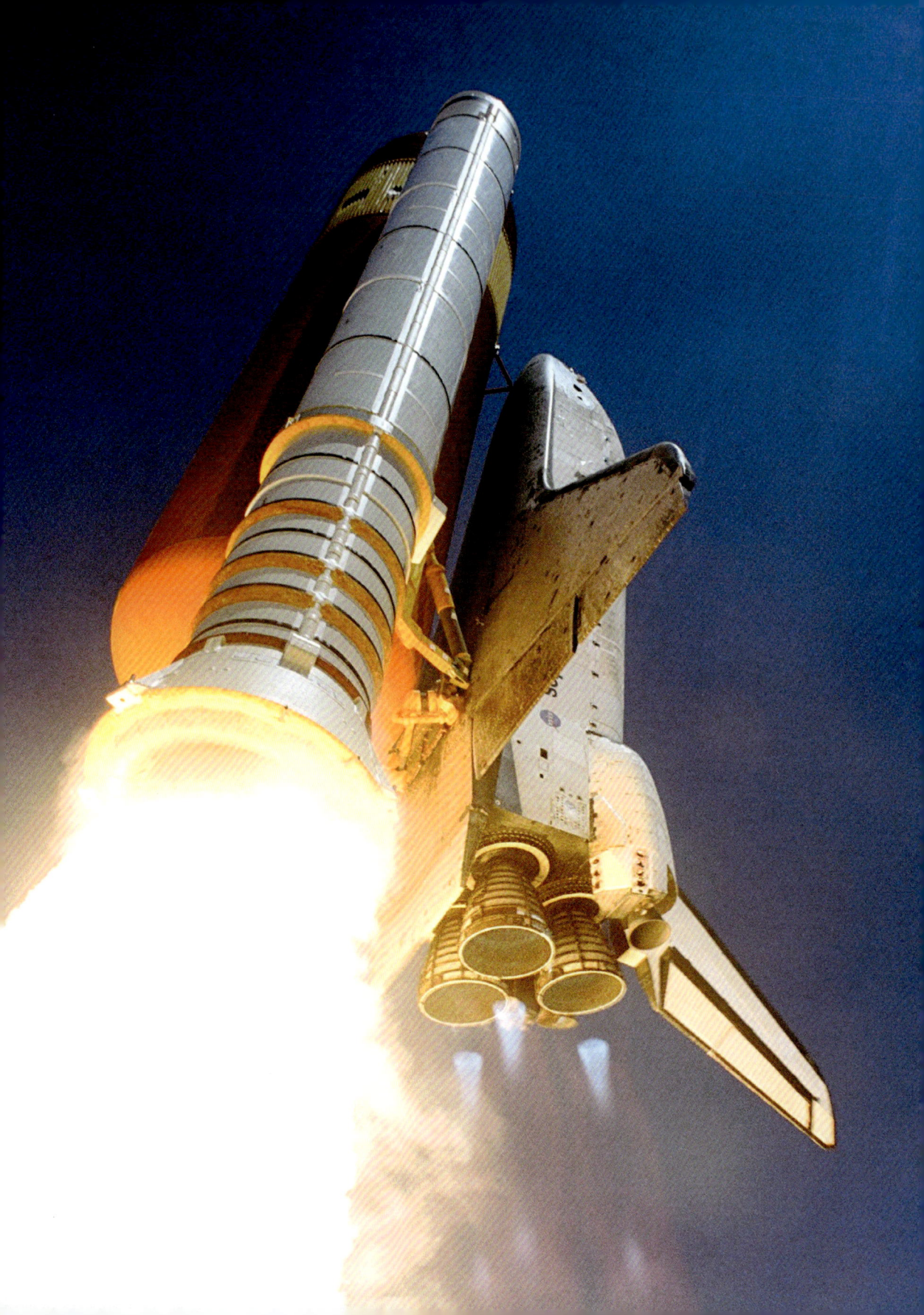

History in the making as Discovery becomes the first piloted spacecraft to launch into space on American Independence Day. As well as providing supplies and repair work to the International Space Station, new equipment and procedures will be tested to improve and verify the shuttle's safety, following the Columbia disaster in February 2003.

16–27 November 2009

The incredible thrust of the solid rocket boosters, along with the shuttle's three main engines, propels Atlantis into the clear, blue Florida sky on its way to the International Space Station. The mission consisted of three Extra Vehicular Activity (EVA) space walks and delivered many large spare components to the International Space Station before the imminent closure of the space shuttle program.

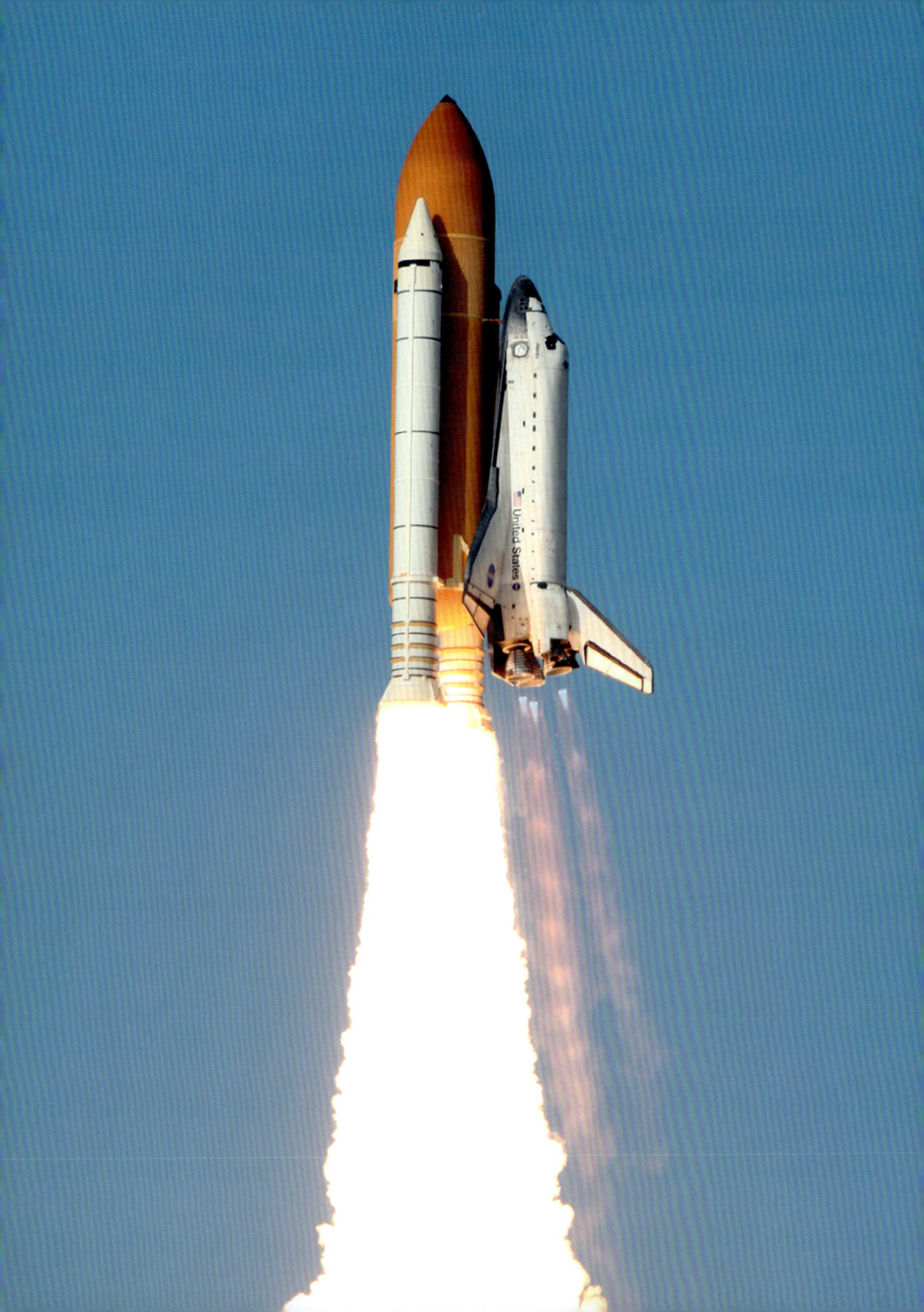

27 May to 6 June 1999

Sunrise at the Banana Creek viewing site, and Discovery sets off for space. Discovery will deliver approximately 4,000 lbs of supplies to the International Space Station for use by future crew members.

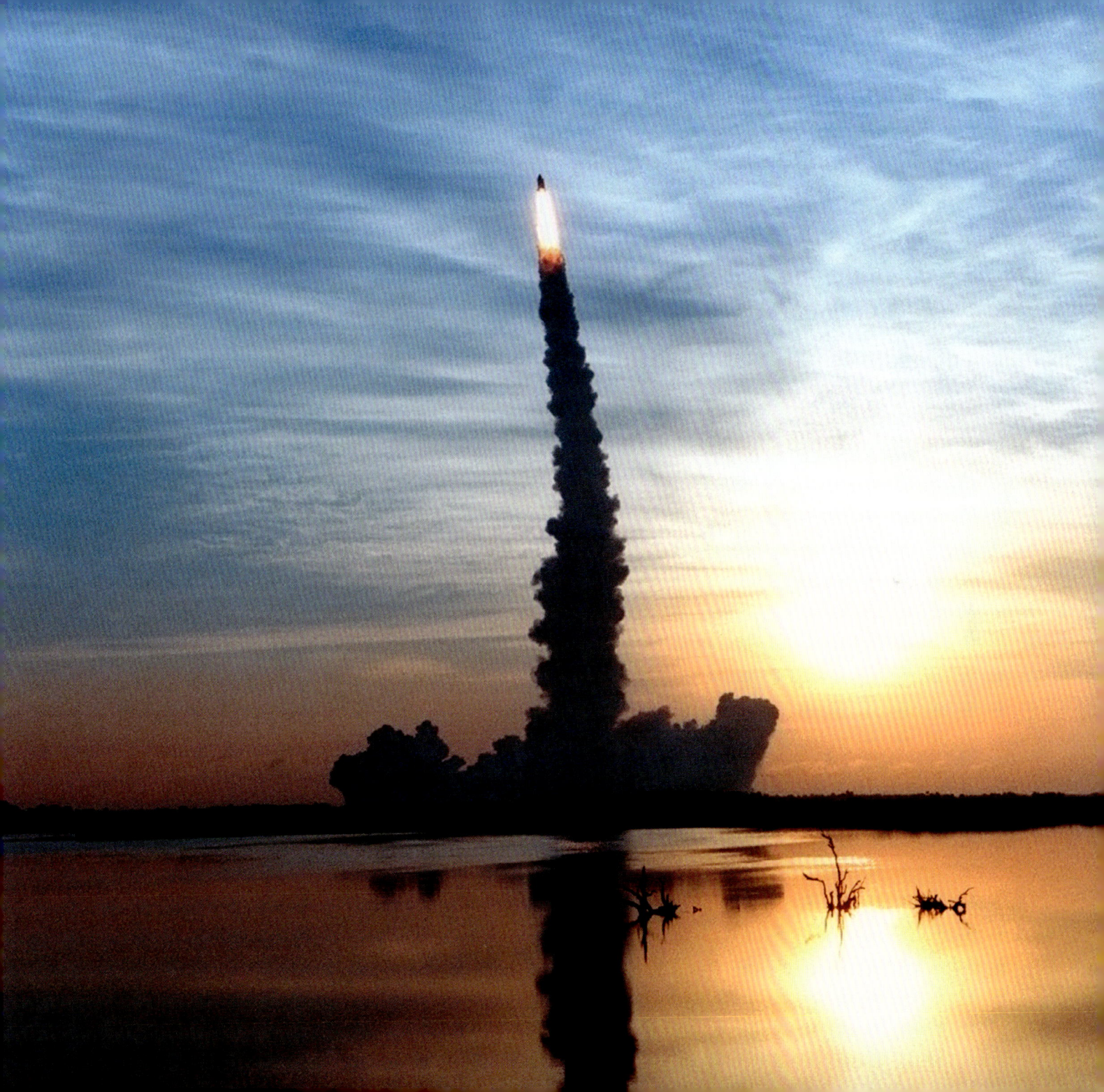

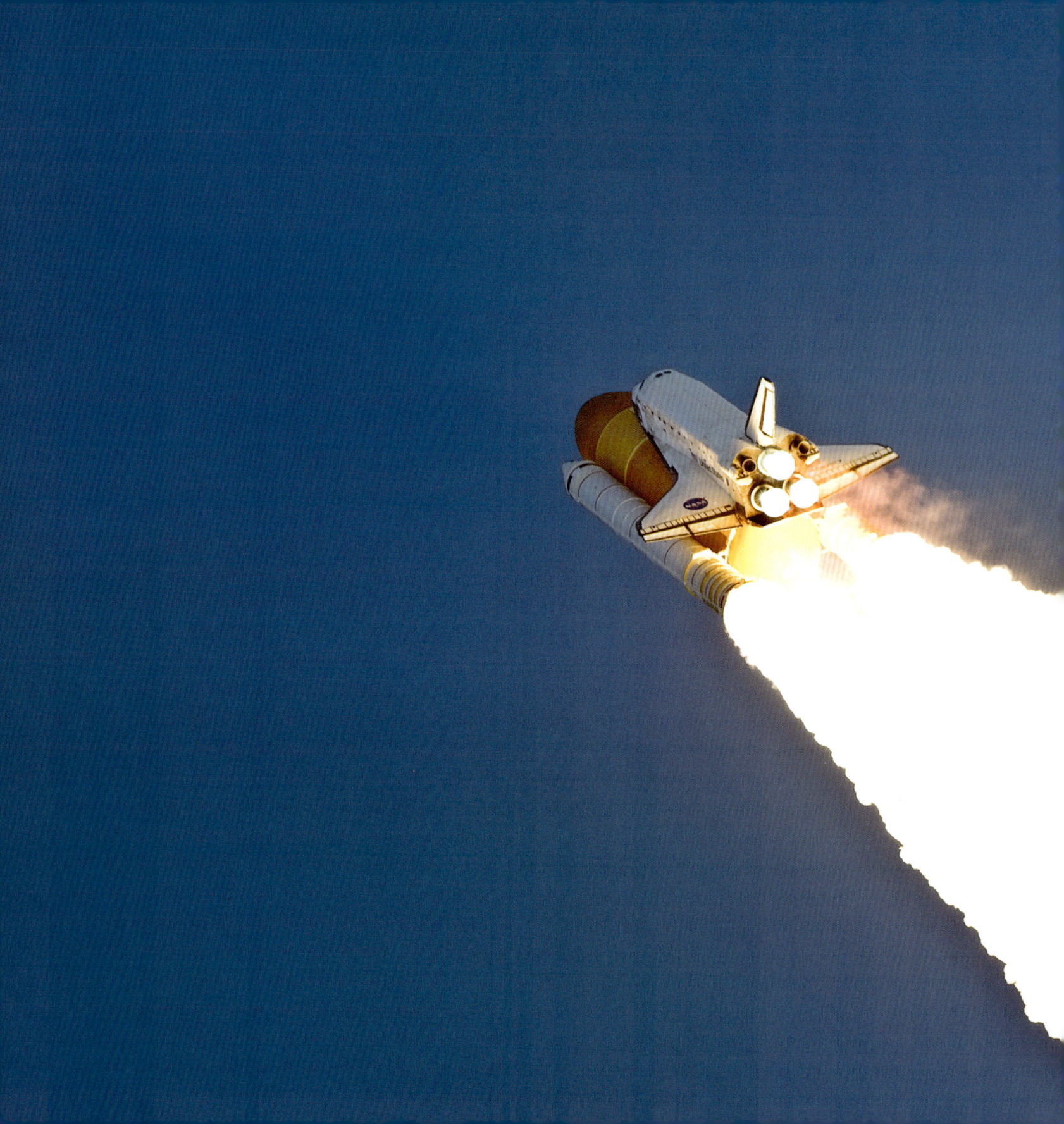

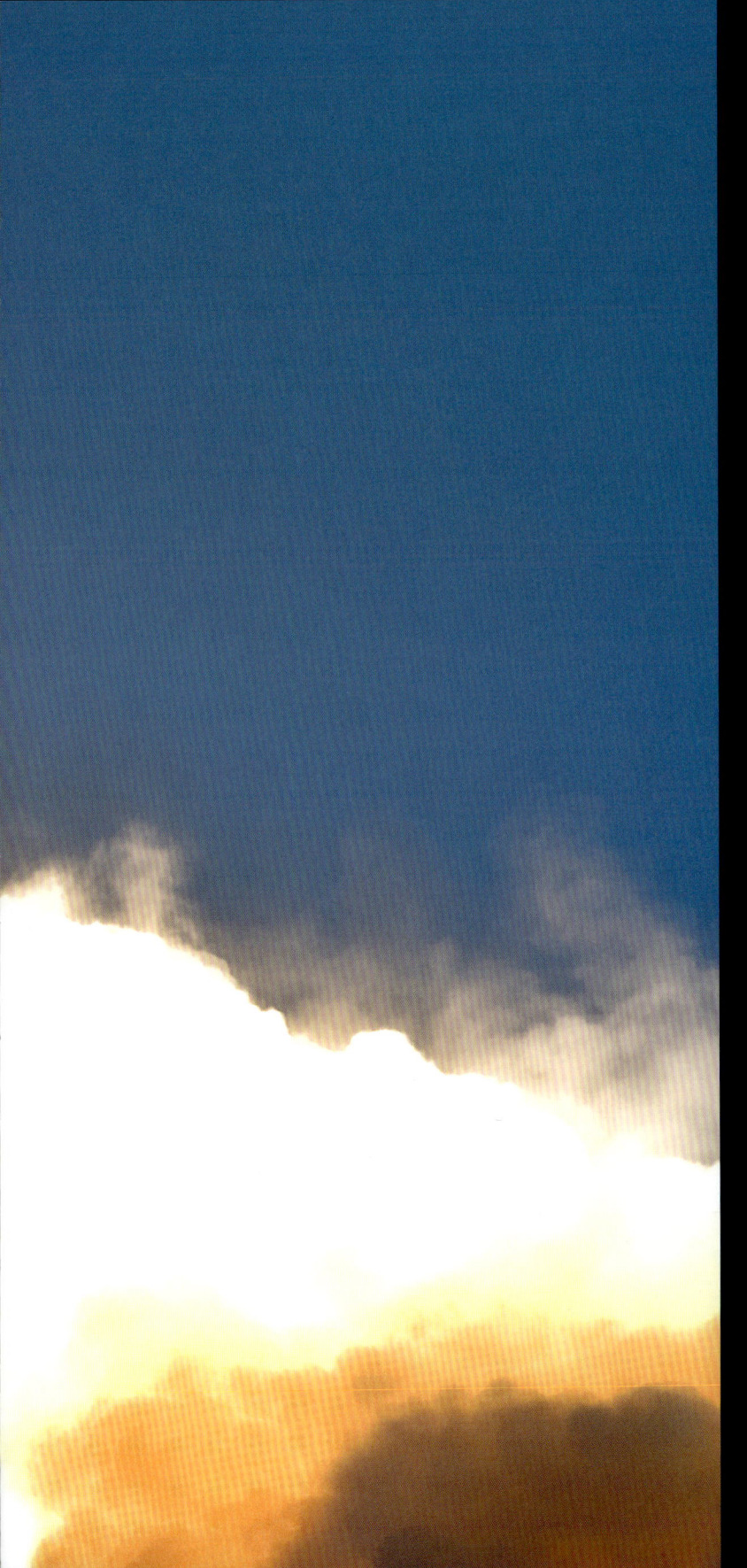

31 May to 14 June 2008

Discovery lifts off from Launch Pad 39A and rockets towards the heavens. The shuttle and her crew delivered the Pressurized Module for the Japanese Experiment Module (JEM) nicknamed Kibo to the International Space Station. The Japanese Remote Manipulator System robotic arm was also delivered and attached to Kibo, as well as replacement parts for a faulty toilet on board the space station.

STS-107
16 January to 1 February 2003

(Right) Columbia rips its way through Earth's blue sky in a triumphant and defiant pose as the white, slender solid rocket boosters and three main orbiter engines leave behind a fiery trail. The mission ended in disaster when the orbiter disintegrated during re-entry into Earth's atmosphere.

STS-2
12–14 November 1981

(Following spread: Left) This photograph, taken by legendary astronaut John W. Young while aboard the shuttle training aircraft, shows an aerial view of Columbia racing towards space and embarking on its second space flight, which lasted just over two days.

STS-7
18–24 June 1983

(Following spread: Right) Space shuttle Challenger soars through the clouds on its journey to Earth orbit, with Robert L. Crippen in command. This photograph was taken from the cockpit of the NASA shuttle training aircraft.

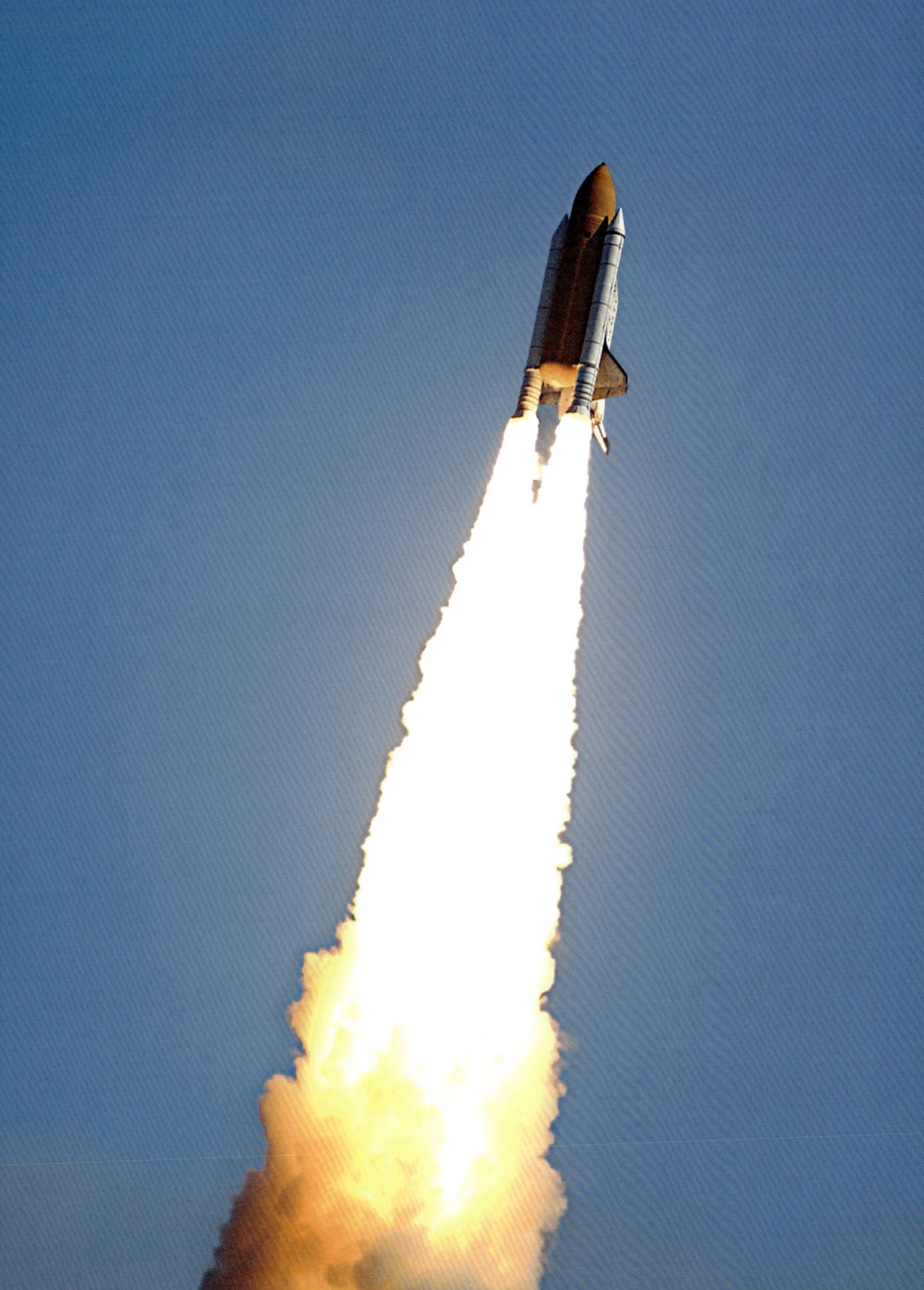

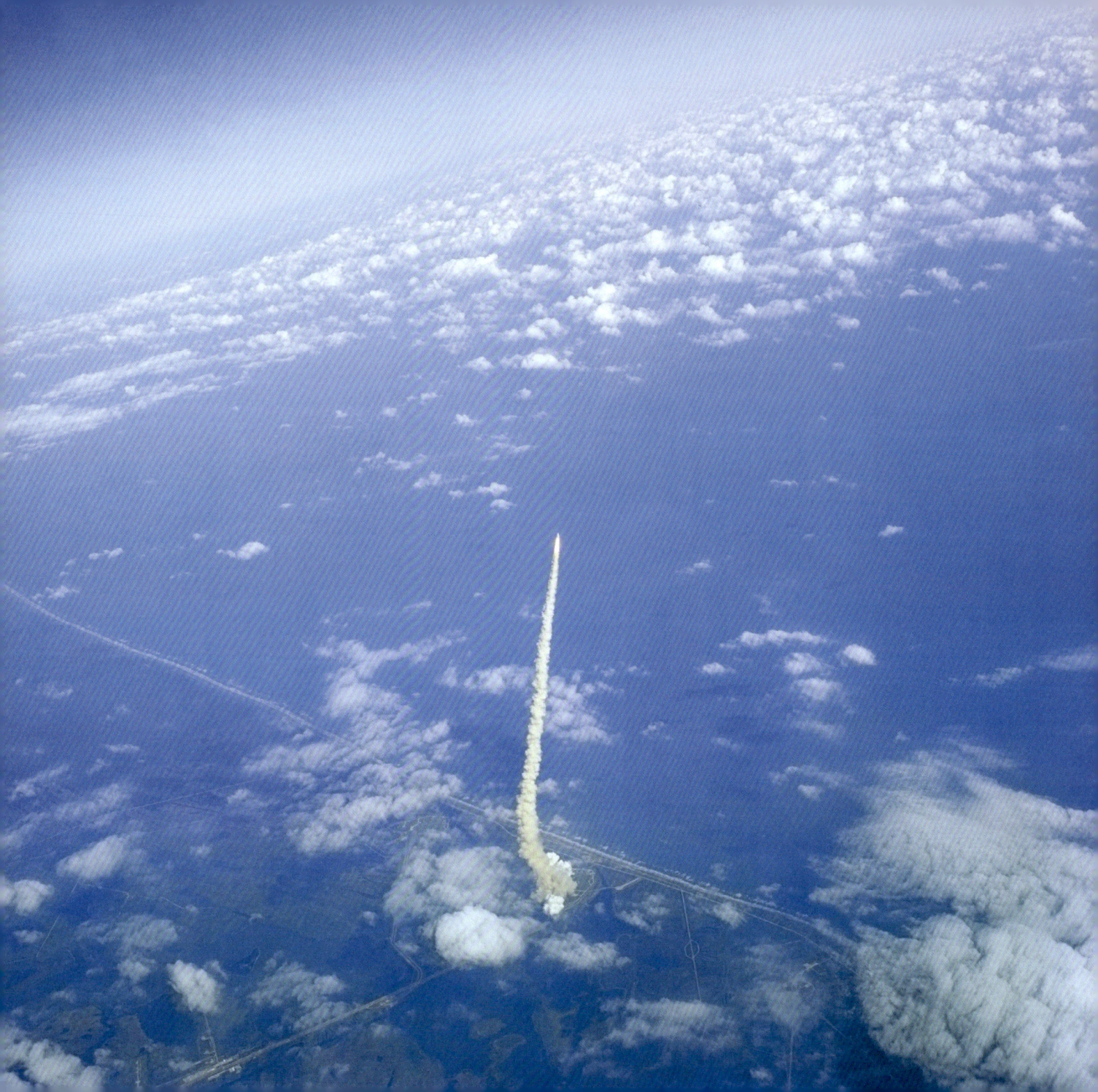

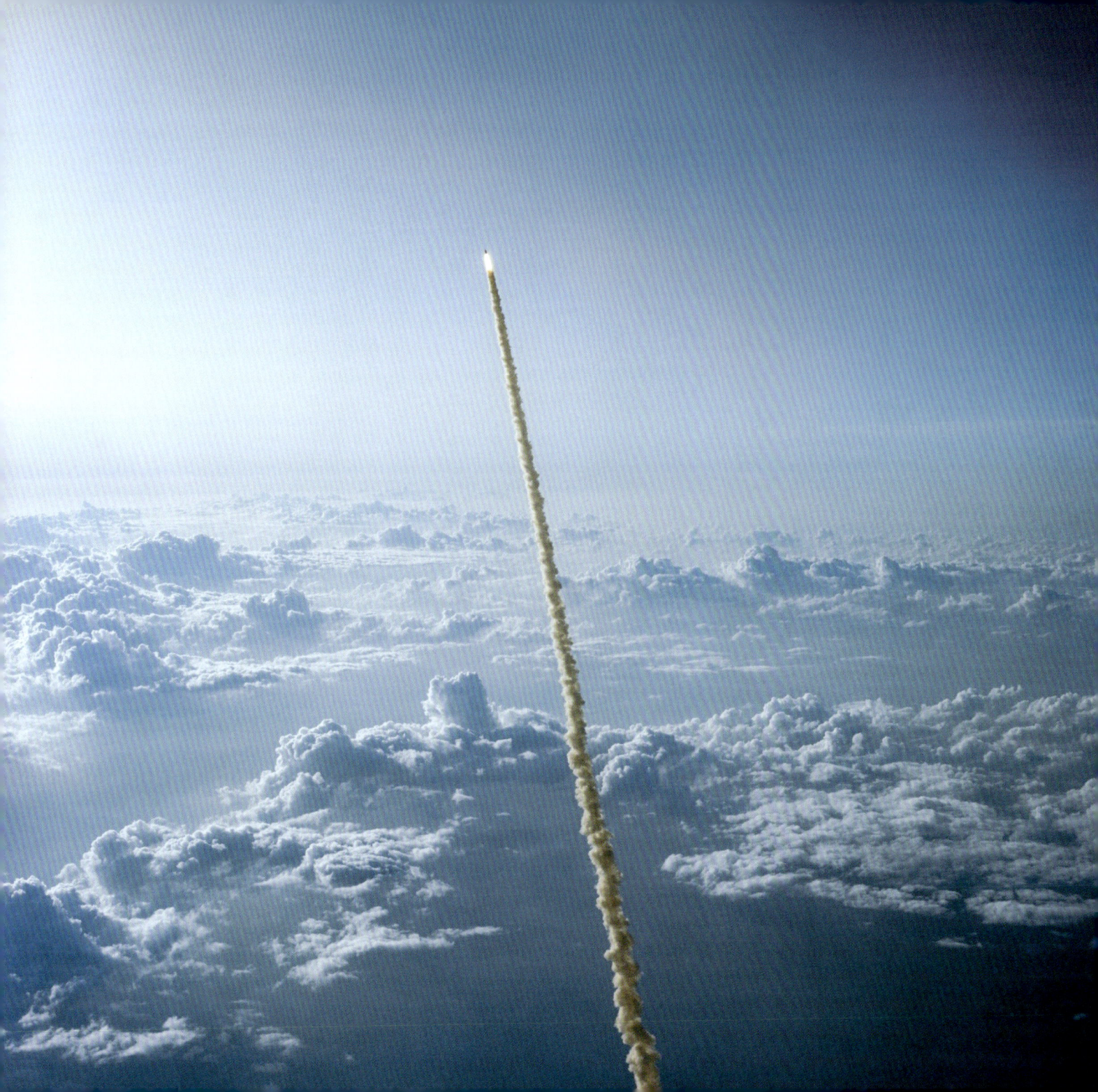

The four-astronaut crew of space shuttle Atlantis rockets high into the Florida sky for a scheduled link-up with the International Space Station. The shuttle will take approximately 8 minutes to achieve orbit, while reaching speeds of 29,000 km/h (18,000 mph). The mission objectives are to deliver supplies and equipment to the space station, including the Raffaello Multi-Purpose Logistics Module and the Lightweight Multi-Purpose Carrier.

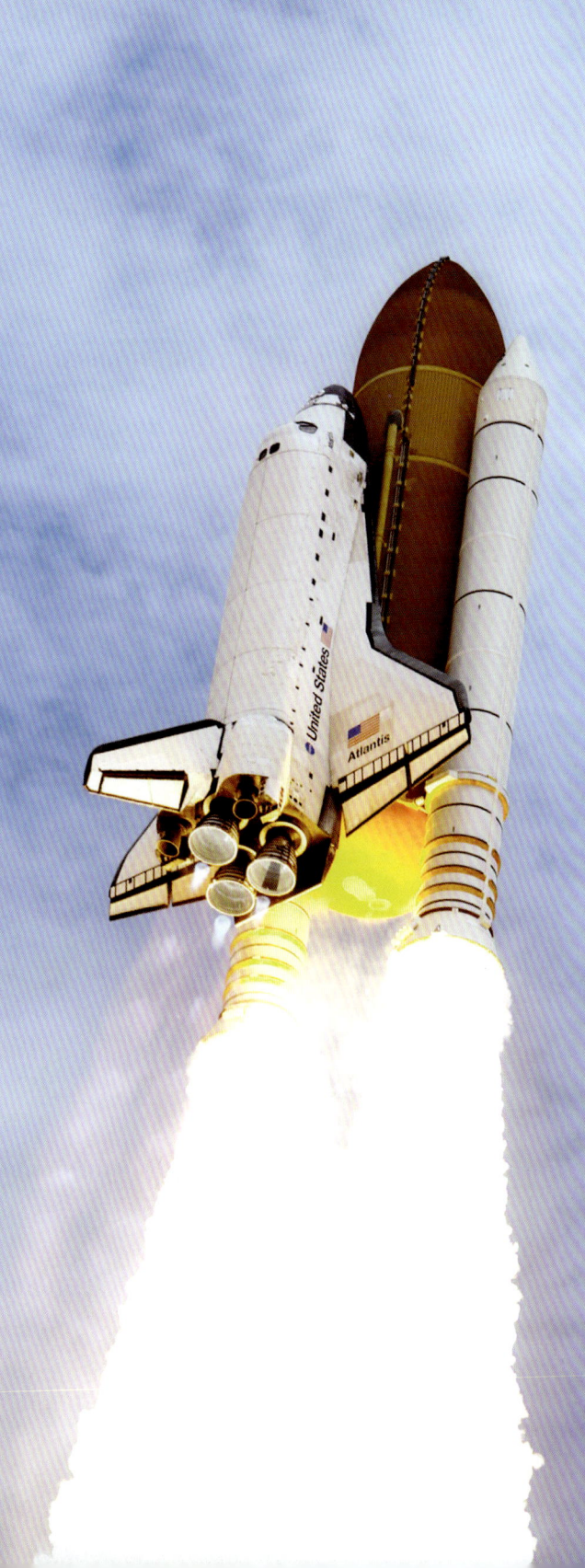

3–11 February 1984

Challenger blasts its way over Florida towards the deep-blue sky above, leaving a perfectly straight trail of smoke in its path. This was the tenth space shuttle mission in total and the fourth for Challenger. The mission was largely celebrated due to the landmark free-flight space walk by Bruce McCandless II, using the Manned Maneuvering Unit jet pack.

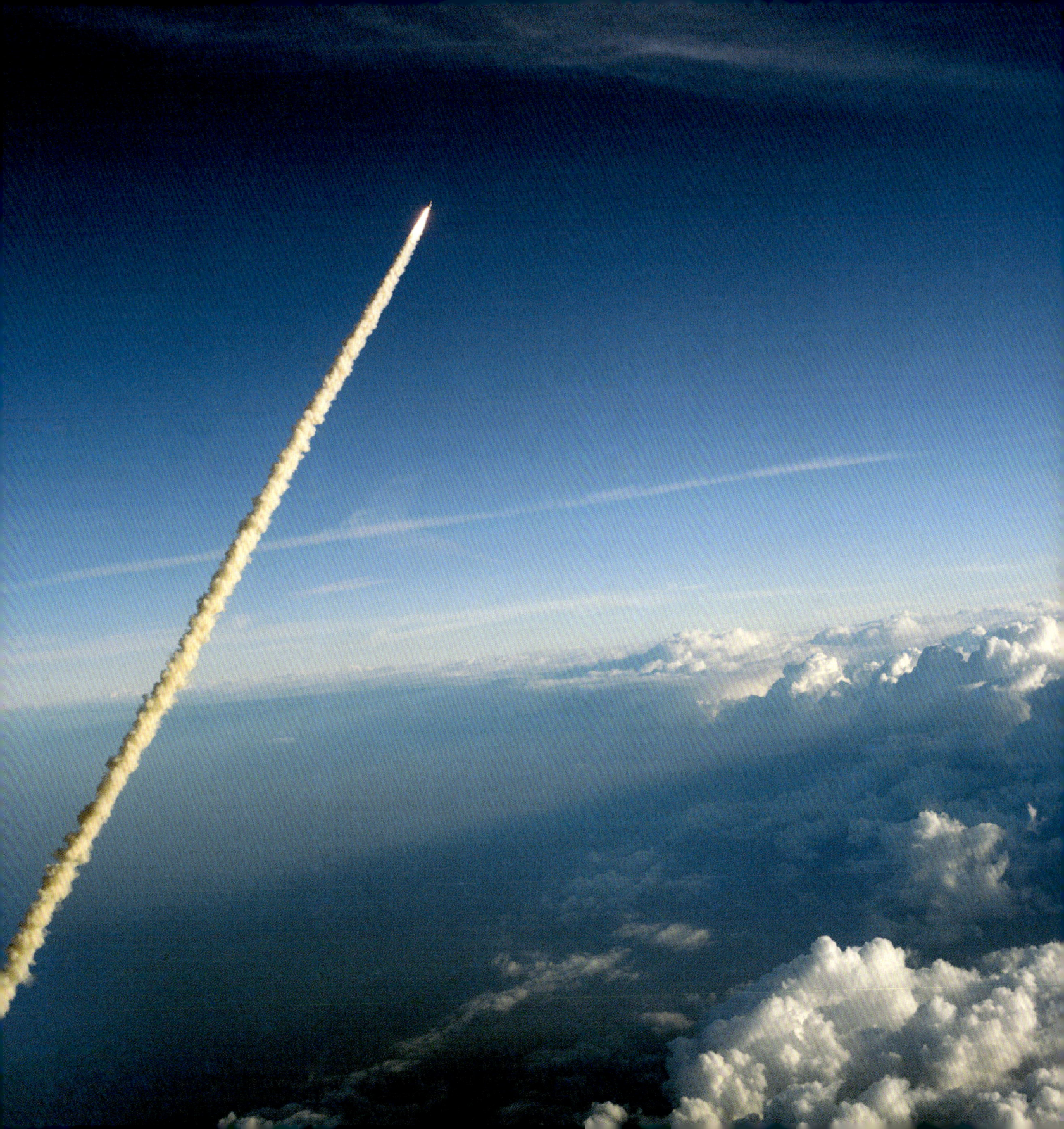

STS-96
27 May to 6 June 1999

(Right) Leaving a twisted trail of smoke in its wake, space shuttle Discovery lifts off over a beautiful summer sunrise in this dramatic photograph.

STS-131
5–20 April 2010

(Following spread: Left) The external tank separates from space shuttle Discovery following a successful launch into space. The external tank will eventually fall back into Earth's atmosphere where it will break up under the extreme temperatures.

STS-29
13–18 March 1989

(Following spread: Right) A distant view of Discovery's external tank falling back to Earth after being jettisoned. The tank carries the liquid hydrogen fuel and liquid oxygen oxidizer needed to propel the orbiter into space.

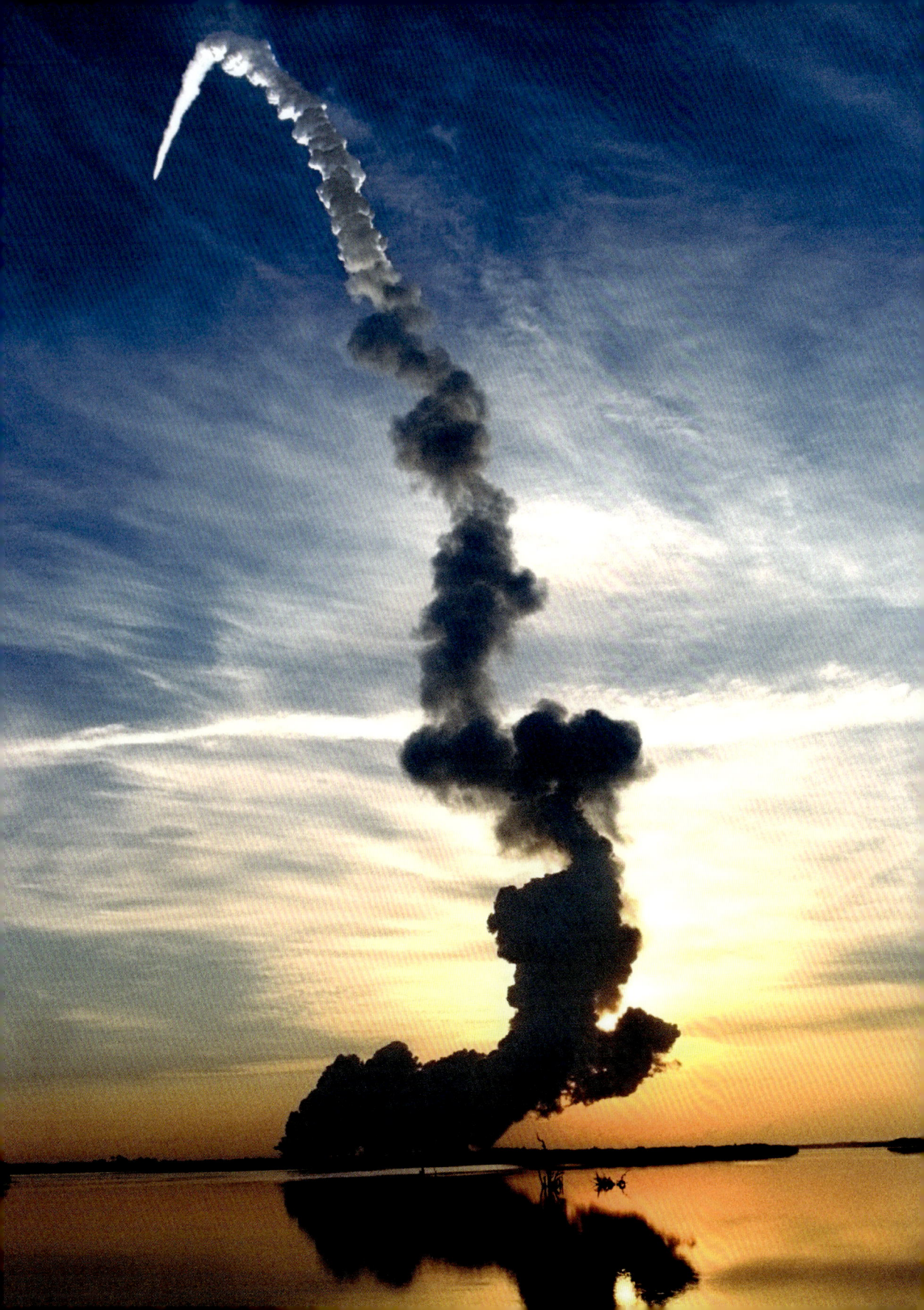

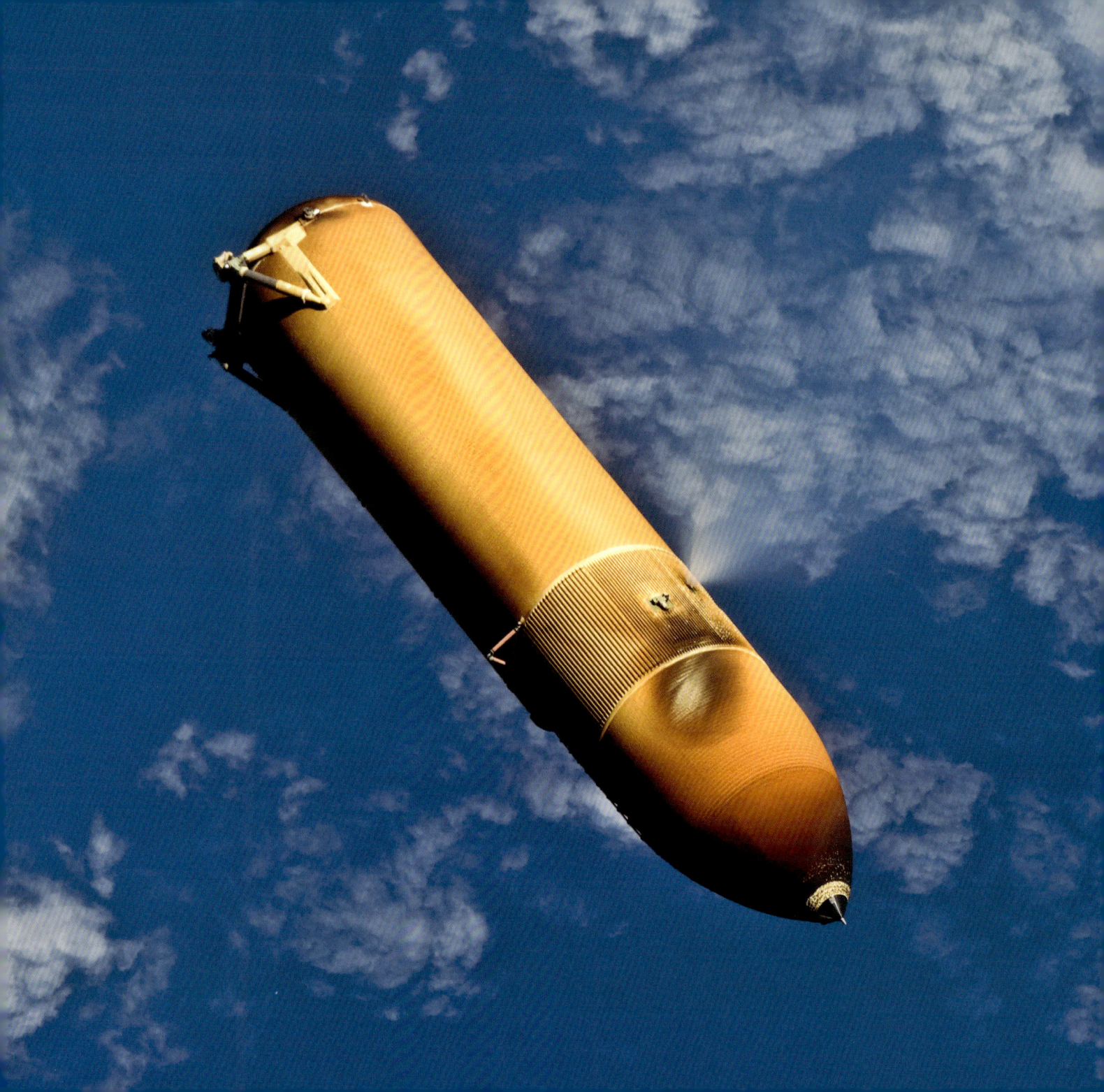

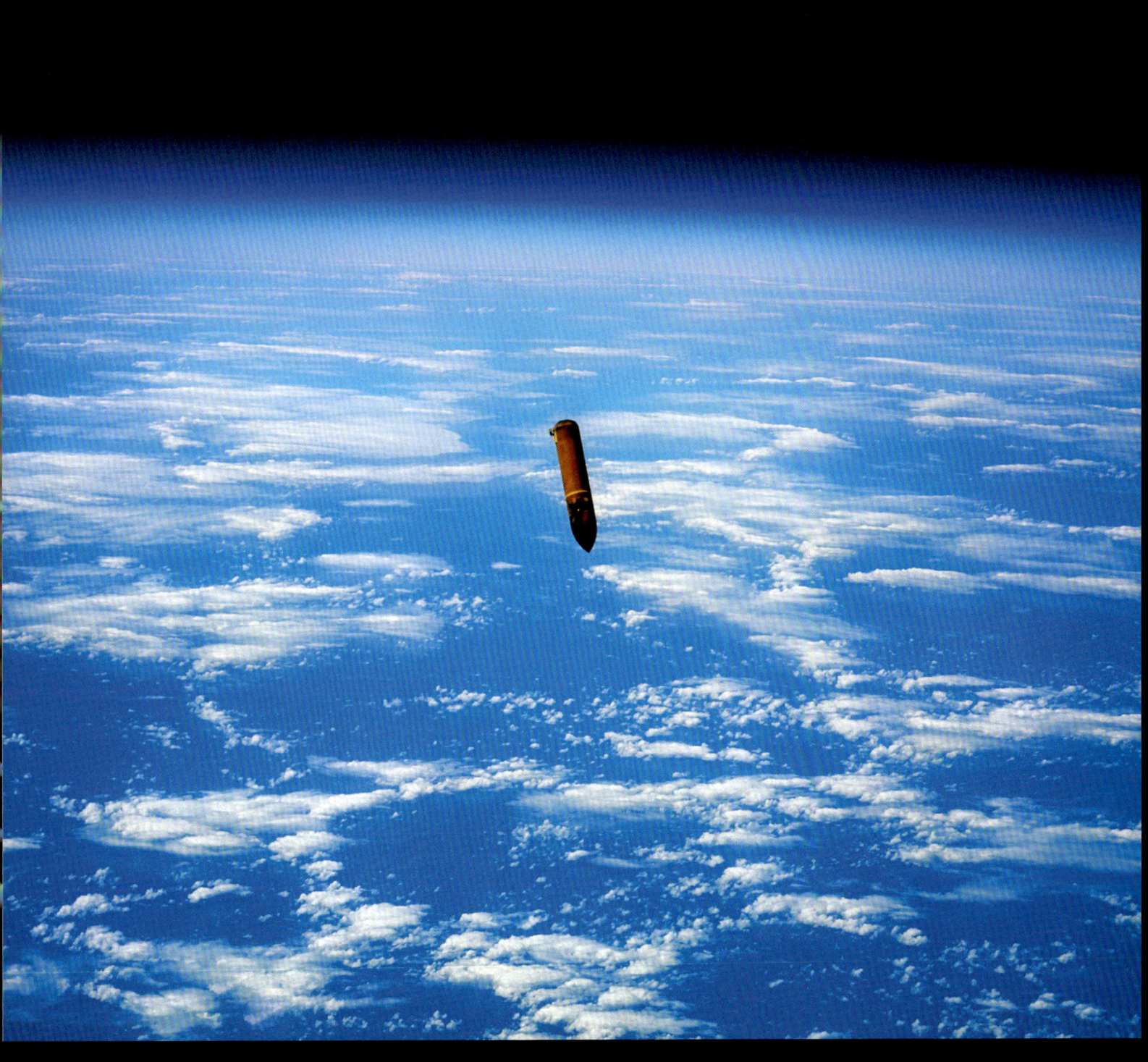

00:00:03

IN SPACE

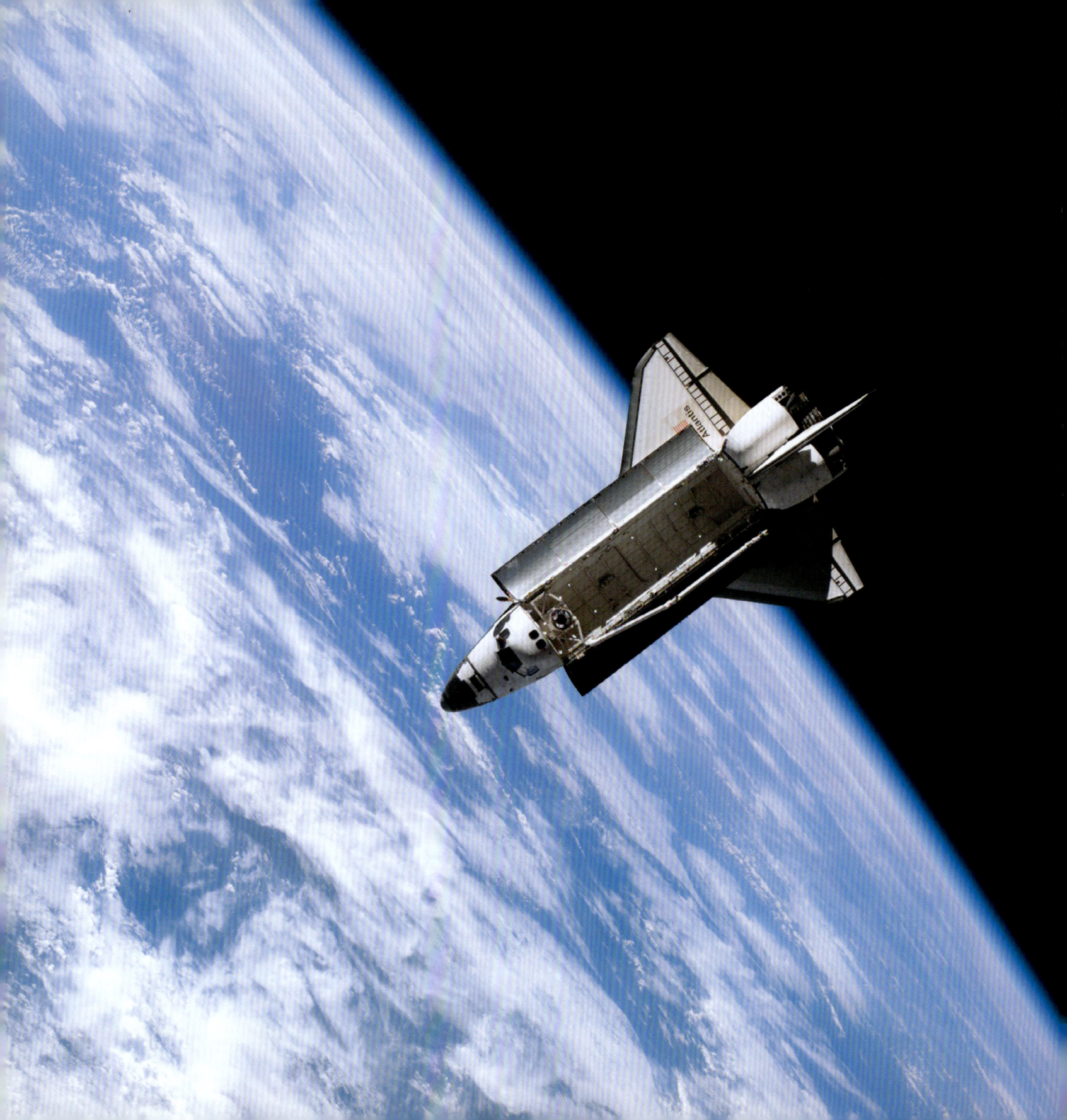

STS-115
9–21 September 2006

After undocking from the International Space Station, Atlantis is photographed beautifully against a cloudy Earth backdrop with the aft creeping into the deep blackness of space. The crew continued assembly work on the space station over a course of three space walks.

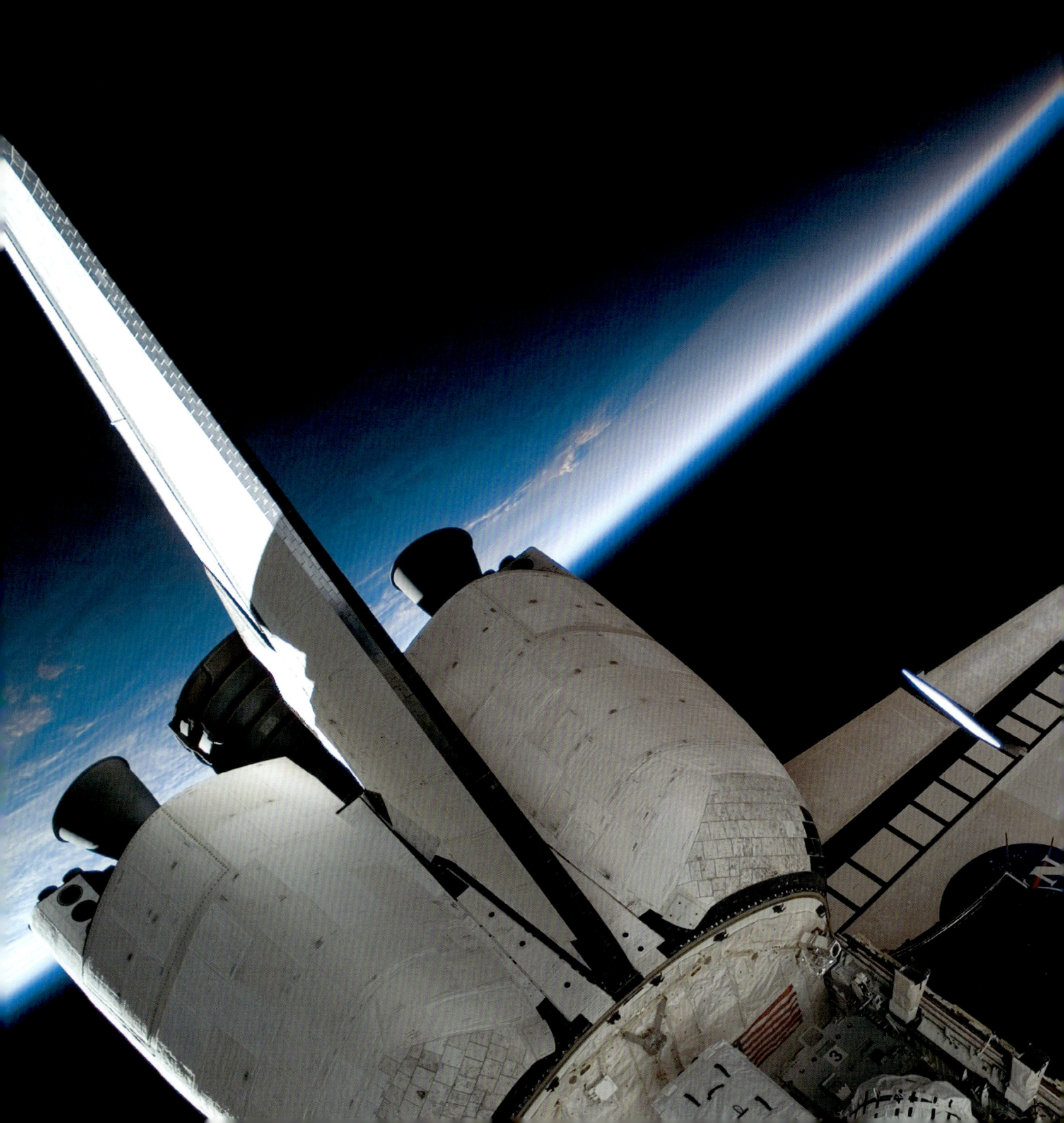

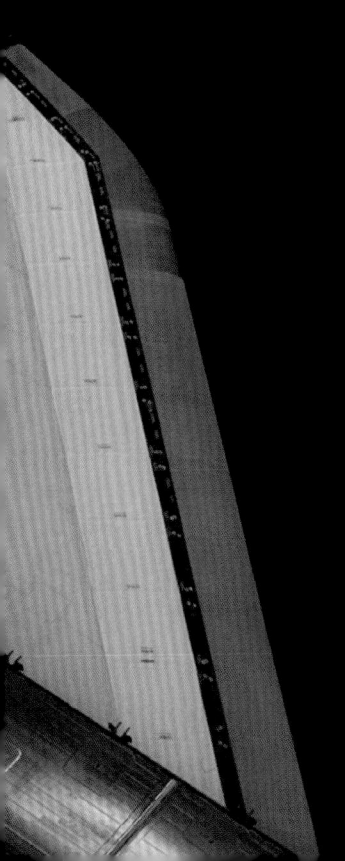

Earth's glowing horizon provides a beautiful backdrop for space shuttle Endeavour while docked with the International Space Station. The shuttle's vertical stabilizer is seen standing proud with the large, white Orbital Maneuvering System (OMS) pods either side. The OMS engines are used to alter the shuttle's orbit in space flight.

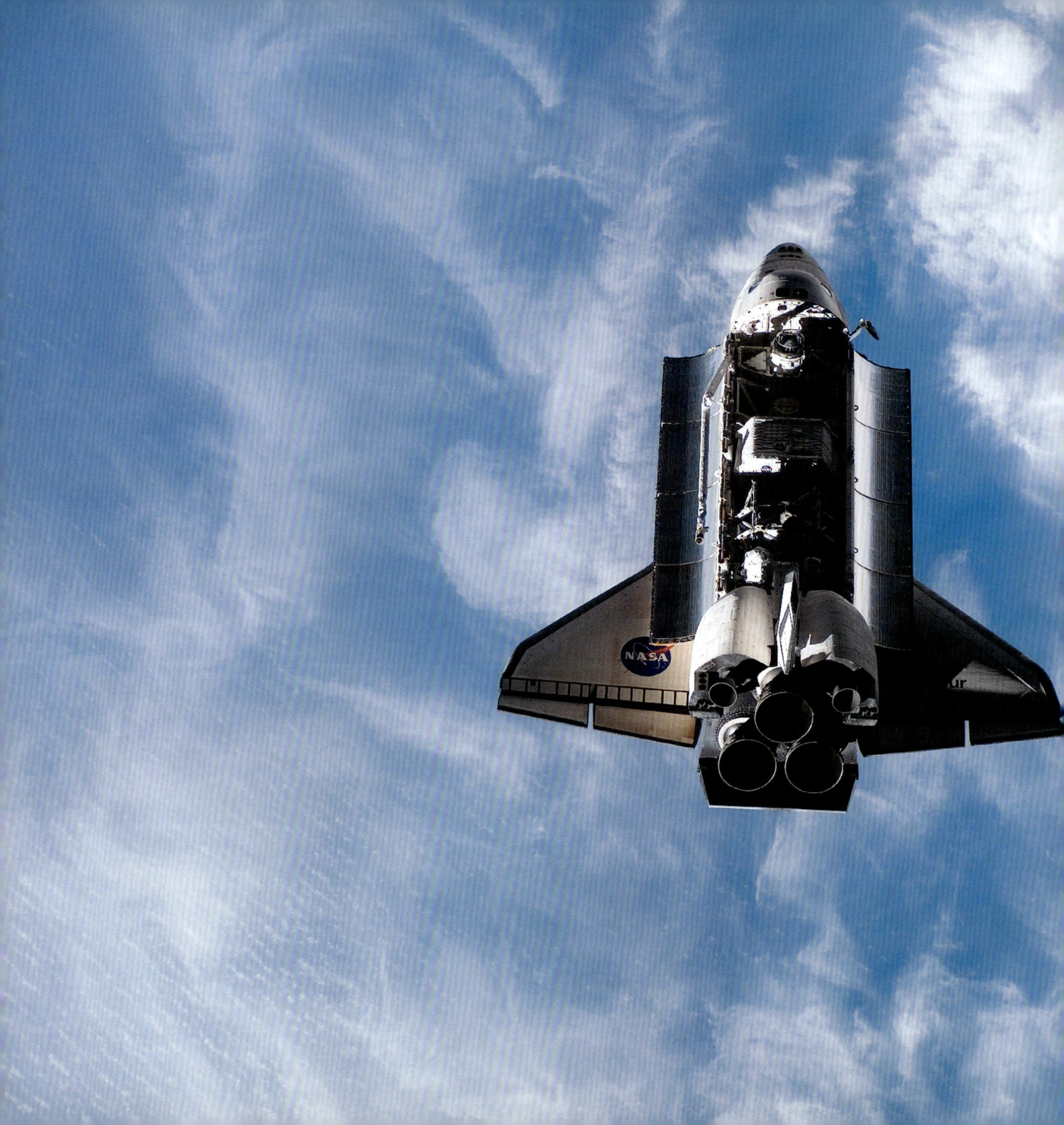

8–21 August 2007

Endeavour suspended against a vibrant, blue Earth atmosphere as viewed from the International Space Station (ISS) in 2007. This mission delivered and constructed the starboard S5 truss segment of the ISS, as well as the External Stowage Platform 3 and a new Control Moment Gyroscope. It would also be the final flight to carry the SPACEHAB Logistics Single Module (LSM) within the shuttle's payload bay. The SPACEHAB LSM is a pressurized habitat designed to carry a variety of cargo and equipment to the ISS. It can be seen here in the middle of Endeavour's payload bay.

The striking contrast between light and shadow is seen beautifully here as the sun beats down on Atlantis while the other side of the orbiter is lost within the blackness of space. The shuttle's open payload bay contains the hardware and equipment that is to be delivered to the International Space Station.

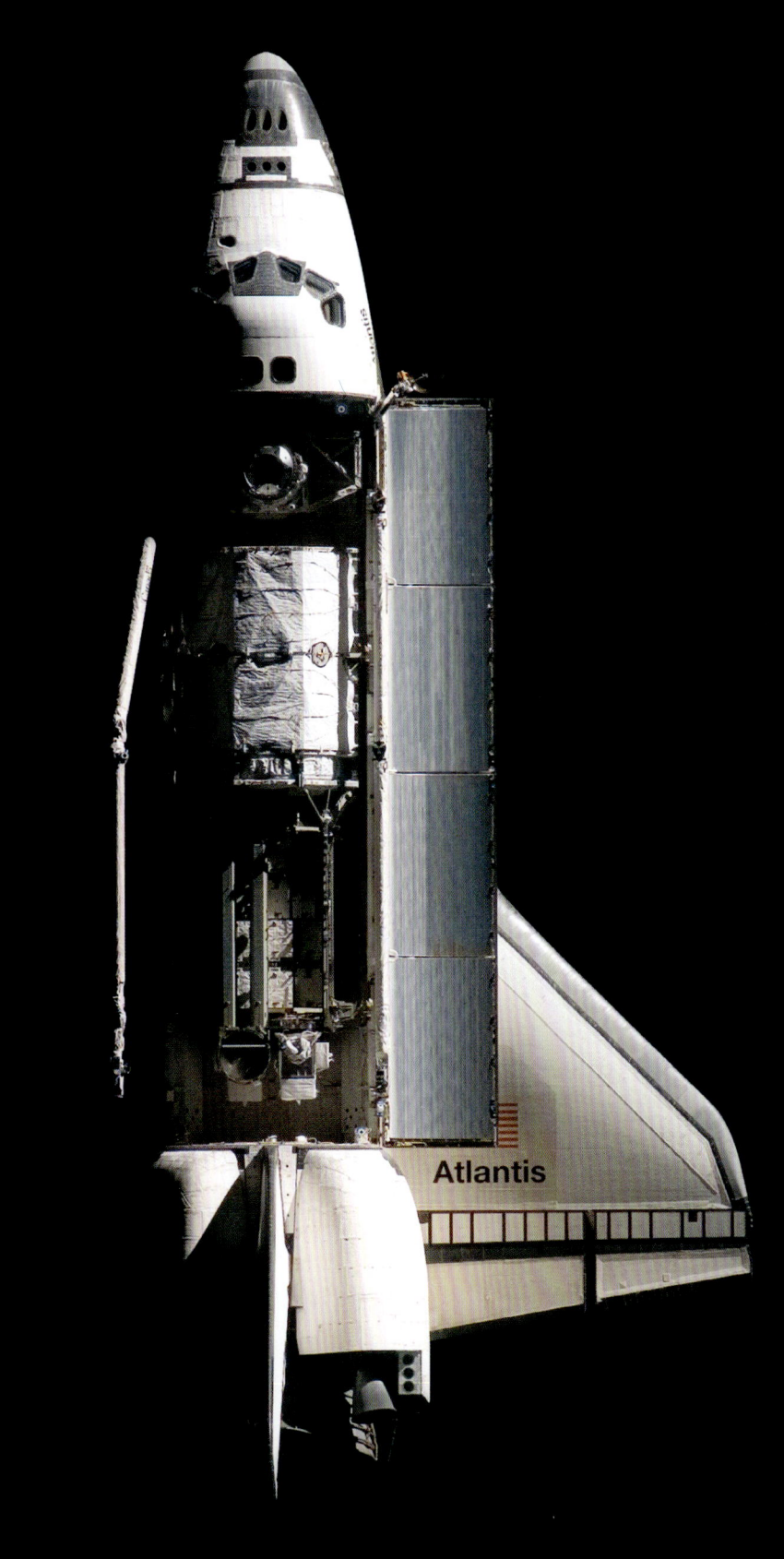

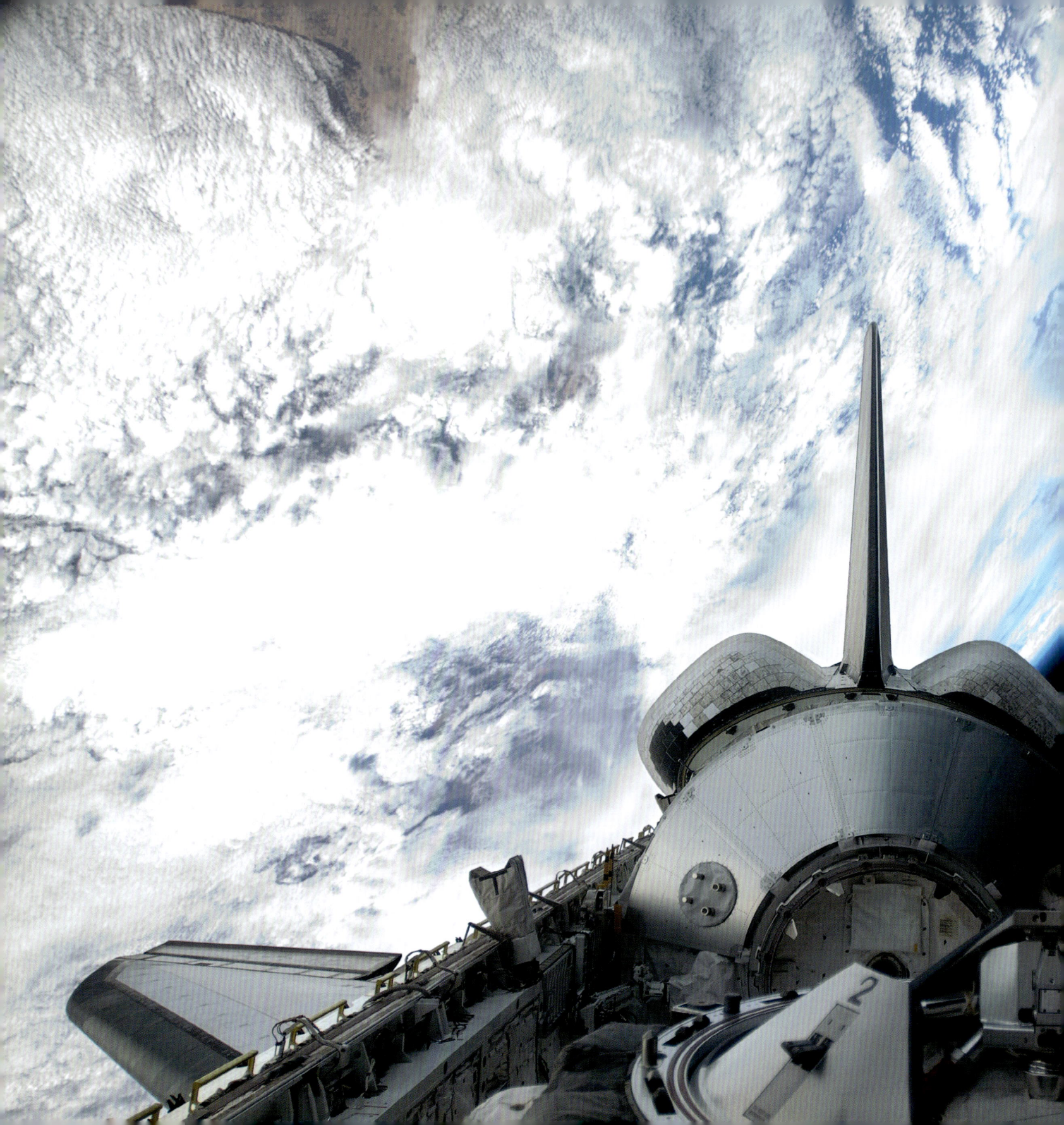

Endeavour's payload bay reveals the Leonardo Multi-Purpose Logistics Module. These modules, built by the Italian Space Agency, were used to transfer equipment and supplies to and from the International Space Station.

This inspiring image of Challenger, taken during its second mission, was captured by a 70mm camera on board an unpiloted and free-flying Shuttle Pallet Satellite. Among the five-person crew for the mission was astronaut Sally Ride, America's first woman in space. Her second and final space flight was also aboard Challenger for STS-41-G in 1984.

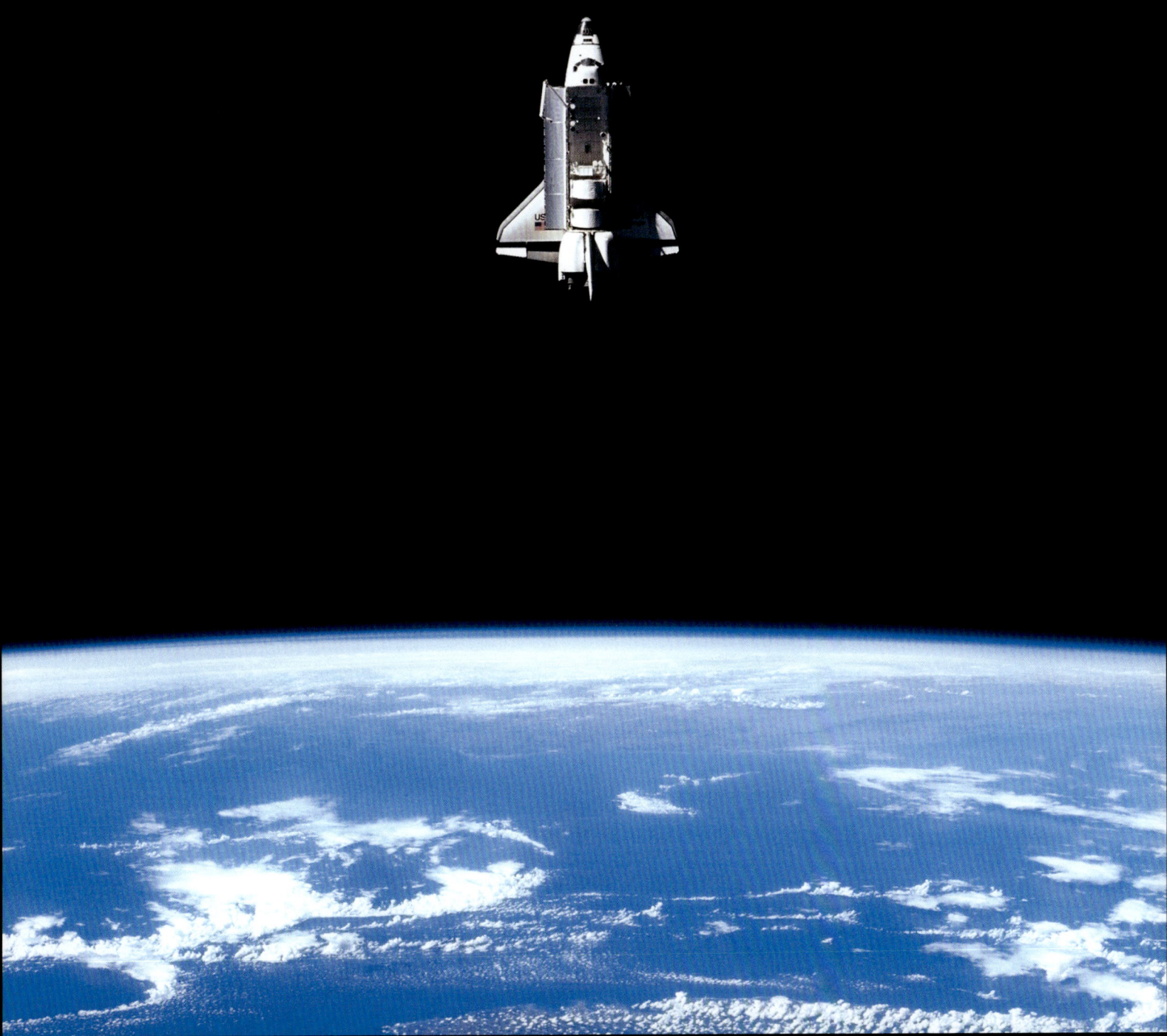

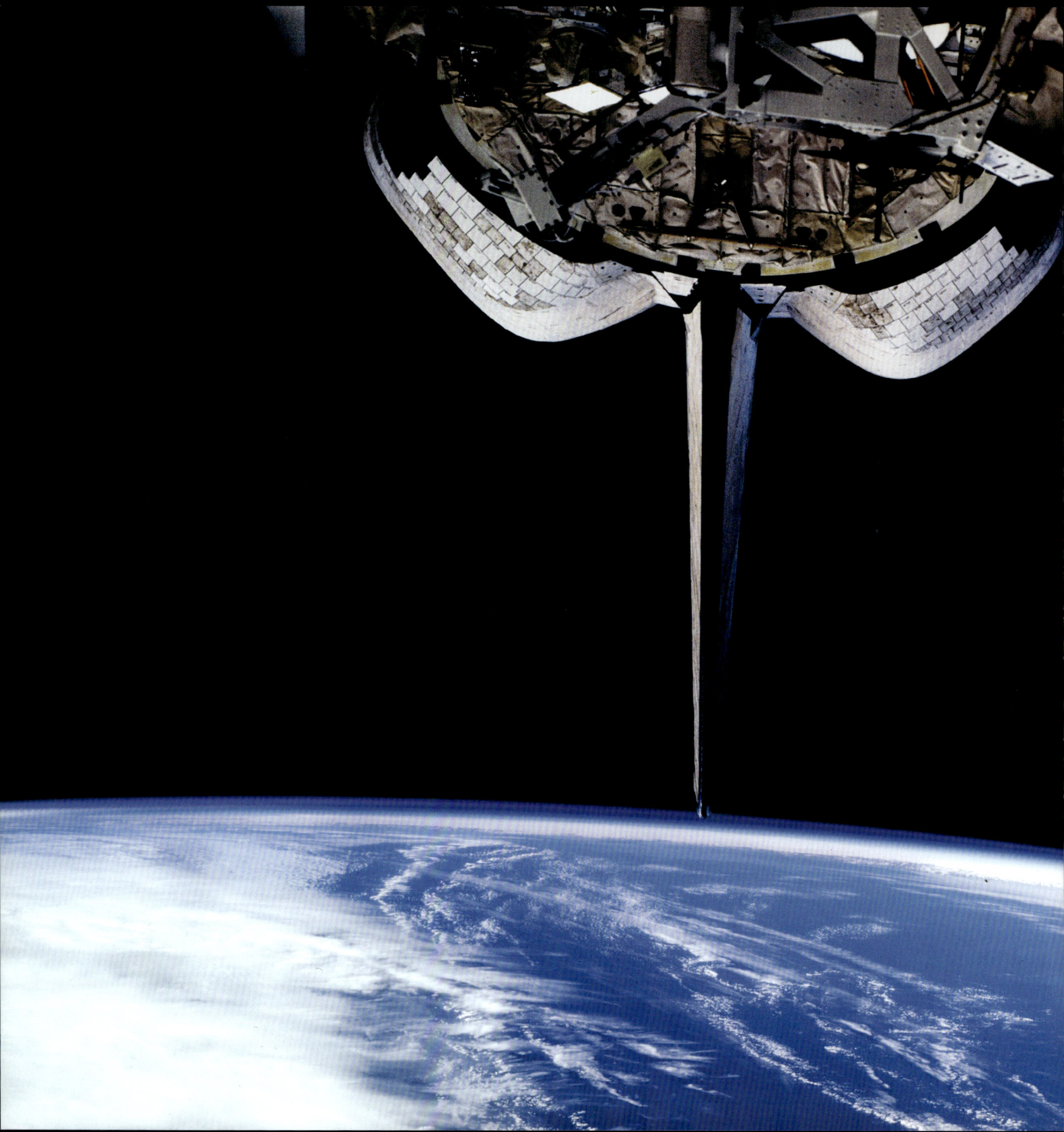

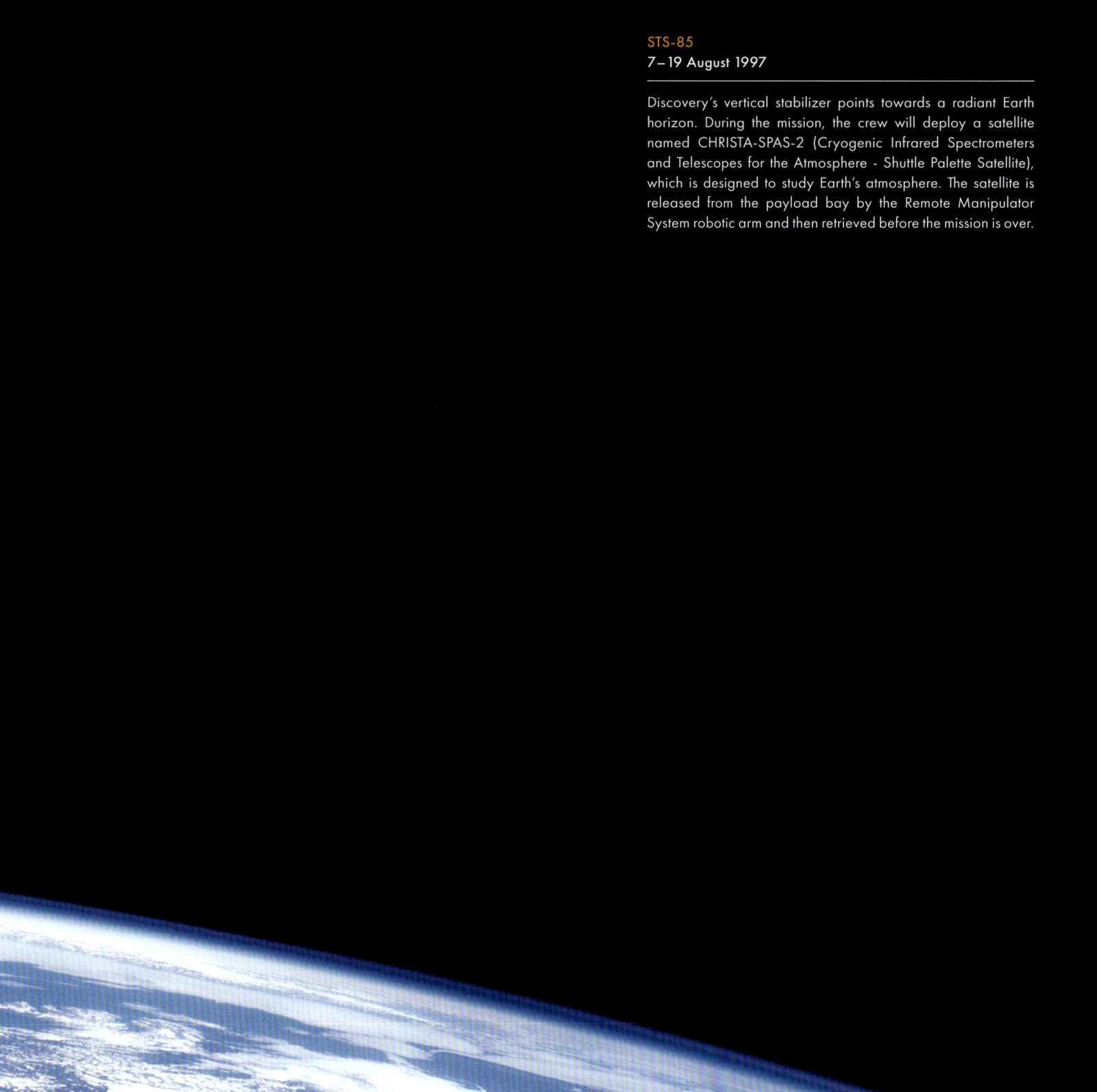

Discovery's vertical stabilizer points towards a radiant Earth horizon. During the mission, the crew will deploy a satellite named CHRISTA-SPAS-2 (Cryogenic Infrared Spectrometers and Telescopes for the Atmosphere - Shuttle Palette Satellite), which is designed to study Earth's atmosphere. The satellite is released from the payload bay by the Remote Manipulator System robotic arm and then retrieved before the mission is over.

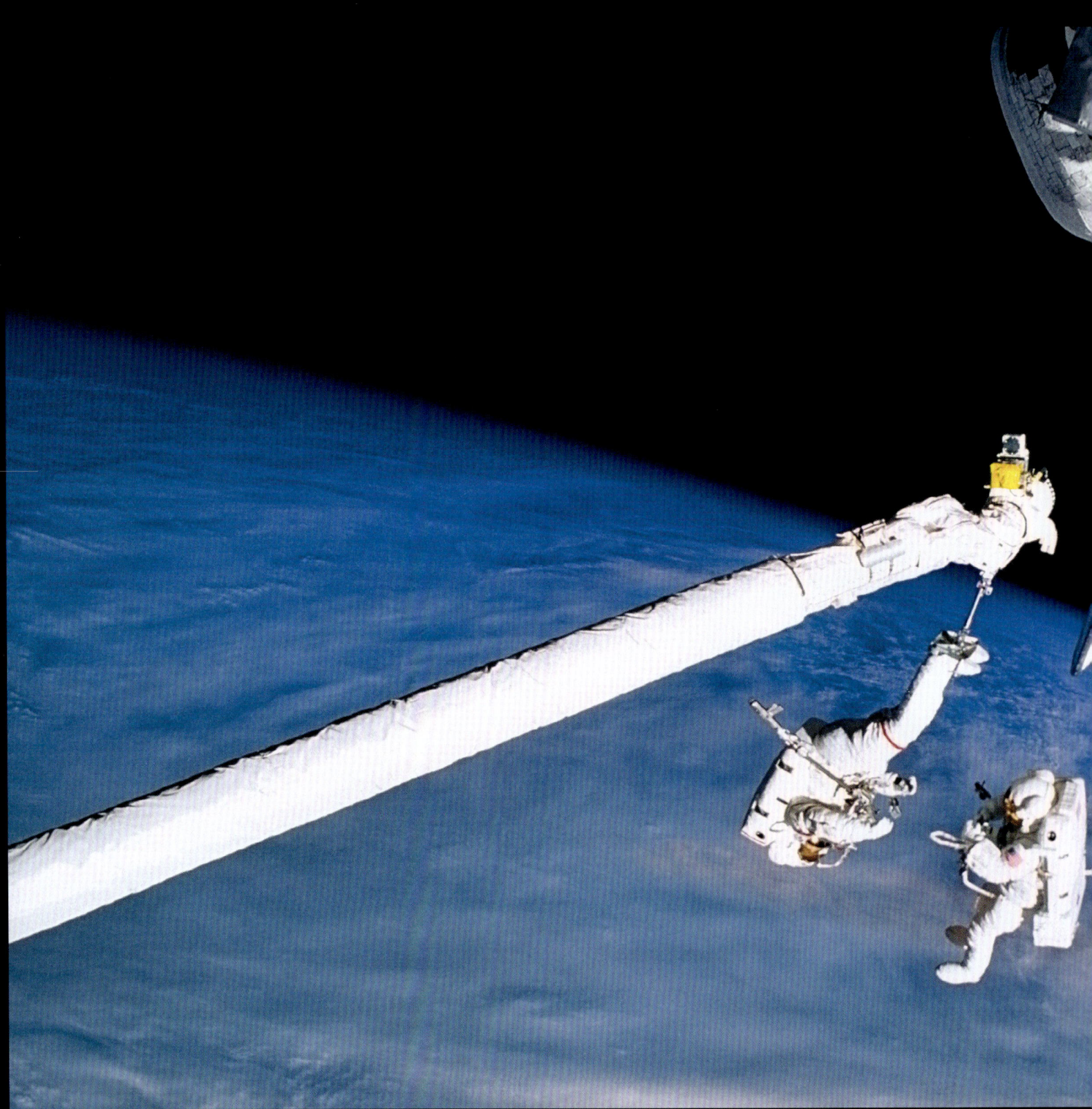

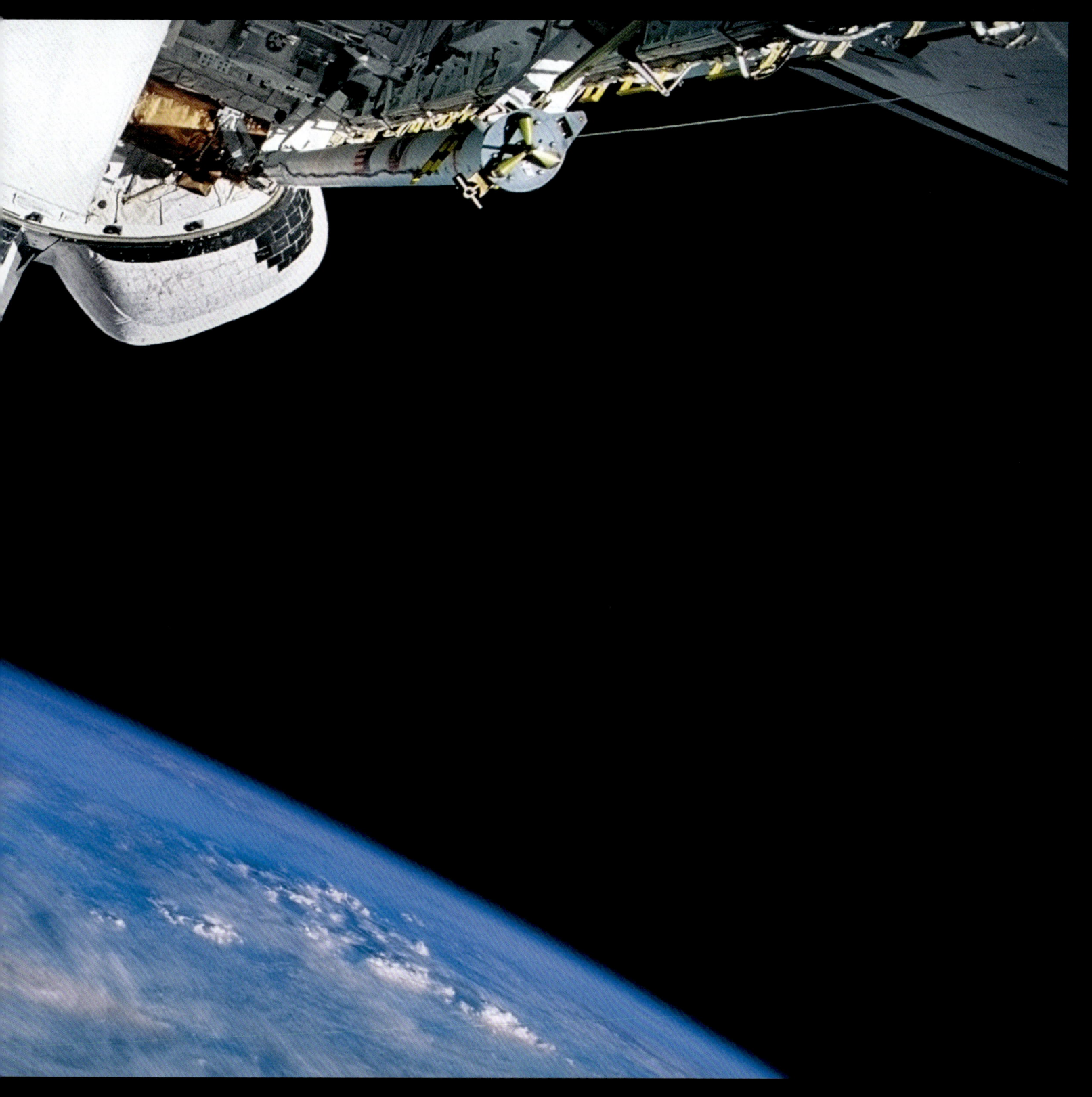

STS-64
9–20 September 1994
_____

(Previous spread) Just another day at the office. Astronauts Carl J. Meade and Mark C. Lee (red stripes on spacesuit) working in space while testing the new SAFER system (Simplified Aid For EVA Rescue). Astronaut Lee can be seen attached to Discovery's Remote Manipulator System robotic arm.

STS-41-B
3–11 February 1984
_____

(Right and Following spread) These inspiring images show astronaut Bruce McCandless II floating alone in the blackness of space. McCandless travelled a distance of 320 feet (98 m) from space shuttle Challenger using the new Manned Maneuvering Unit jet pack. He became the first person to make an untethered space walk and ventured further from his craft than any other astronaut before him.

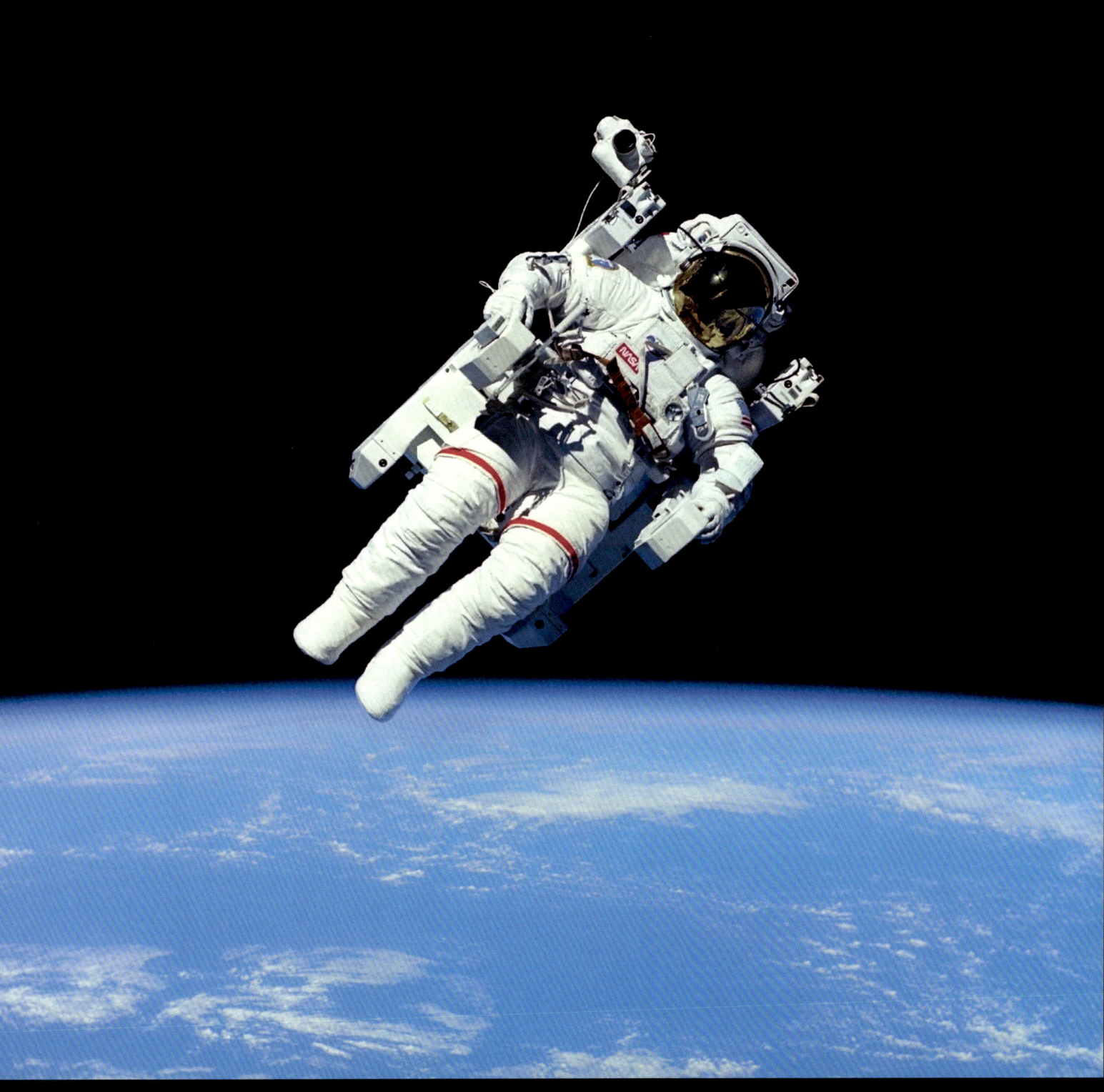

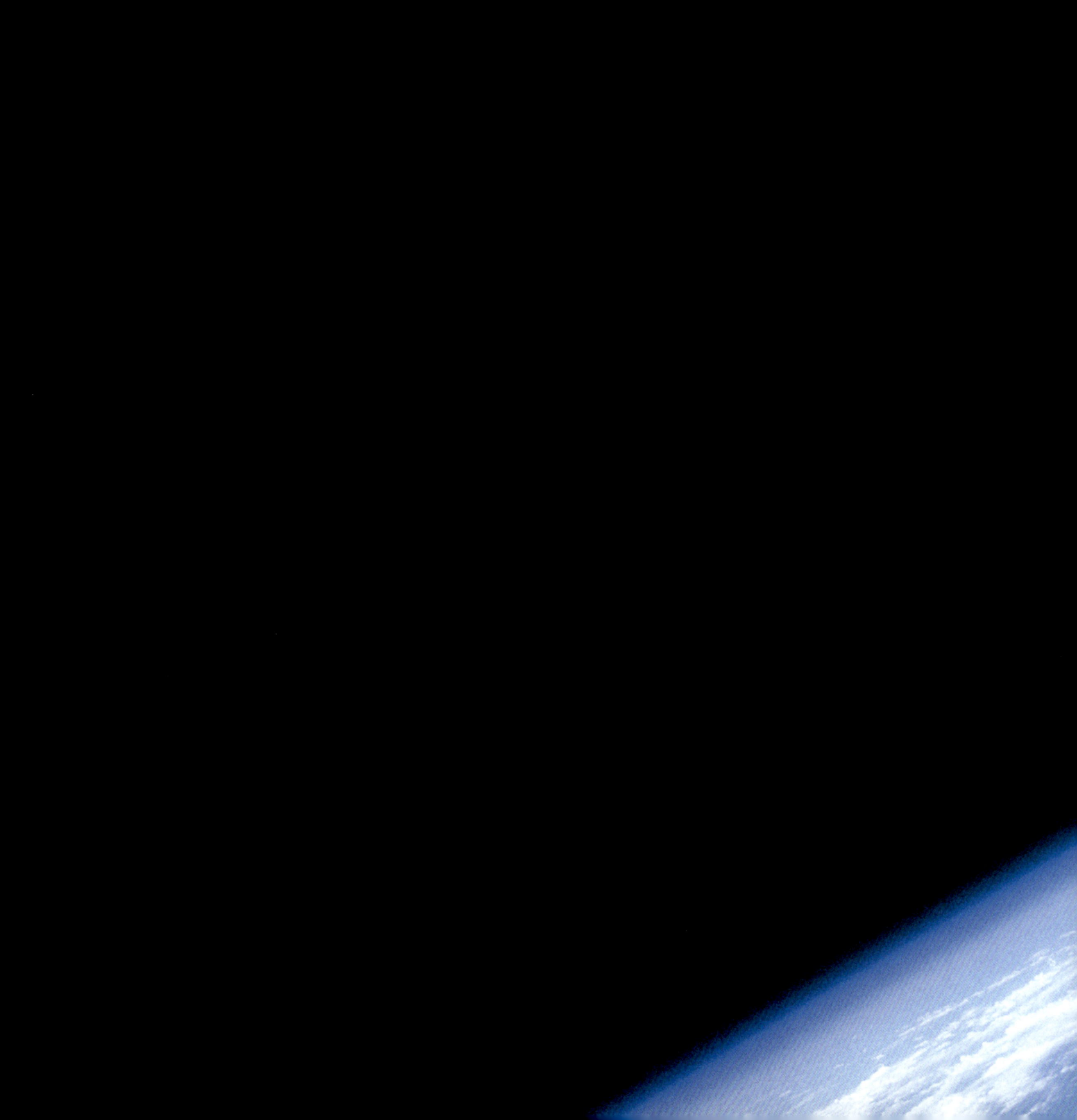

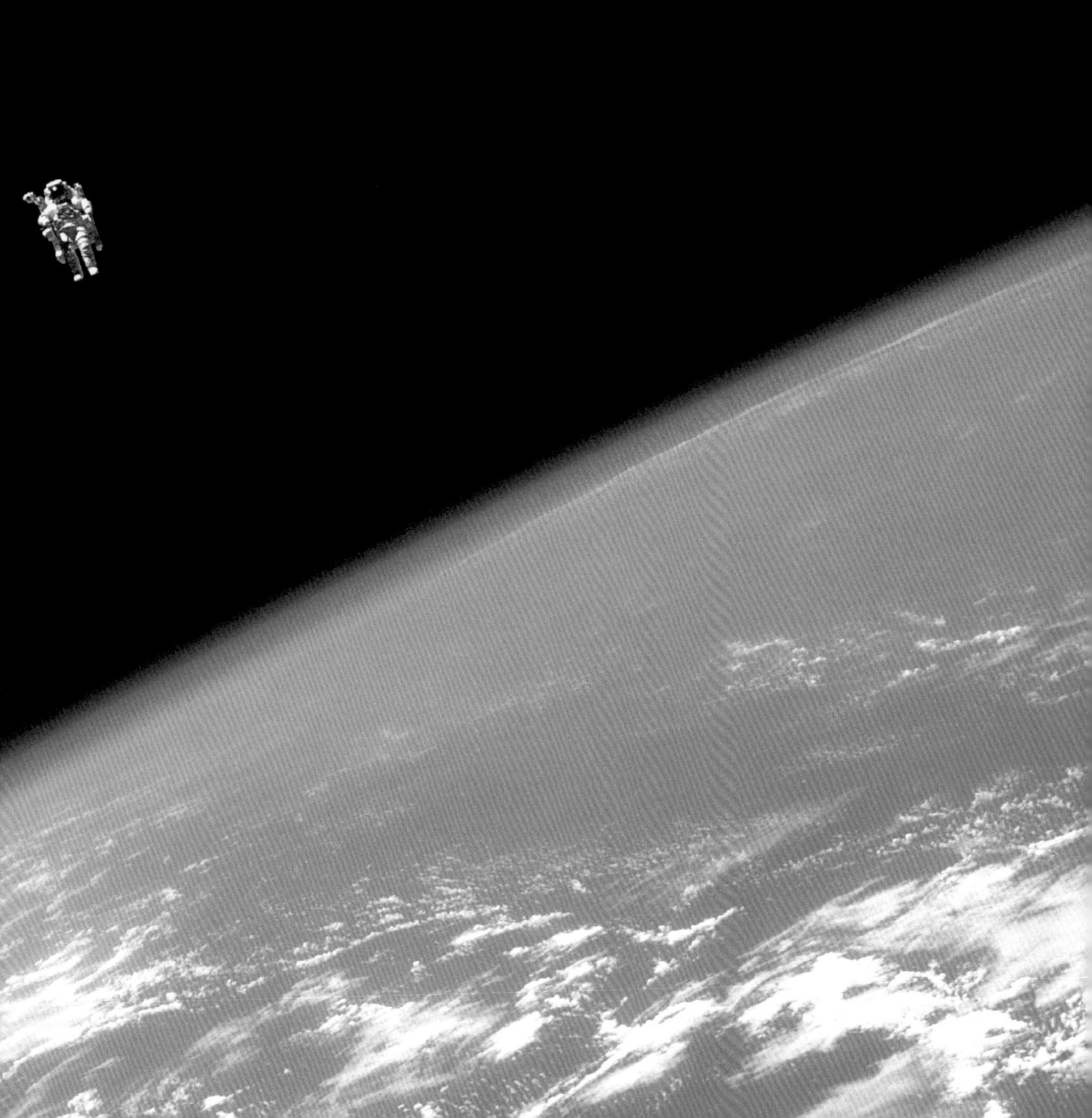

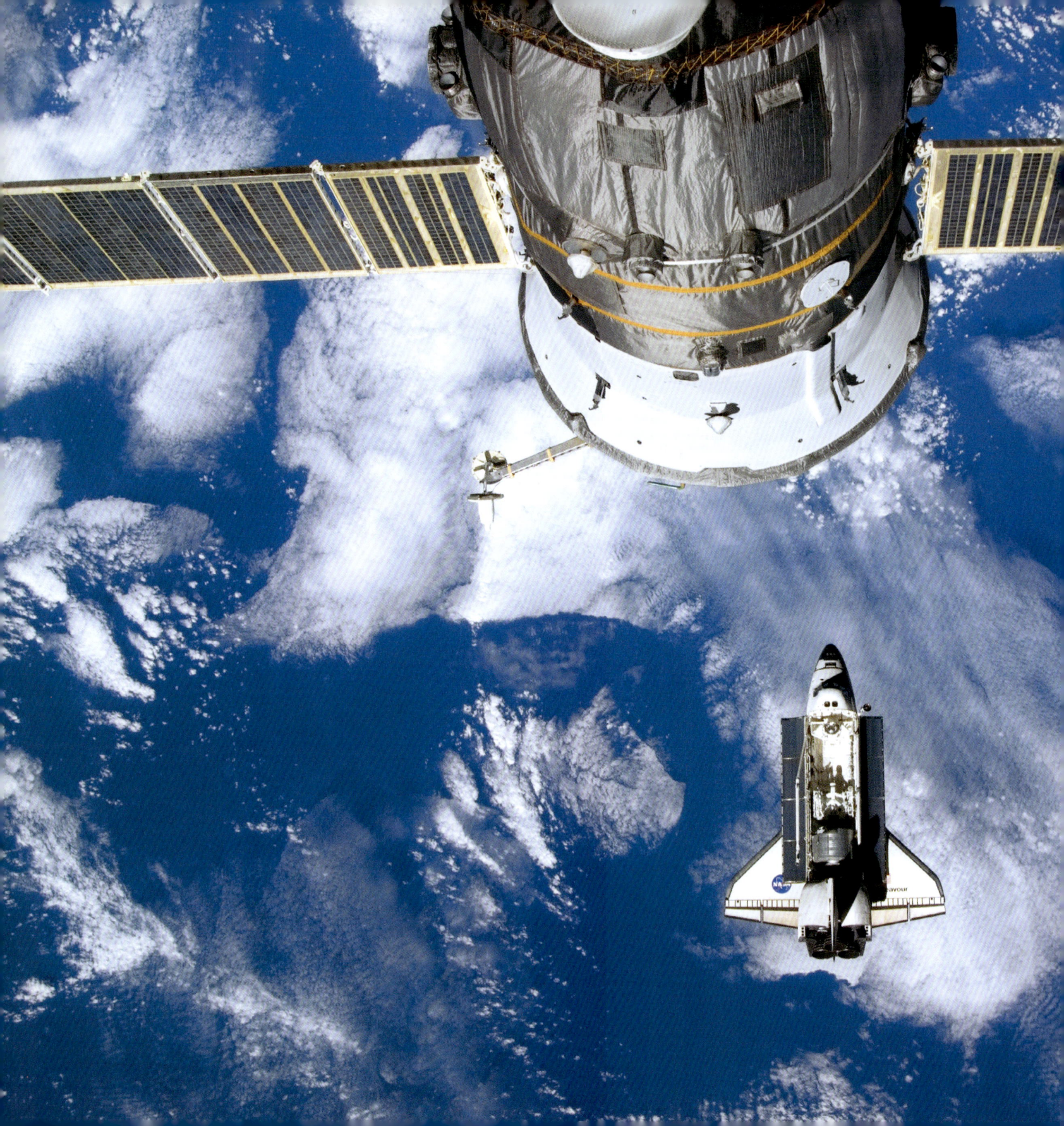

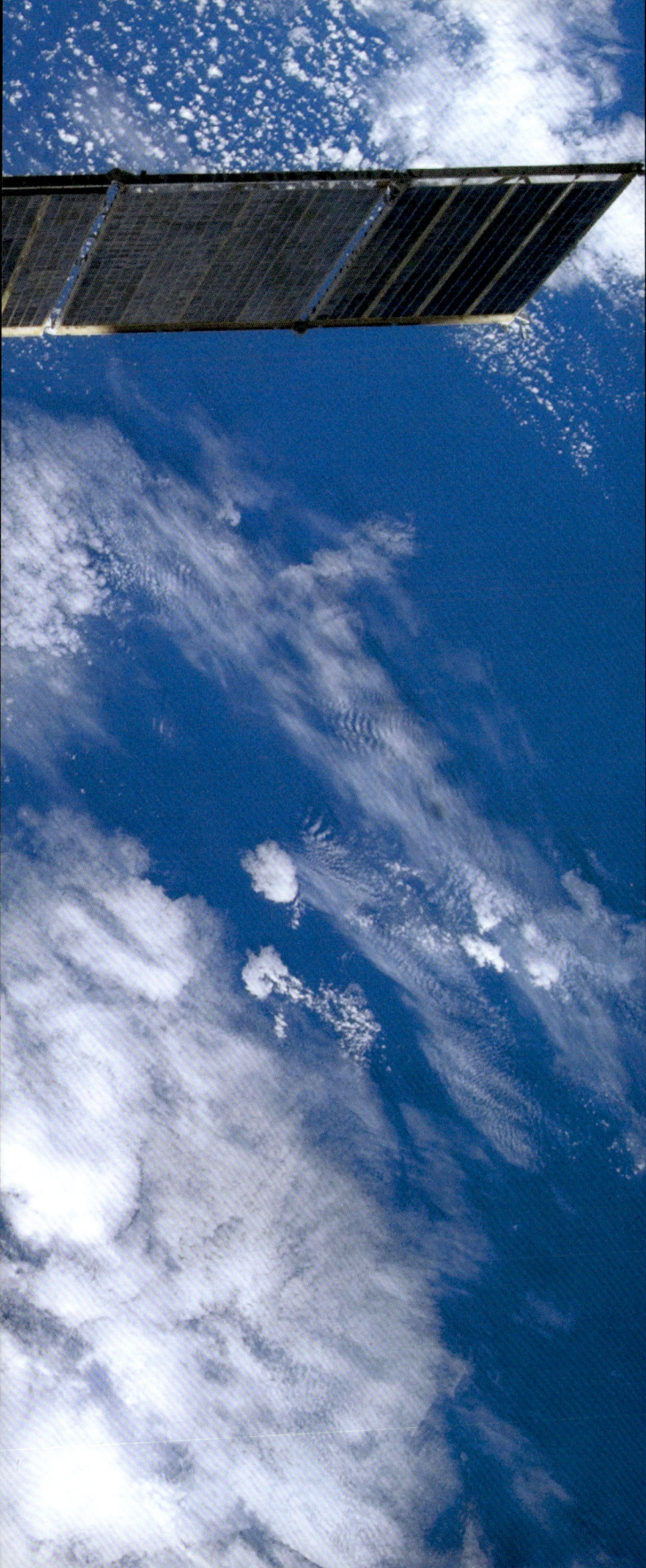

STS-123
11–26 March 2008

(Left) A beautiful view from the International Space Station (ISS). Endeavour prepares for rendezvous and docking at the beginning of a 16-day mission. A Russian spacecraft can be seen docked to the ISS at the top of the frame.

STS-132
14–26 May 2010

(Following spread: Left) In this exquisite and unusual photograph, we see space shuttle Atlantis peacefully floating above the bright-white Andes Mountain range near the border of Chile and Argentina. This image was taken during the approach to the International Space Station.

STS-120
23 October to 7 November 2007

(Following spread: Right) Against the backdrop of a cloudy, blue-and-white Earth, space shuttle Discovery approaches the International Space Station (ISS) during rendezvous and docking operations. The large Harmony Node 2 module, which will be connected to the ISS can be seen in Discovery's payload bay.

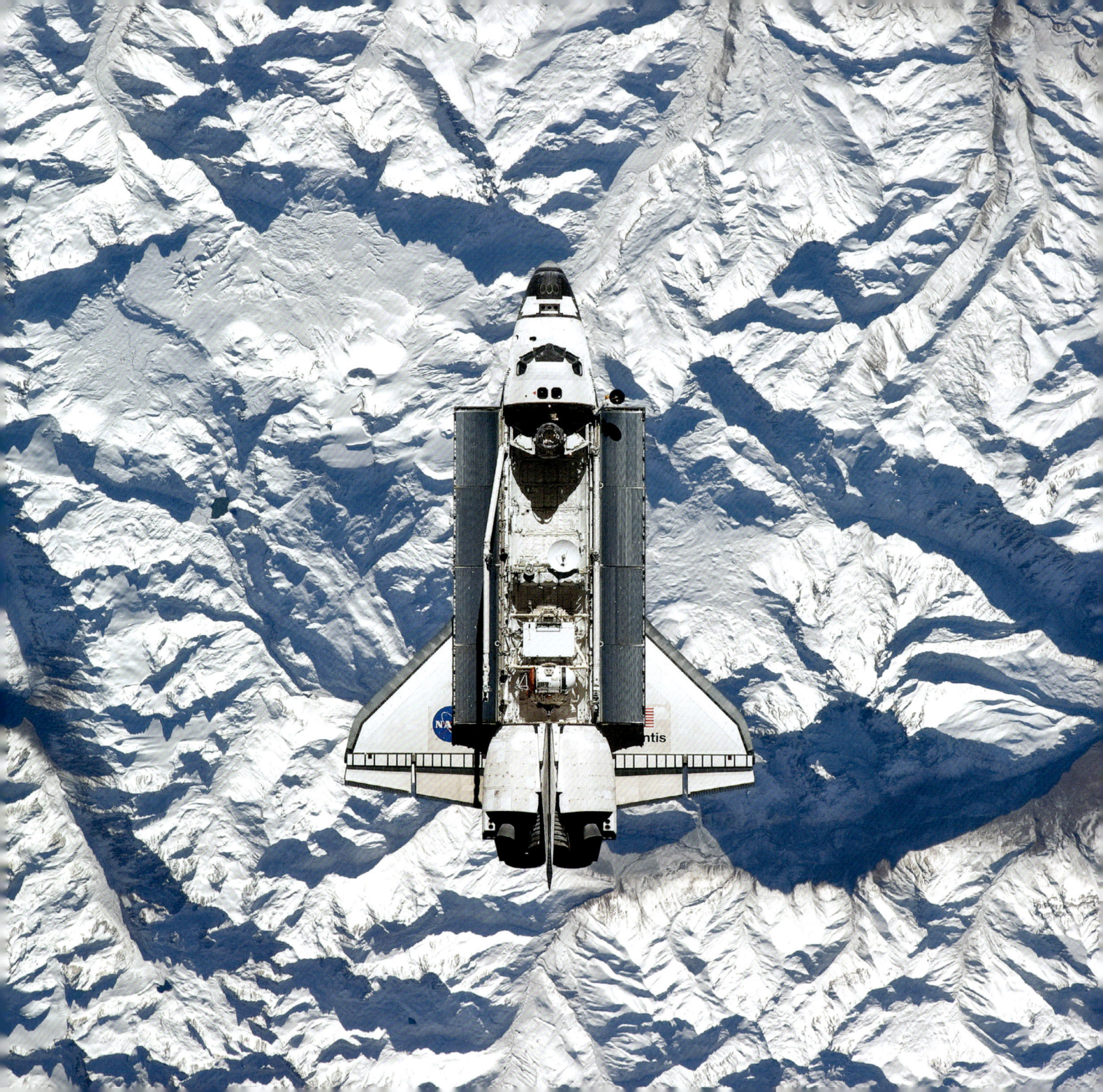

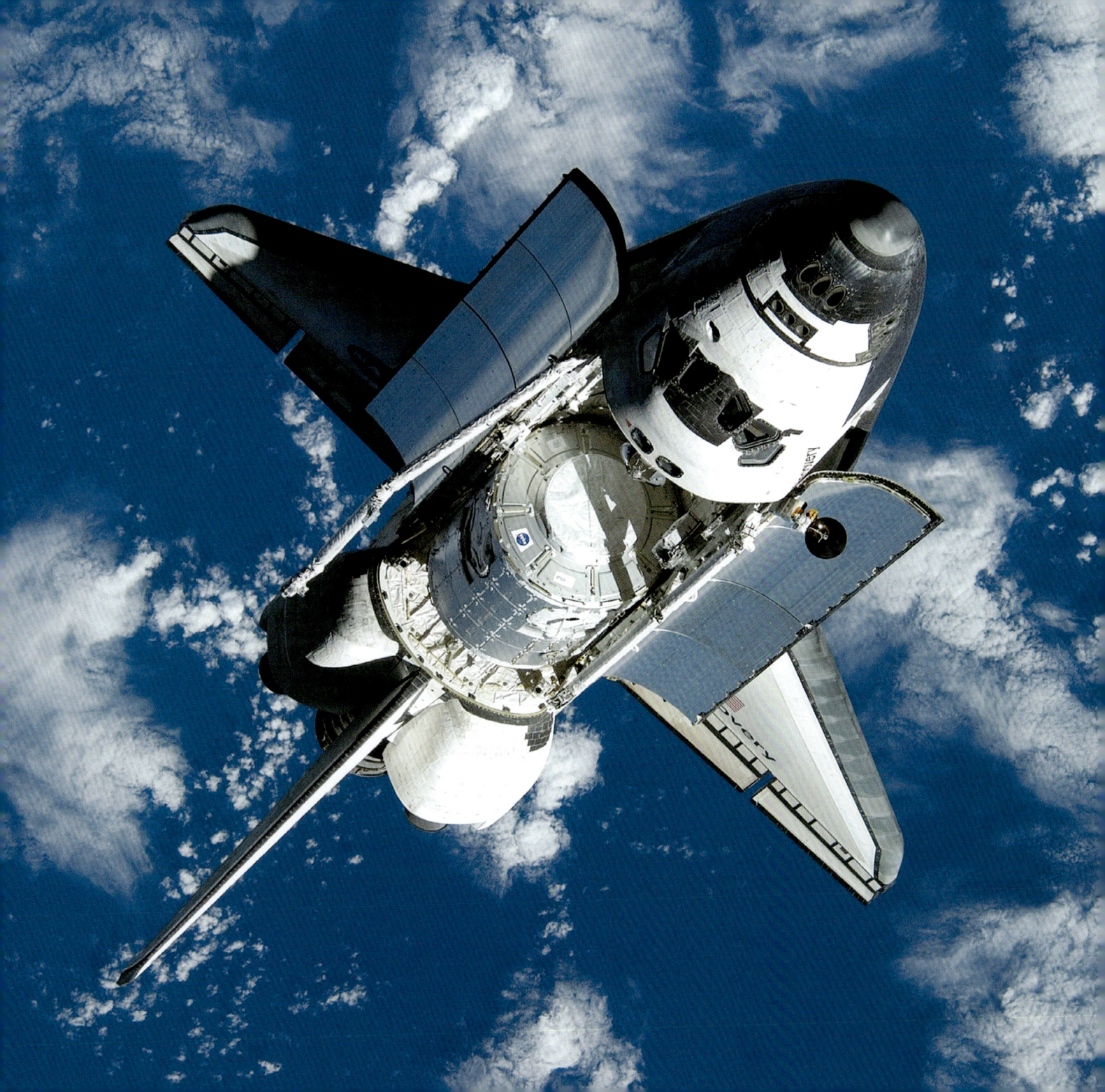

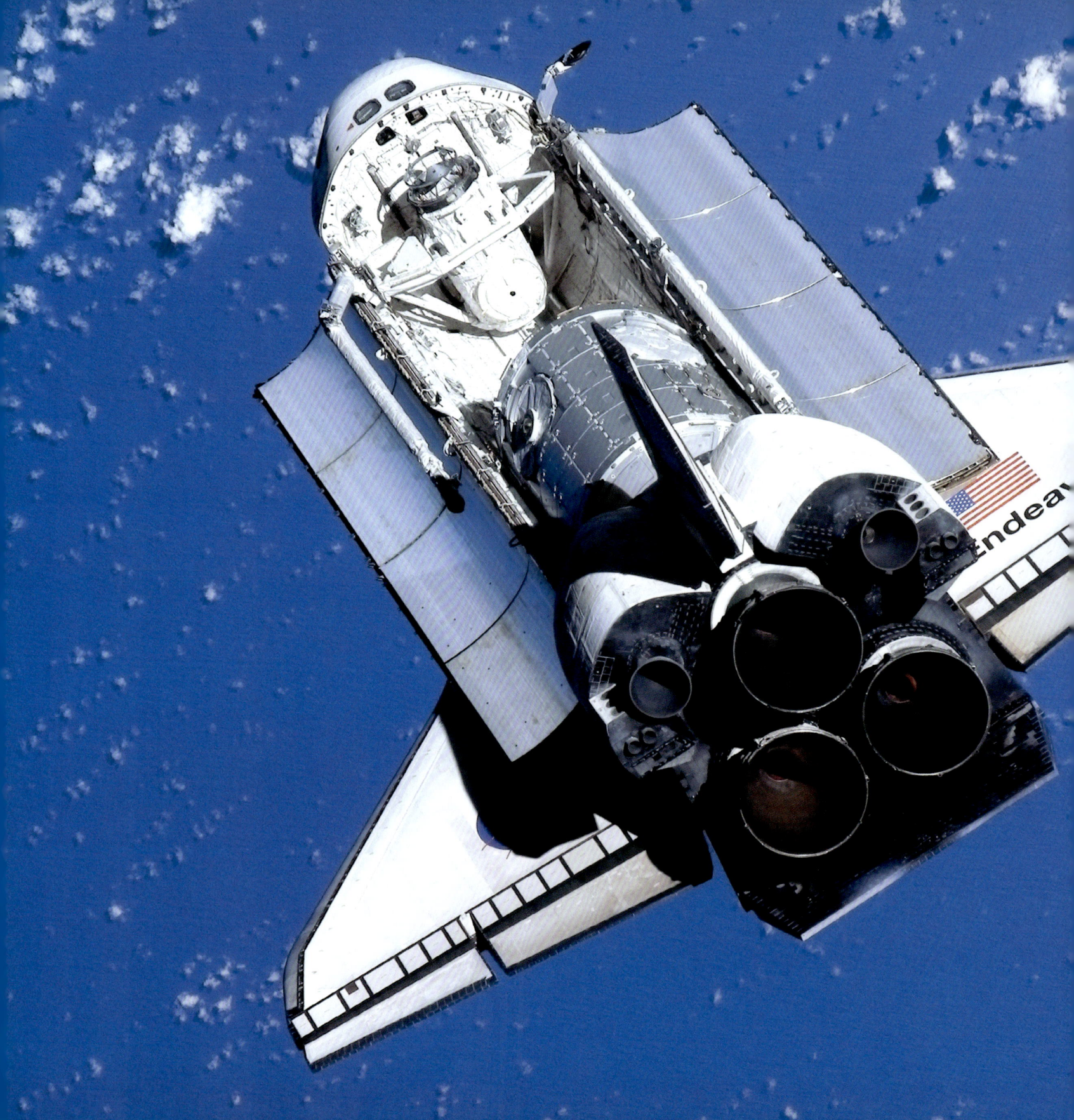

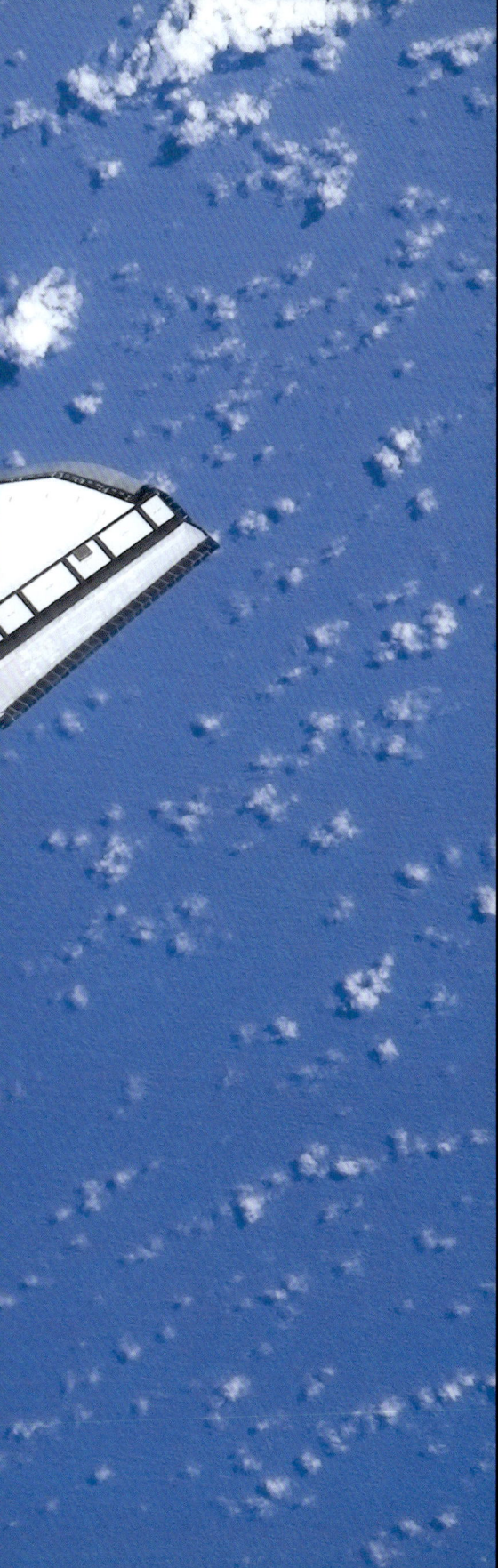

8–21 February 2010

As Endeavour approaches the International Space Station (ISS) for rendezvous operations, the shuttle is put through a series of manoeuvres, allowing crew members aboard the ISS to observe the shuttle's condition.

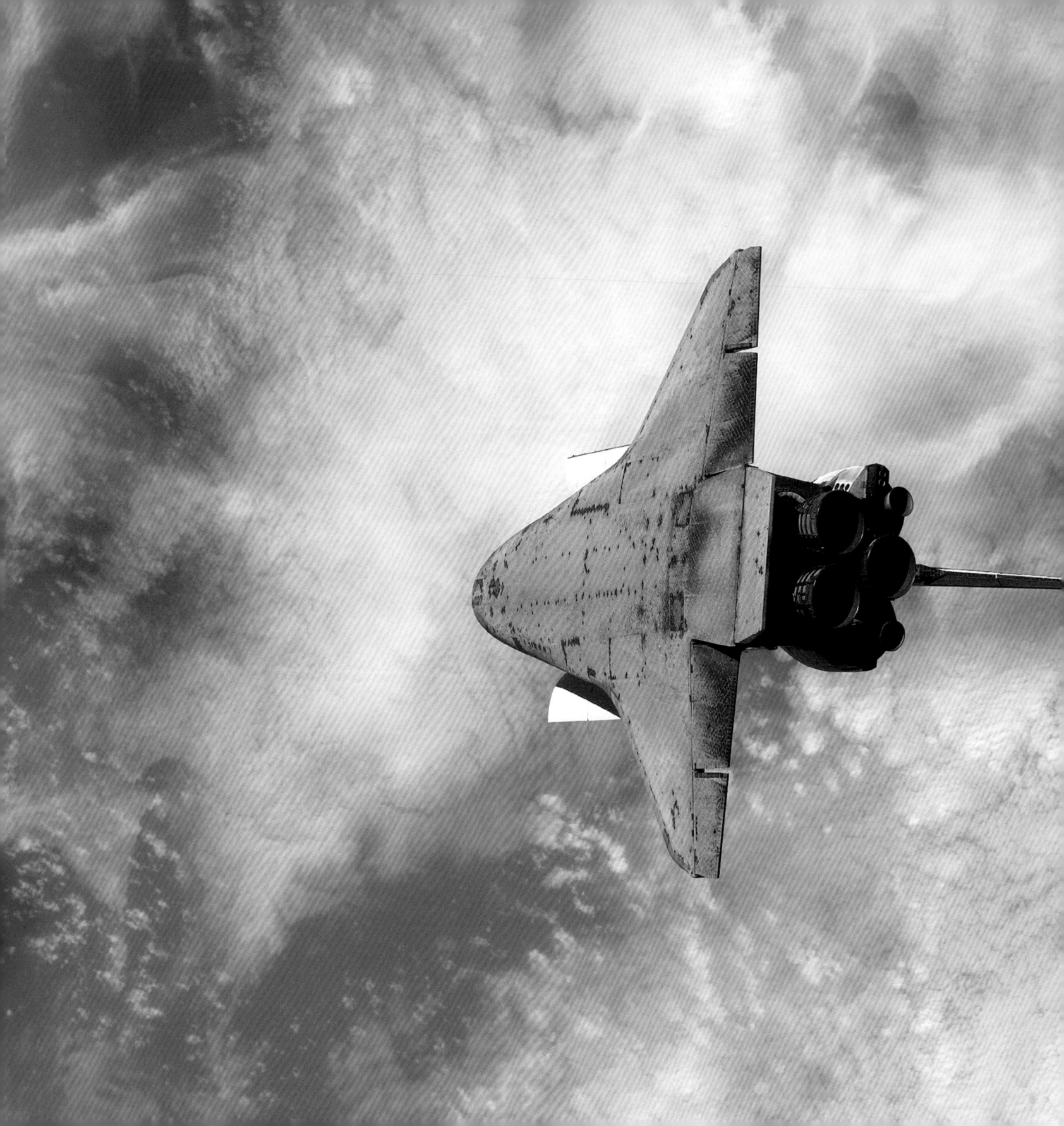

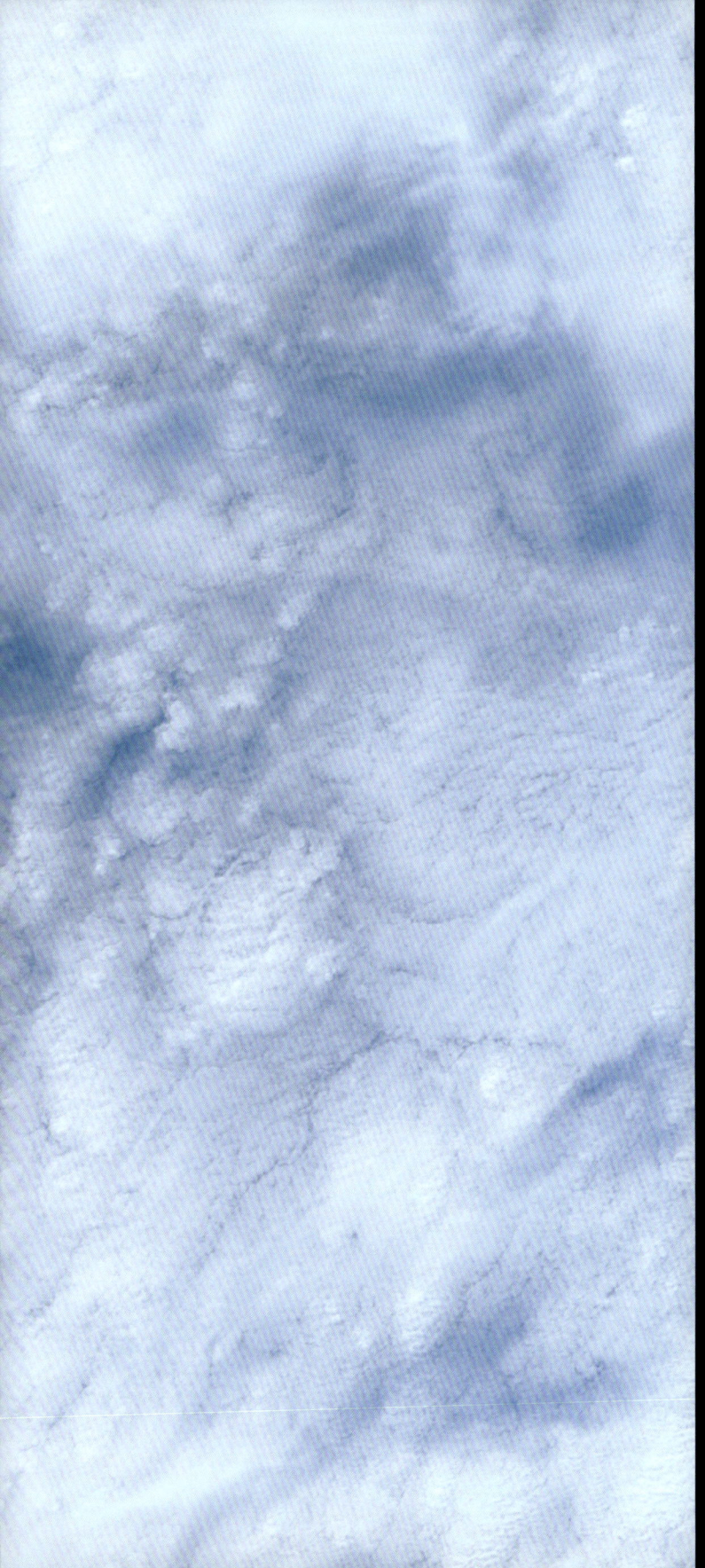

8–21 August 2007

Endeavour performs a rendezvous pitch maneouvre during its approach to the International Space Station, exposing the shuttle's underside thermal protection system. The shuttle's heat shield suffered a puncture caused by debris from the external fuel tank during launch. NASA officials, as well as the crew, decided Endeavour was safe to perform the dangerous Earth re-entry procedure, due to the small size and location of the damage. There were concerns that an attempt to repair the shuttle could have caused more damage to the heat shield. After more than 12 days in space, Endeavour landed safely at Kennedy Space Center, Florida.

STS-120
23 October to 7 November 2007

(Right) Space shuttle Discovery floats above Earth's bright-blue oceans on its approach to the International Space Station (ISS). STS-120 delivered the Harmony module: a central utility hub that provides electrical power and acts as a connection point for various other modules on the ISS.

STS-123
11–26 March 2008

(Following spread) Equipped with digital cameras using 400mm and 800mm lenses, crew members aboard the ISS take a series of detailed photographs of Endeavour's thermal protection system. These images are relayed back to mission control for analysis at the start of the mission. The shuttle's commander manoeuvres Endeavour, allowing the ISS crew to inspect the heat shield thoroughly.

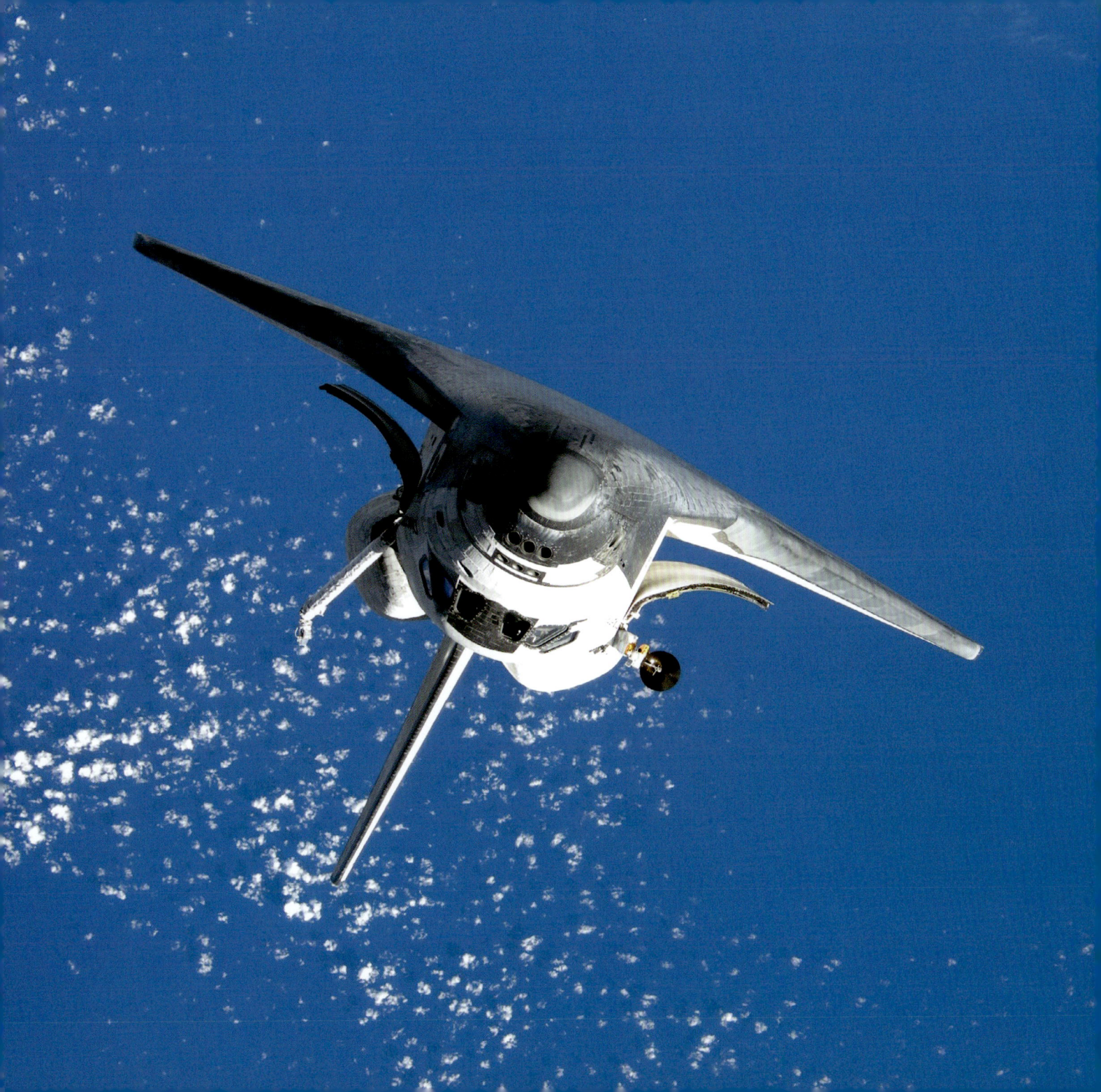

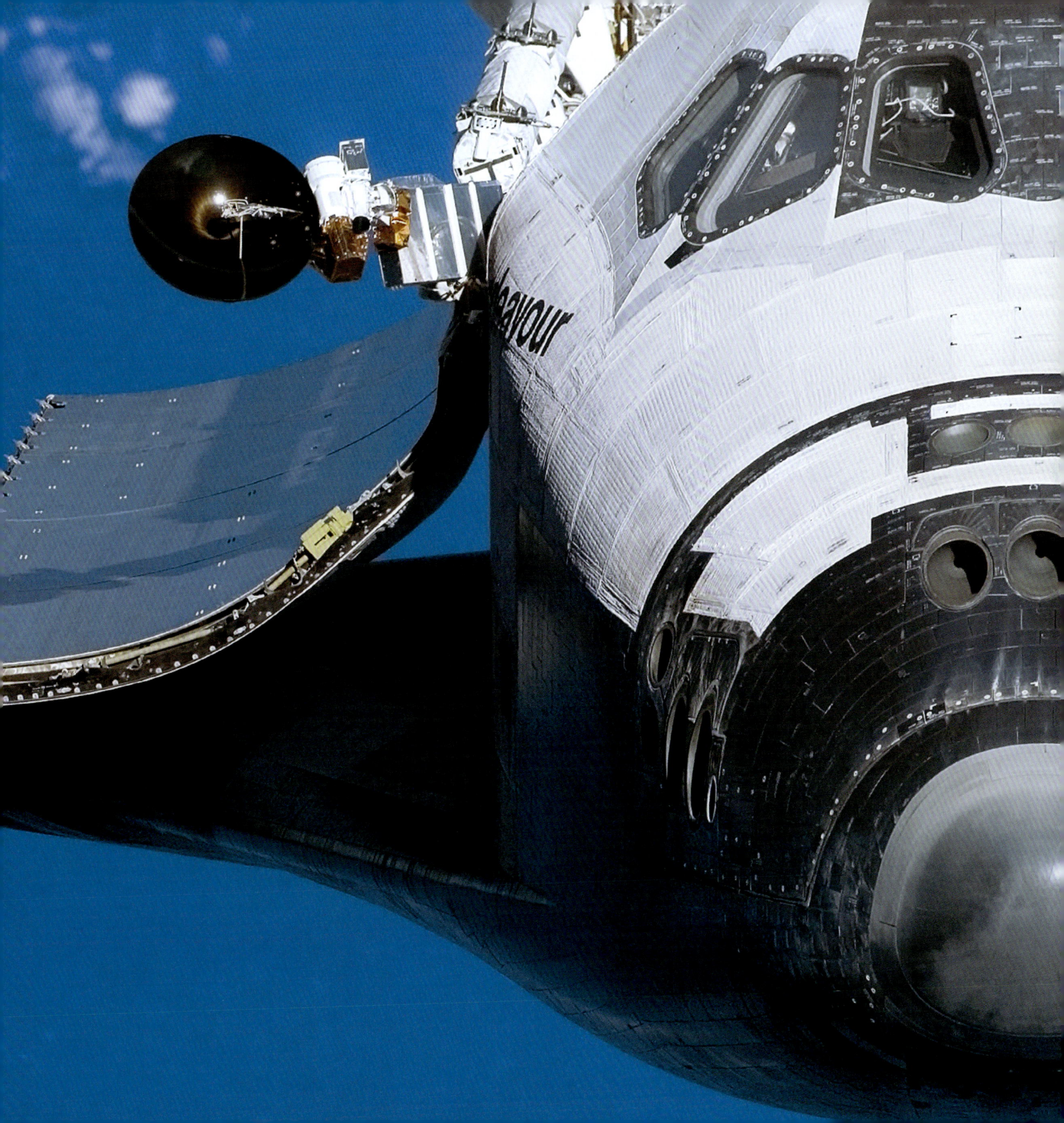

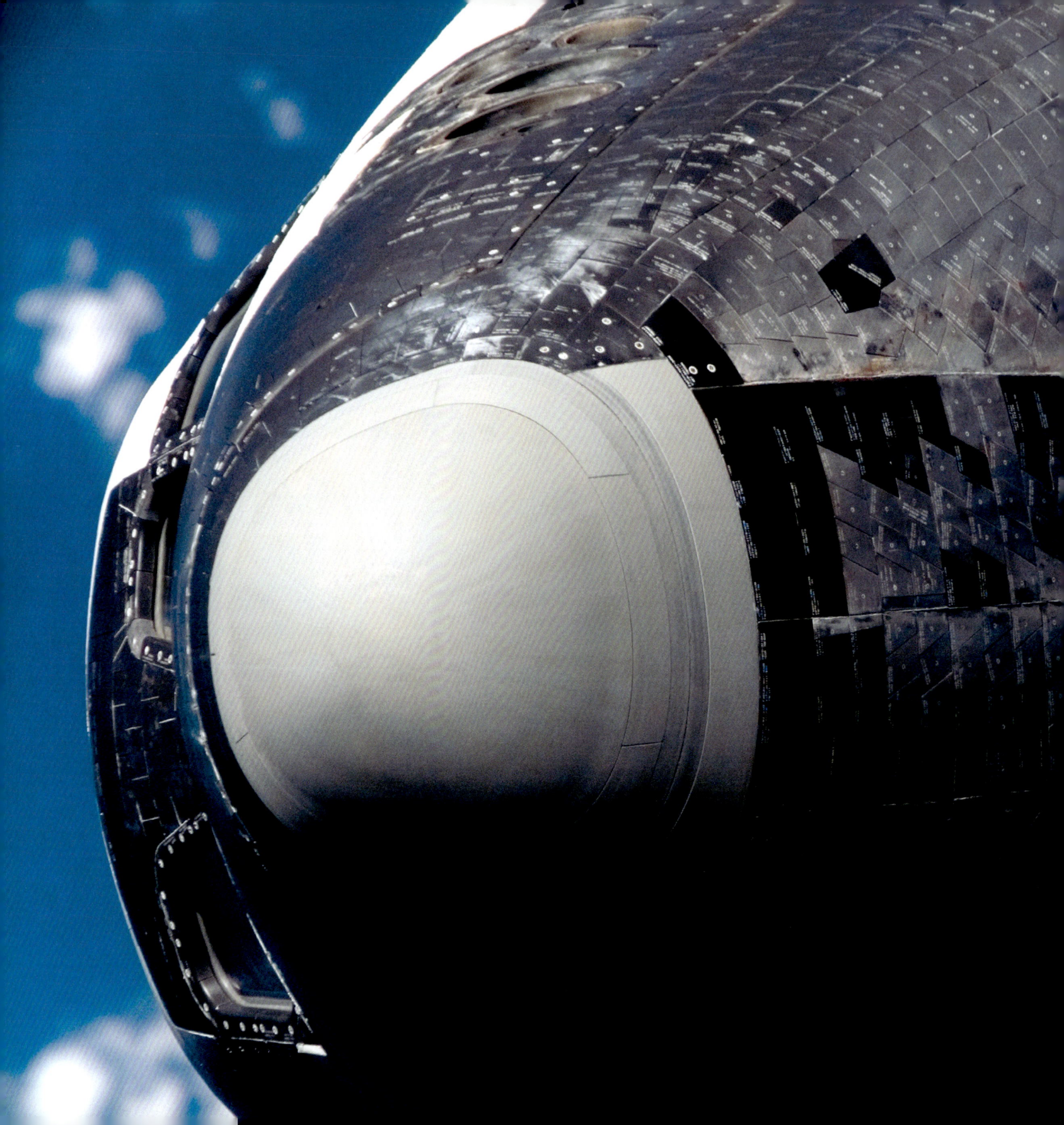

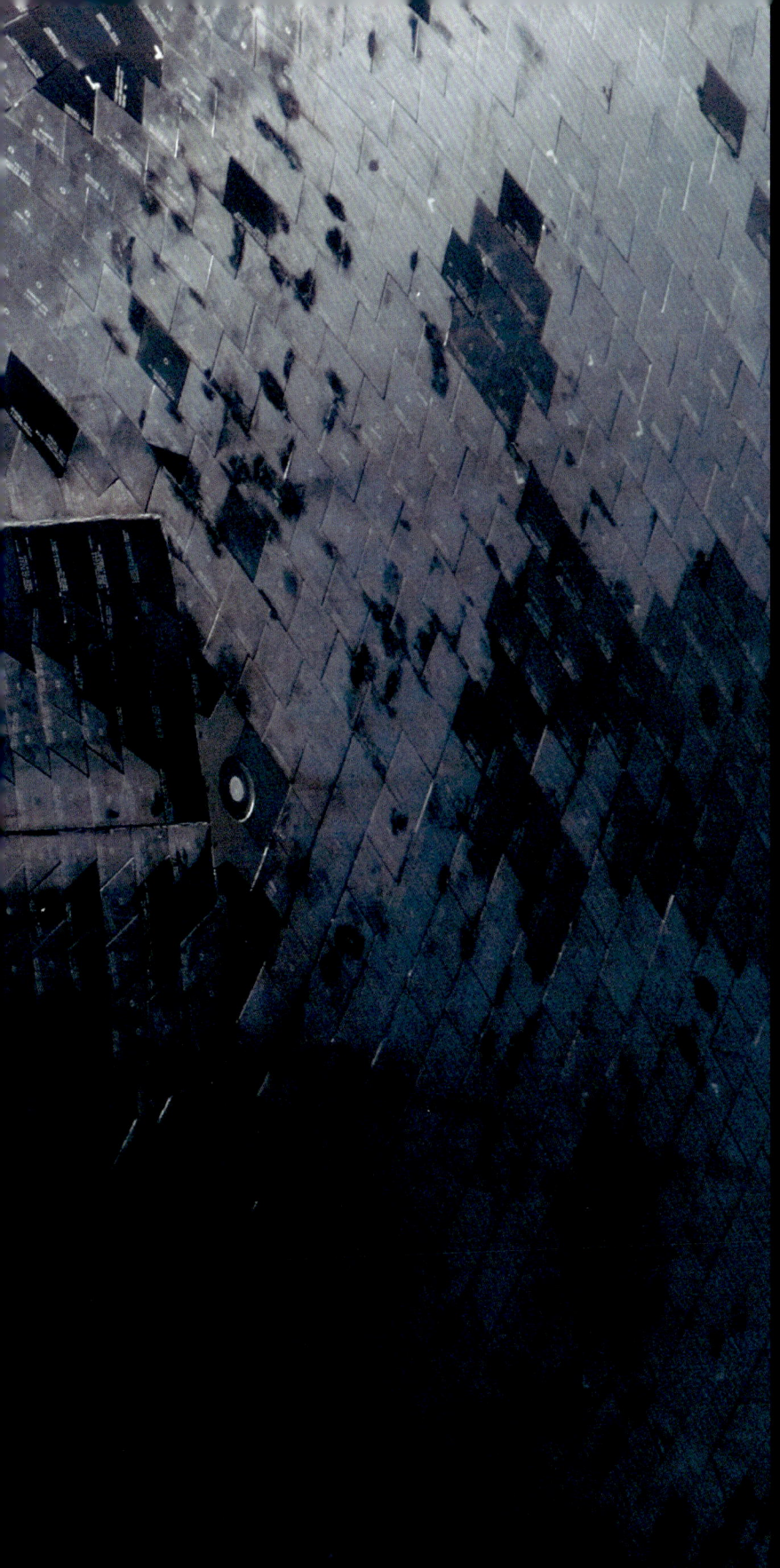

STS-118
8–21 August 2007

Space shuttle Endeavour on approach to the International Space Station. A series of digital images are taken to inspect the shuttle's condition before returning to Earth. This photograph shows a close-up of Endeavour's nose, while offering a detailed view of the heat shield tiles.

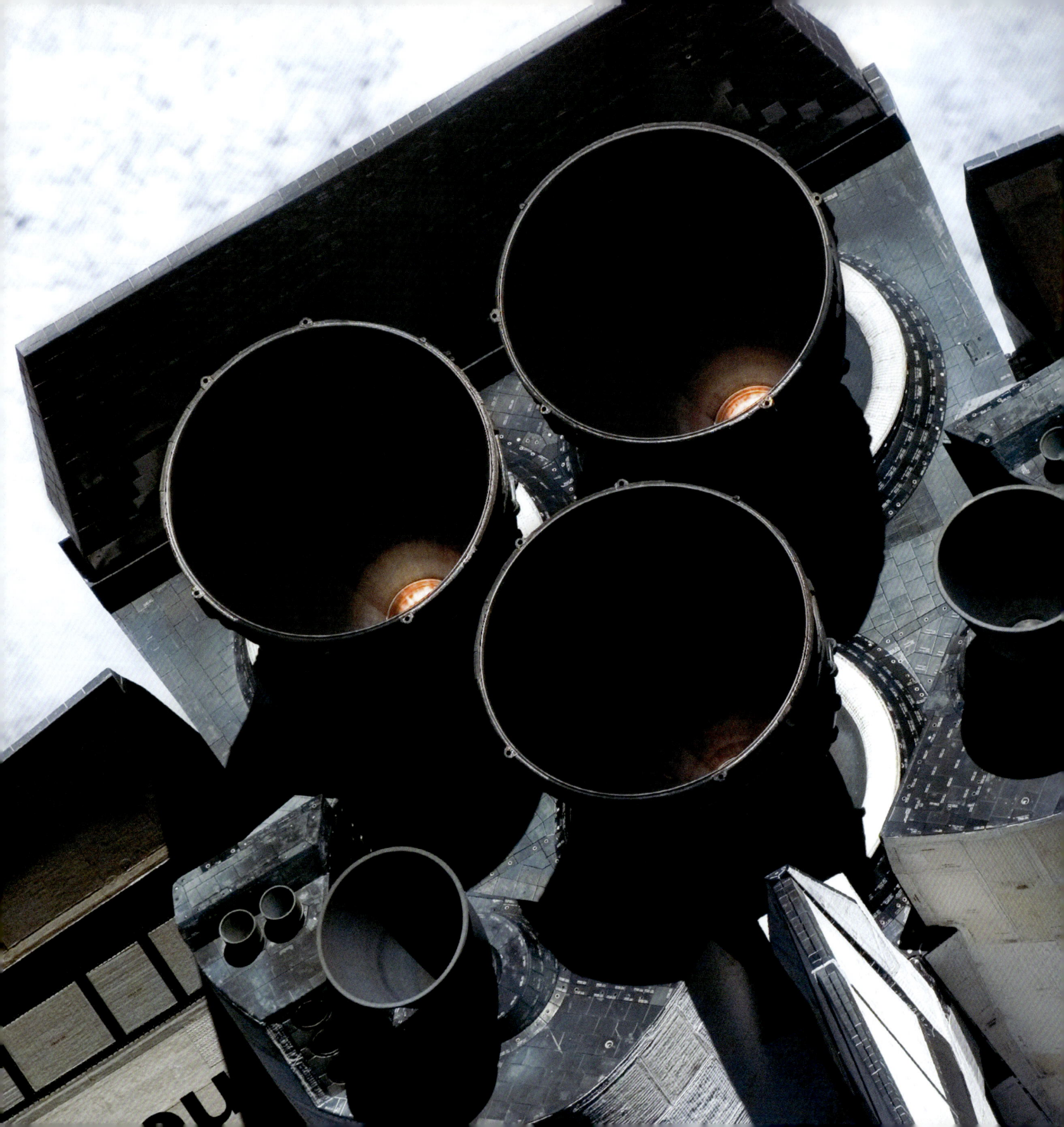

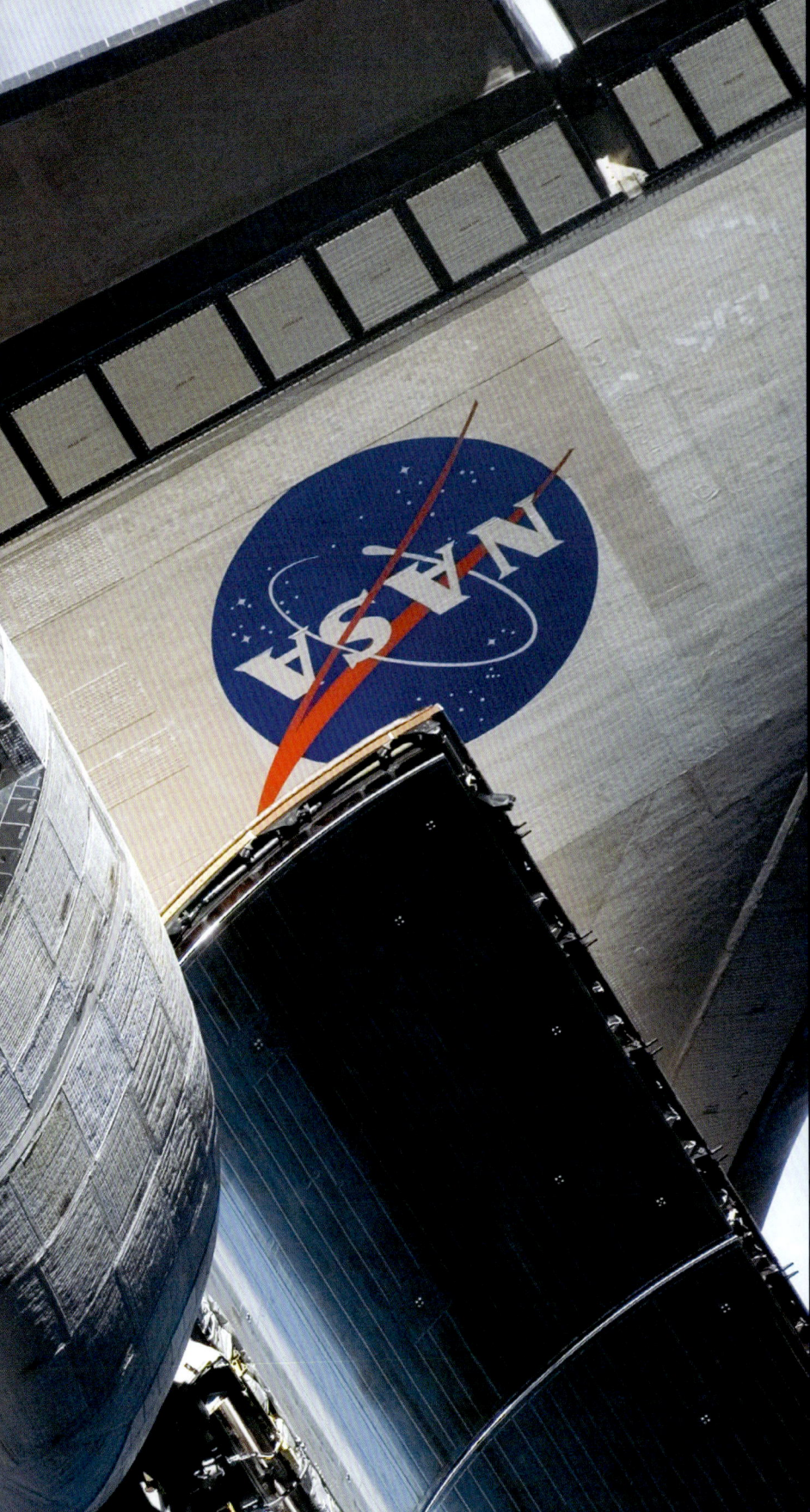

11 – 26 March 2008

This awesome view of Endeavour's tail section shows the shuttle's three main engines and the Orbital Maneuvering System pods, located either side of the shuttle's vertical stabilizer. A digital camera with a 400mm lens was used to capture this image, taken by a crew member aboard the International Space Station.

26 July to 9 August 2005

Commander Eileen M. Collins manoeuvres Discovery into a back-flip, which allows crew members aboard the International Space Station to capture detailed photographs of the shuttle's heat shield. This was the highly anticipated return-to-flight mission after the tragic loss of Columbia in 2003.

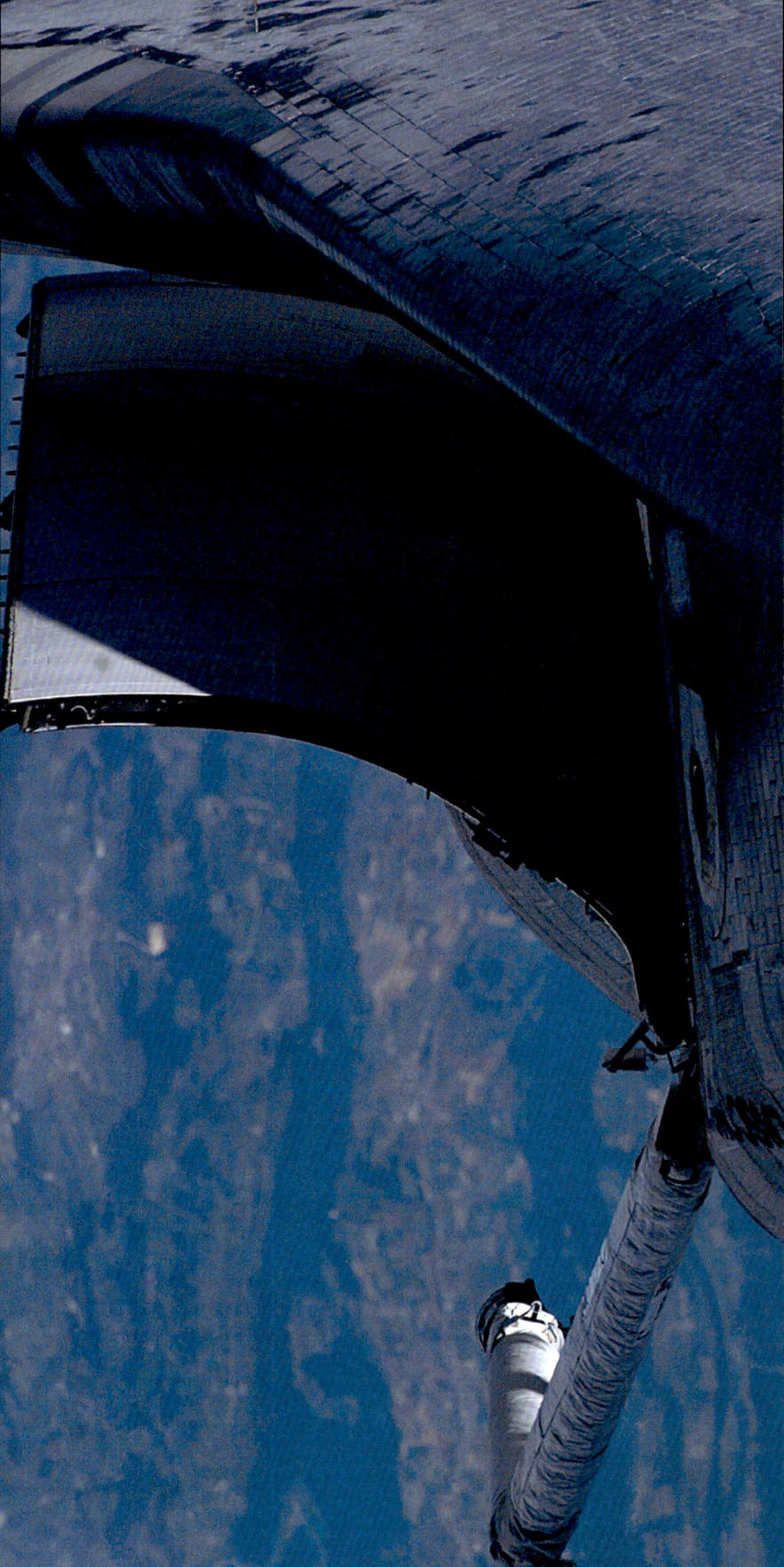

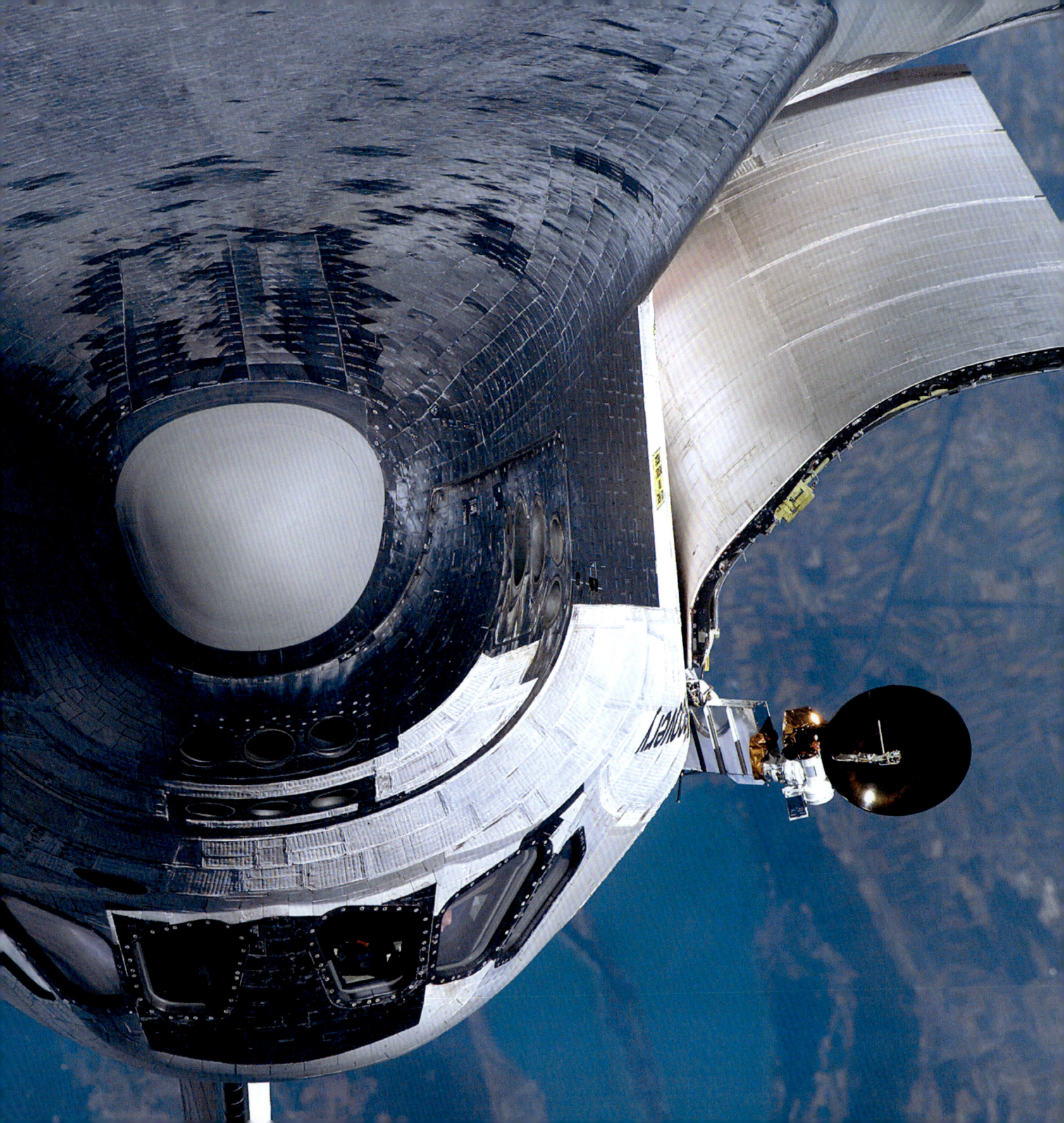

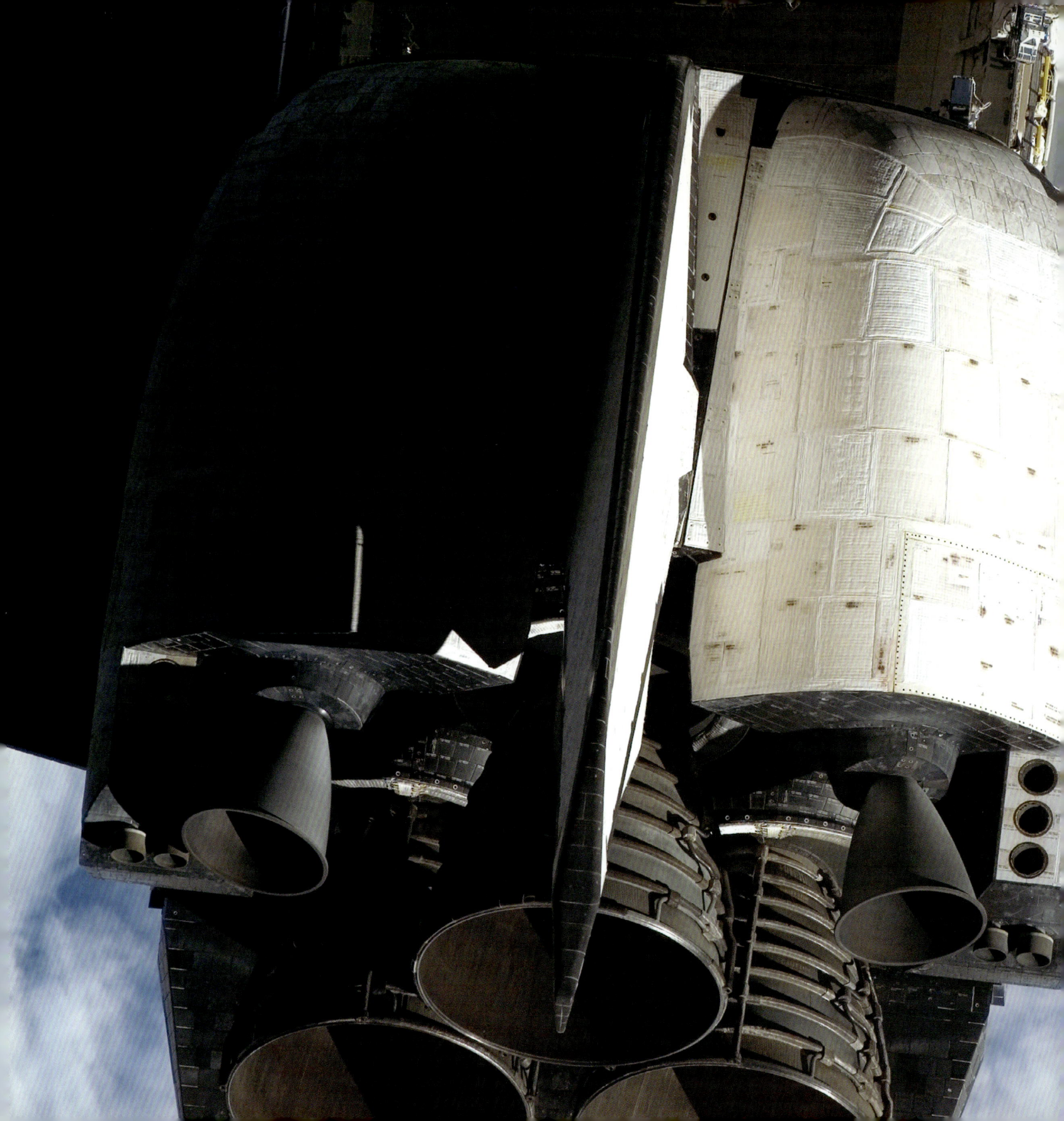

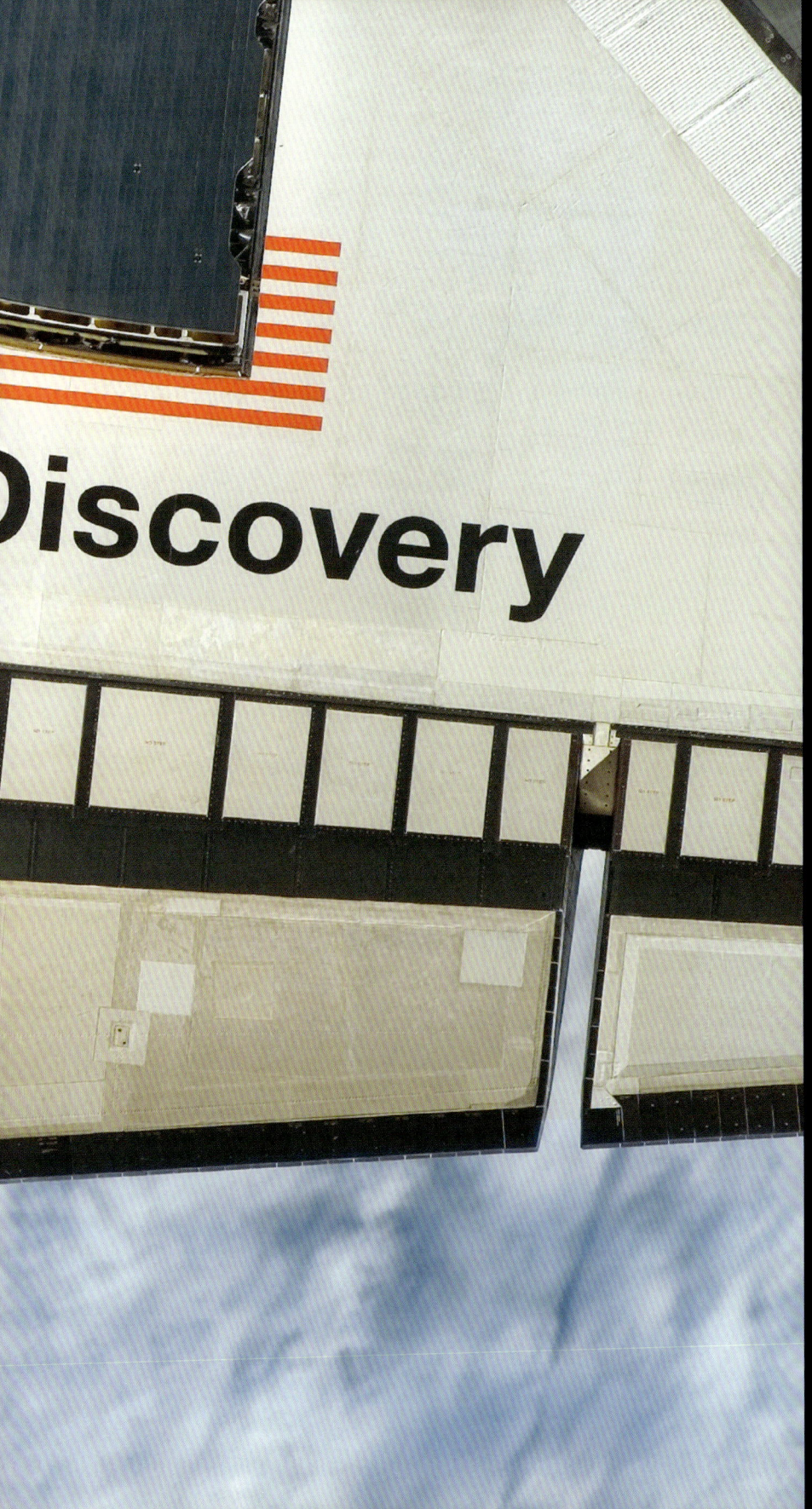

STS-120
23 October to 7 November 2007

(Left) Discovery performs a back-flip manoeuvre on approach to the International Space Station. Crew members aboard the space station are treated to a great show, as the shuttle flaunts its main engines and Orbital Maneuvering System pods, just as sunlight illuminates the name Discovery.

STS-134
16 May to 1 June 2011

(Following spread) This beautiful and rare image, taken by Expedition 27 crew member Paolo Nespoli from a Russian Soyuz spacecraft, shows Endeavour docked with the International Space Station for the final time during its last ever mission.

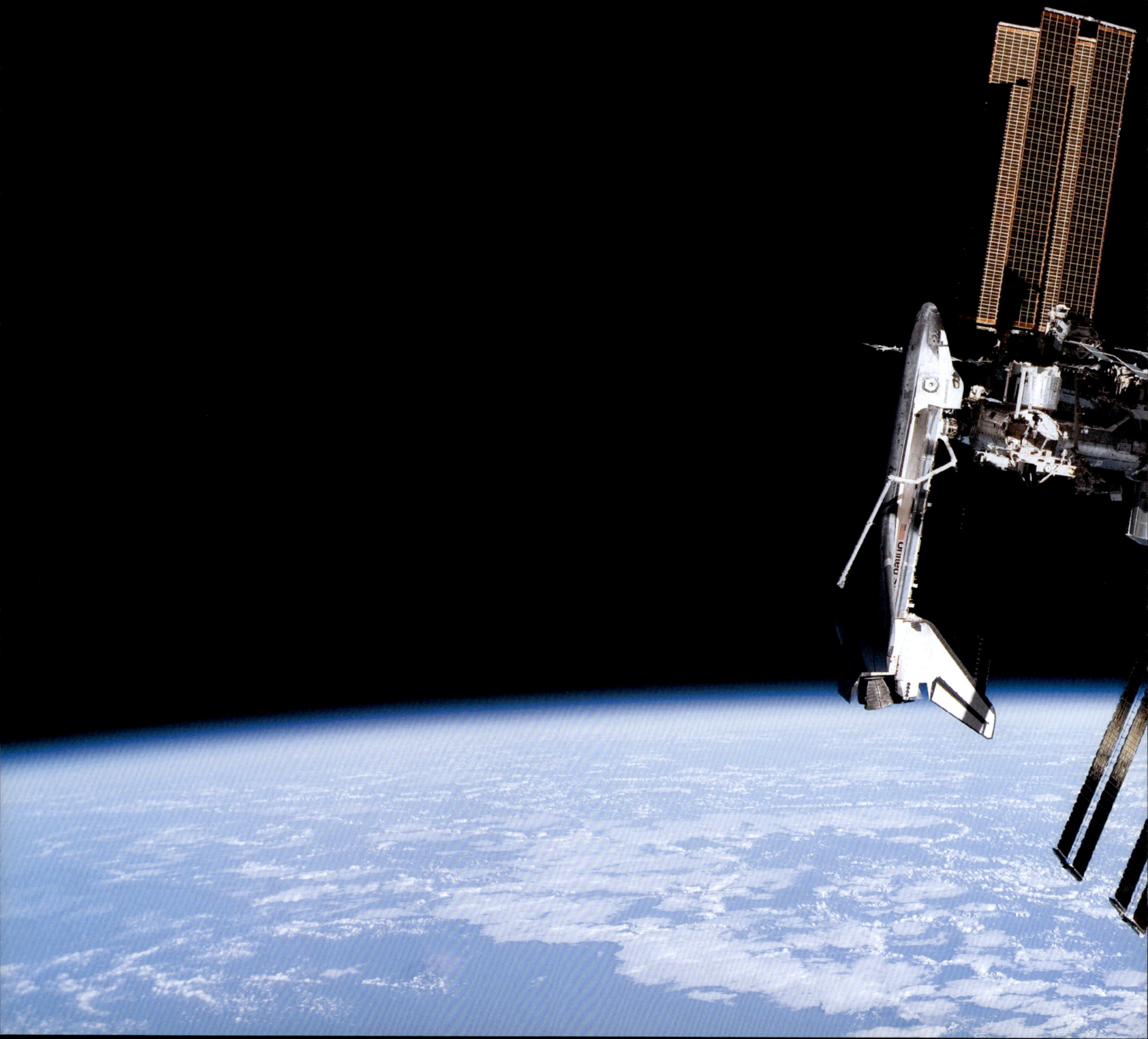

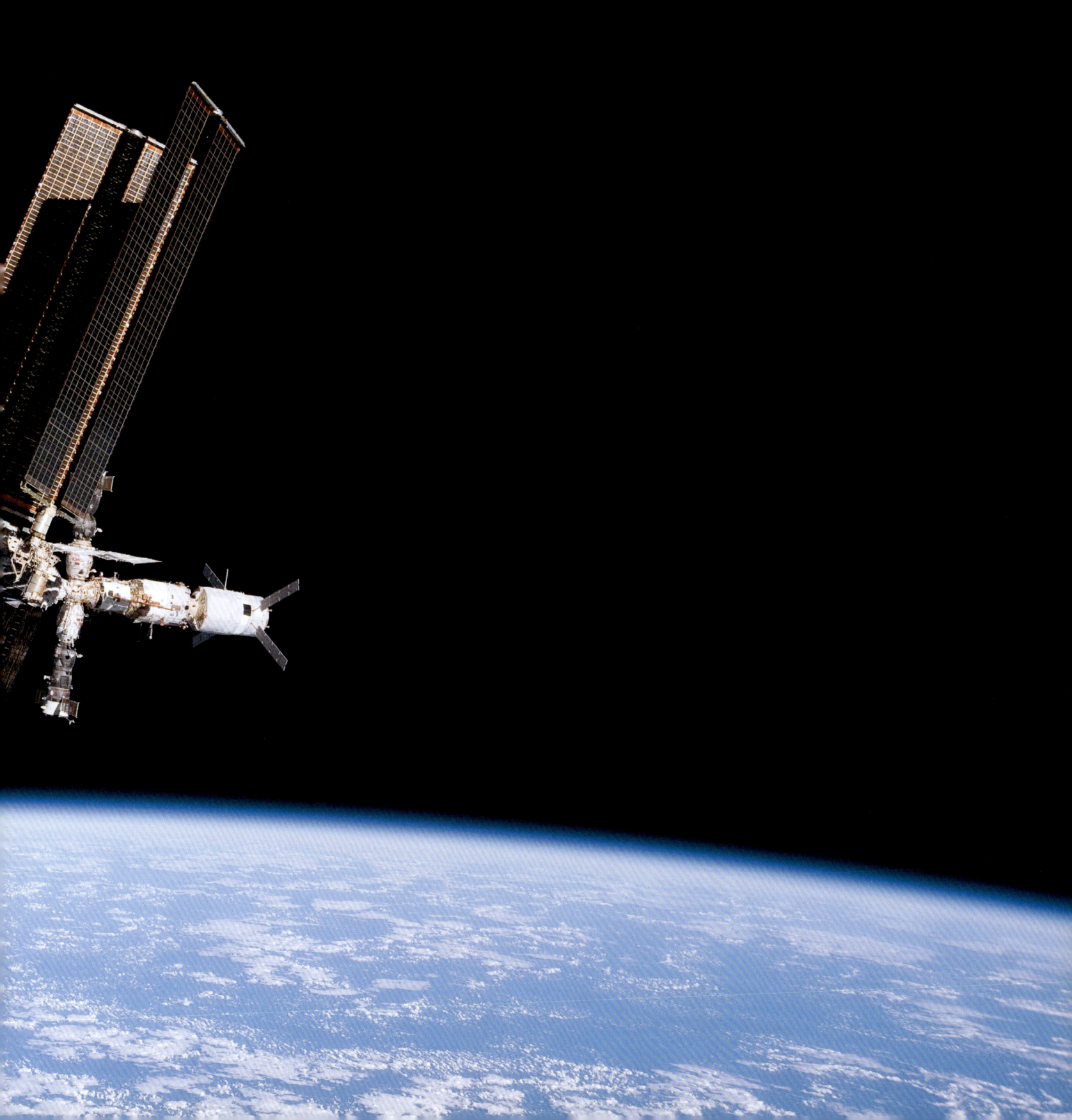

Discovery clings to the International Space Station, while Earth's horizon provides a stunning contrast with the deep blackness of space. This photograph was taken during one of three space walks, where the crew carried out repairs to the space station and shuttle.

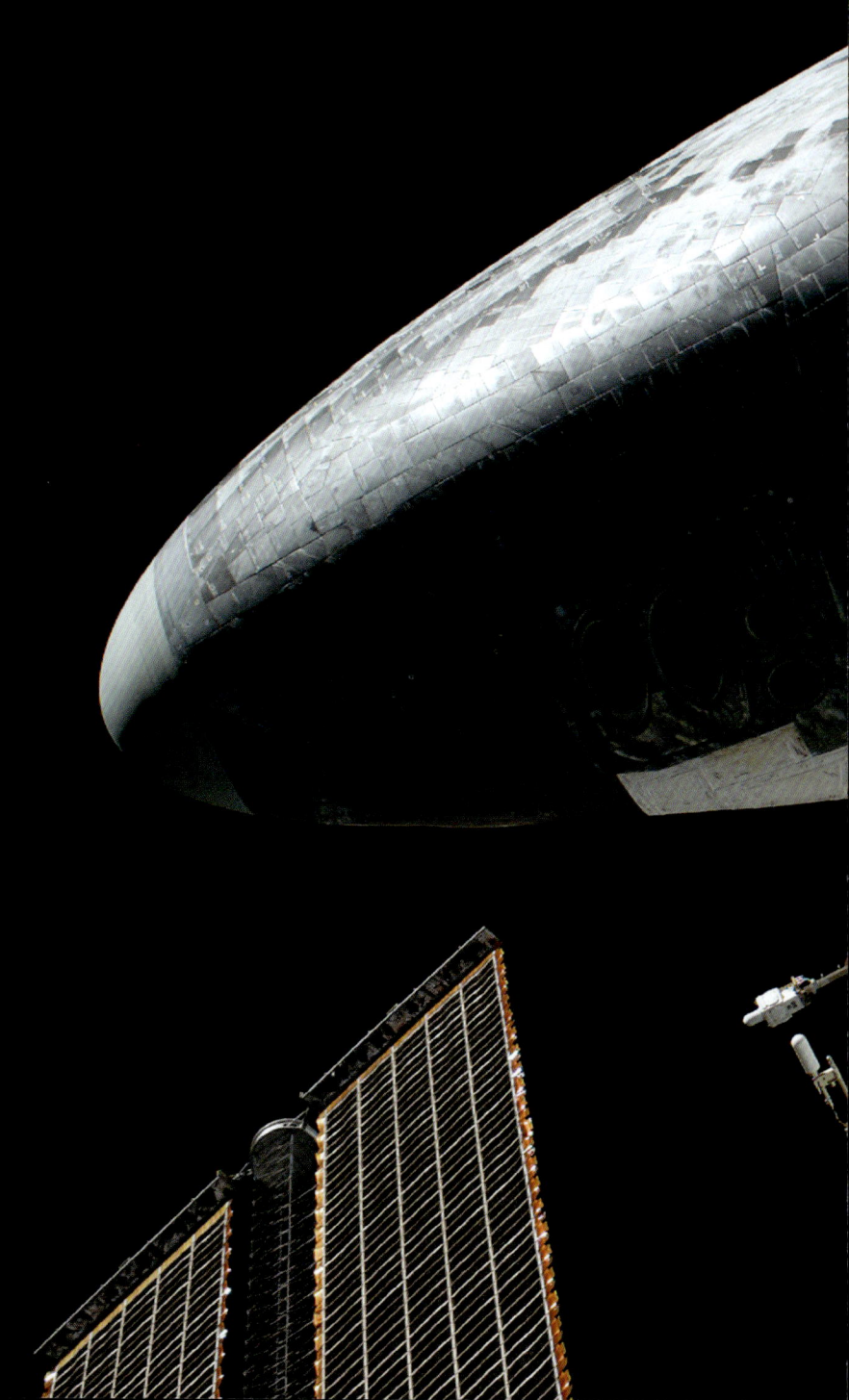

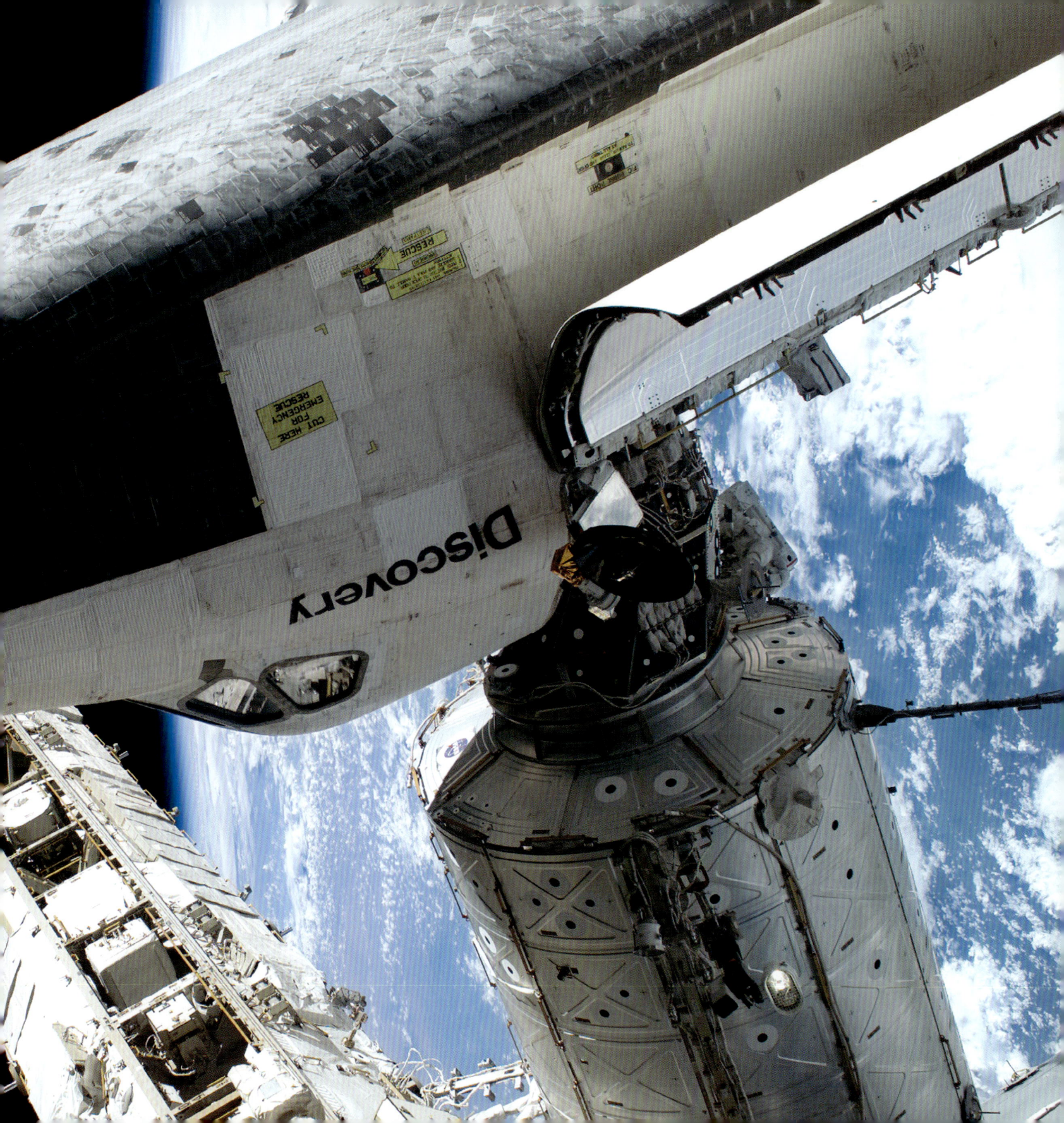

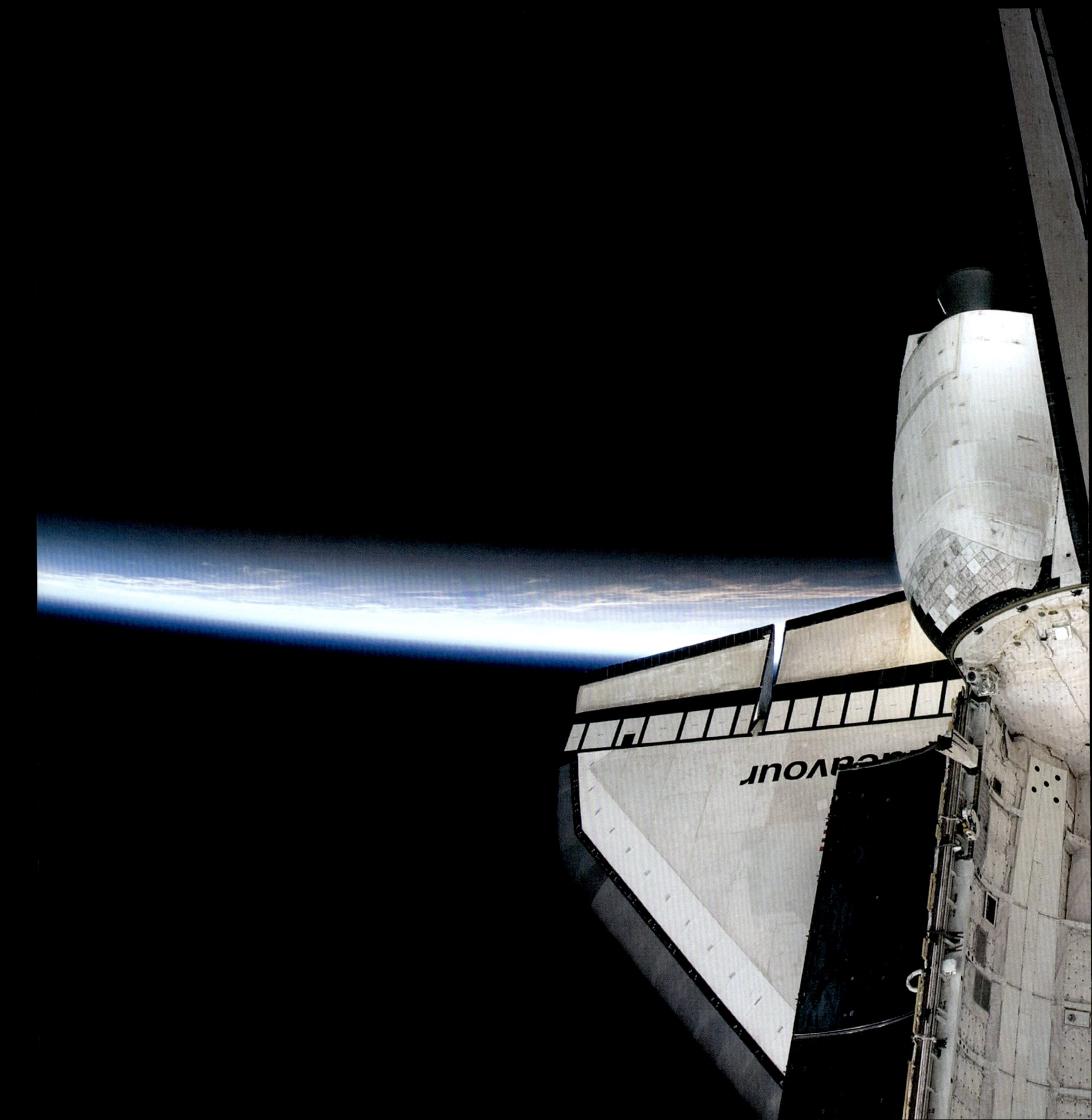

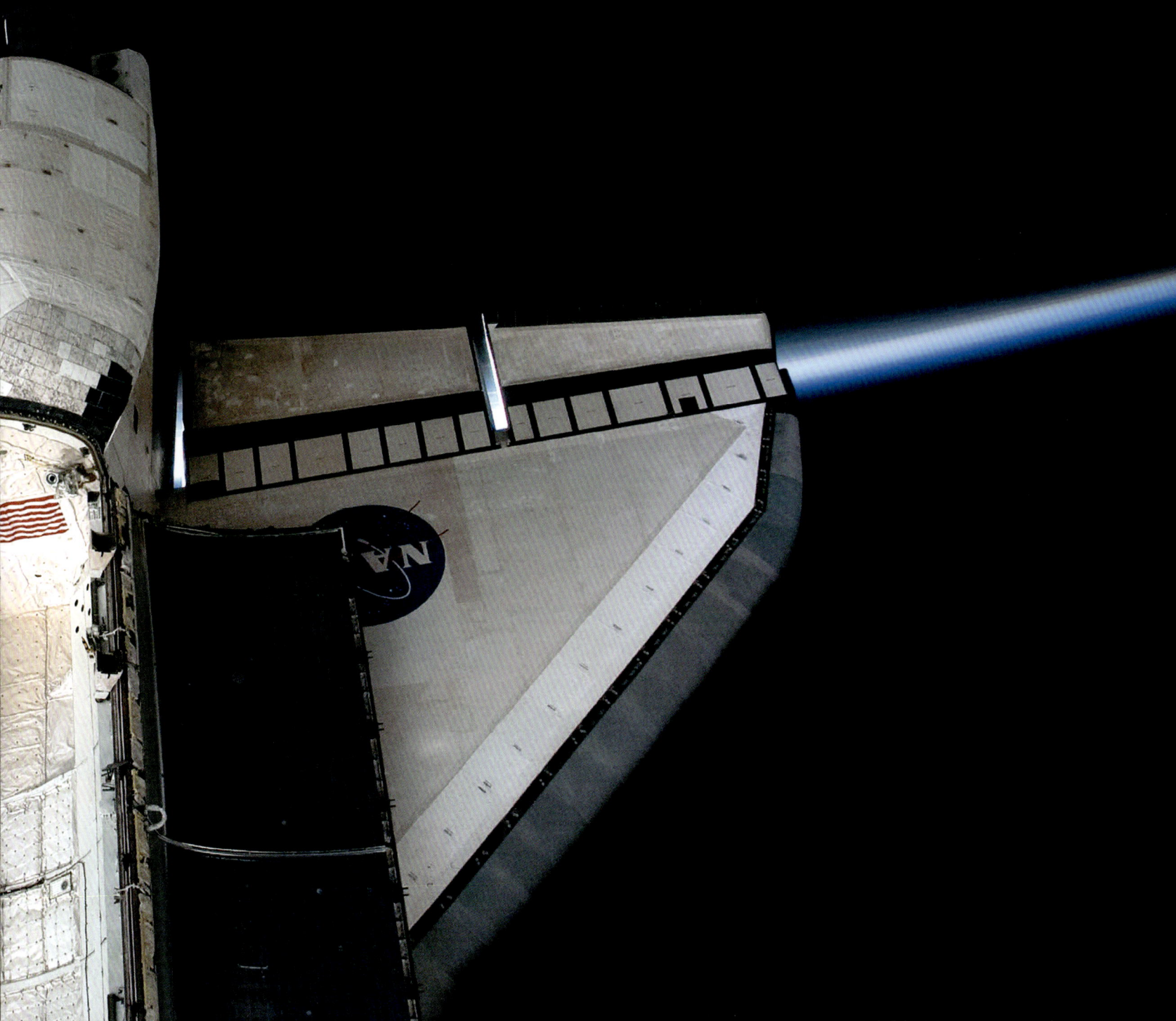

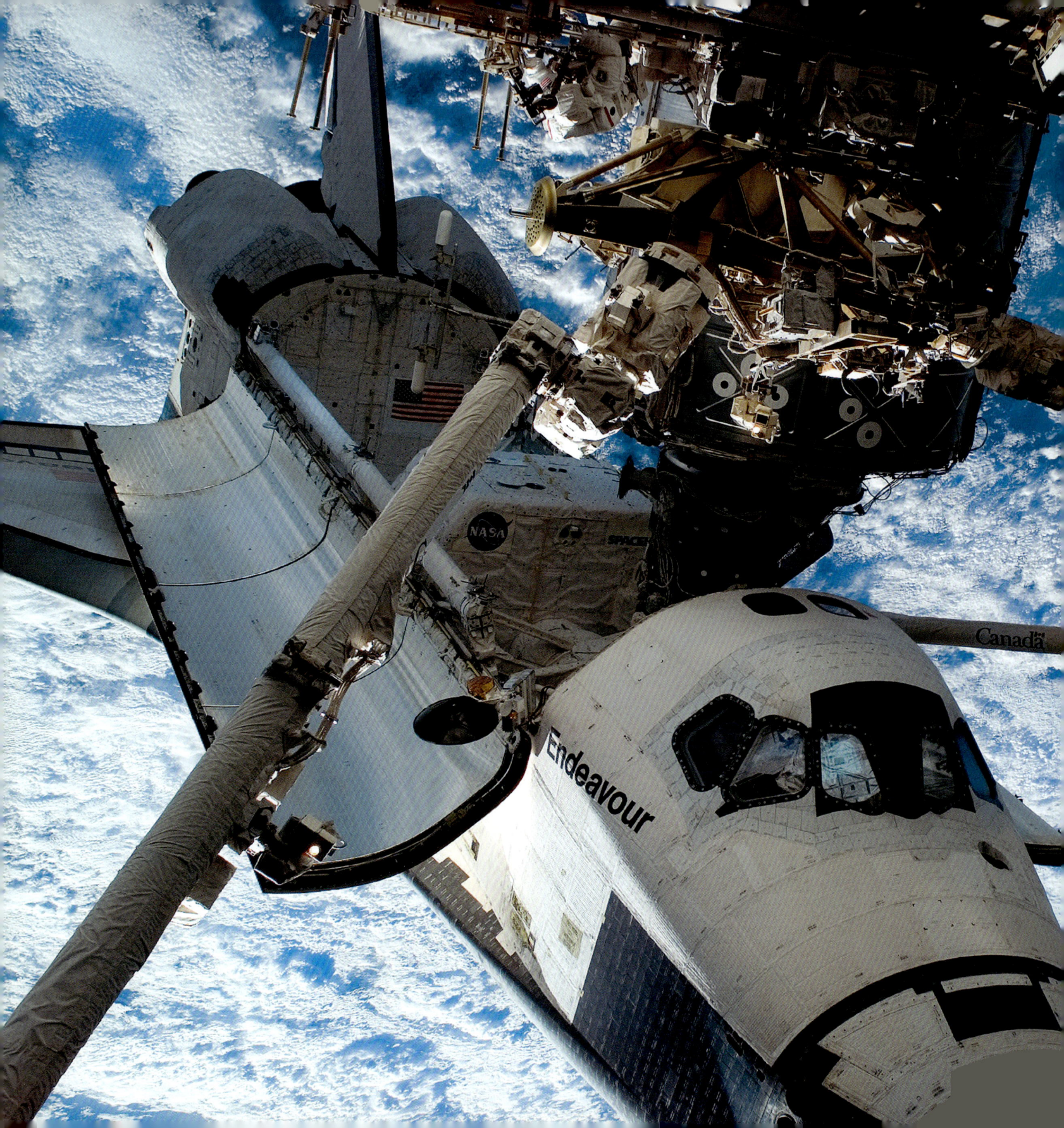

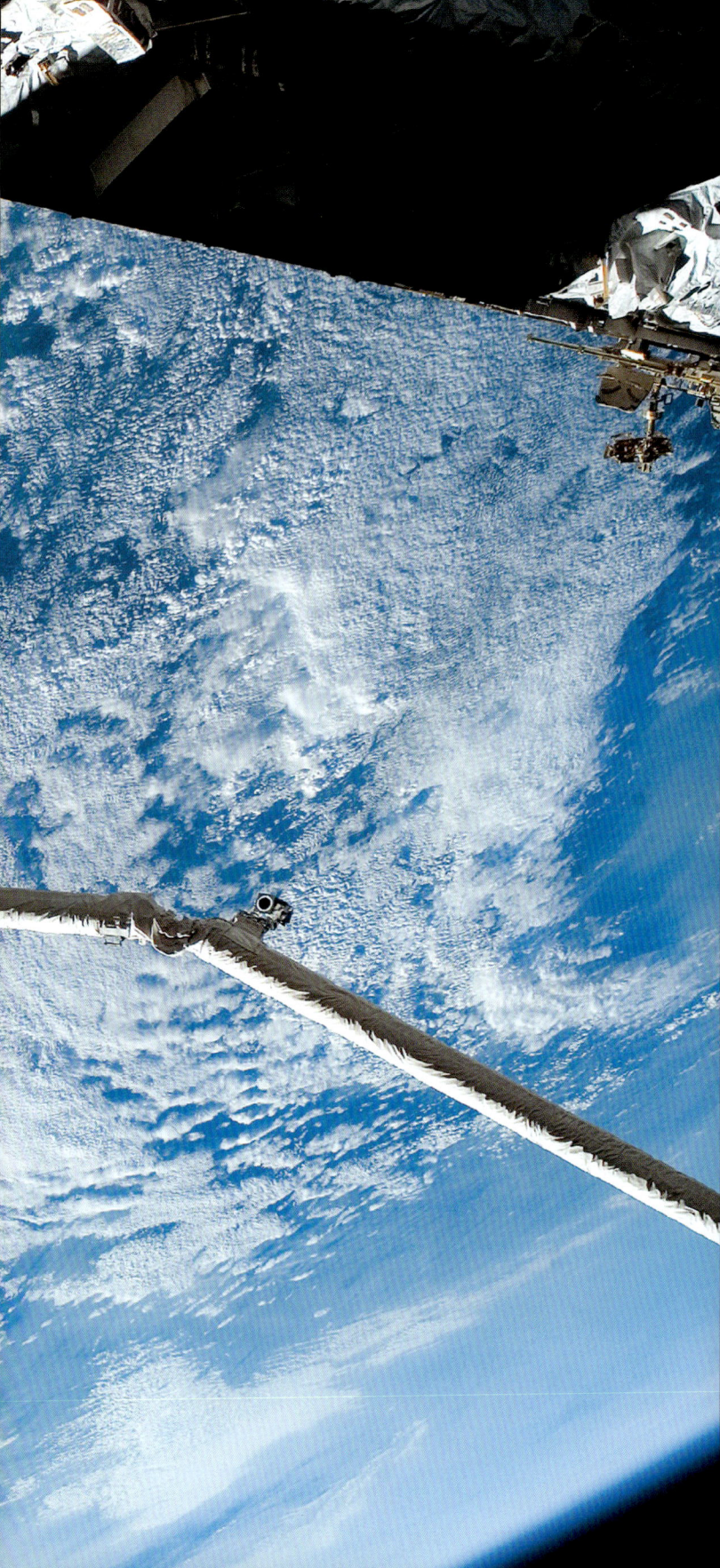

STS-130
8–21 February 2010

(Previous spread) The thin auroral horizon of planet Earth provides a simple yet striking setting, while Endeavour's aft section is photographed with the payload bay doors open wide, boasting nothing but the proud stars and stripes.

STS-118
8–21 August 2007

(Left) Taken during an astronaut space walk, Endeavour can be seen docked to the International Space Station (ISS) with a dramatic Earth backdrop. The shuttle delivered fresh supplies and vital components, including a new starboard S5 truss segment to continue construction of the ISS.

STS-132
14–26 May 2010

With nearly twelve full days in space and seven days docked to the International Space Station, Atlantis was due to retire after this mission. However, NASA later announced that Atlantis would return to space one last time on STS-135, bringing an end to the space shuttle program for good. This photograph was captured by an STS-132 astronaut during the first of three Extra Vehicular Activity (EVA) space walks.

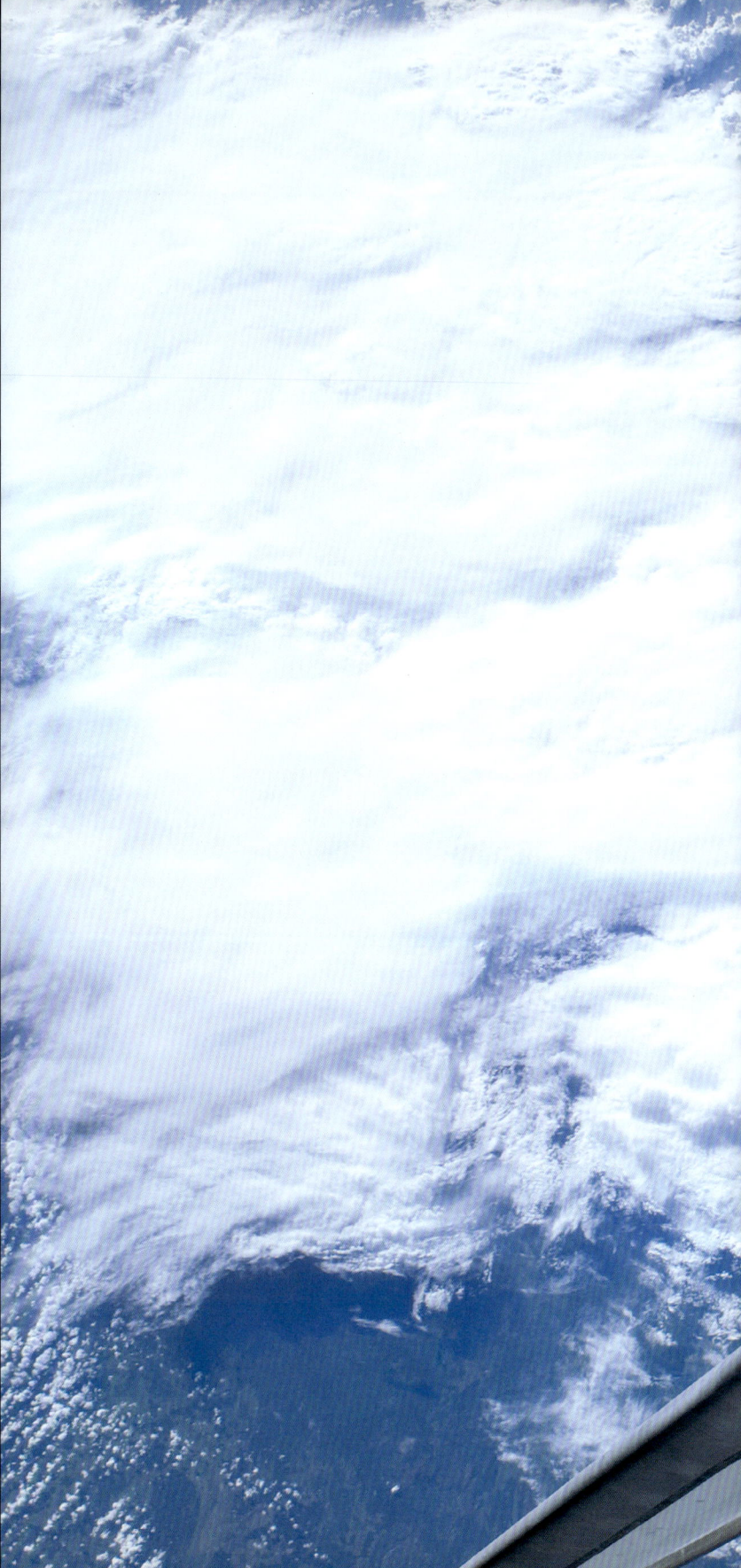

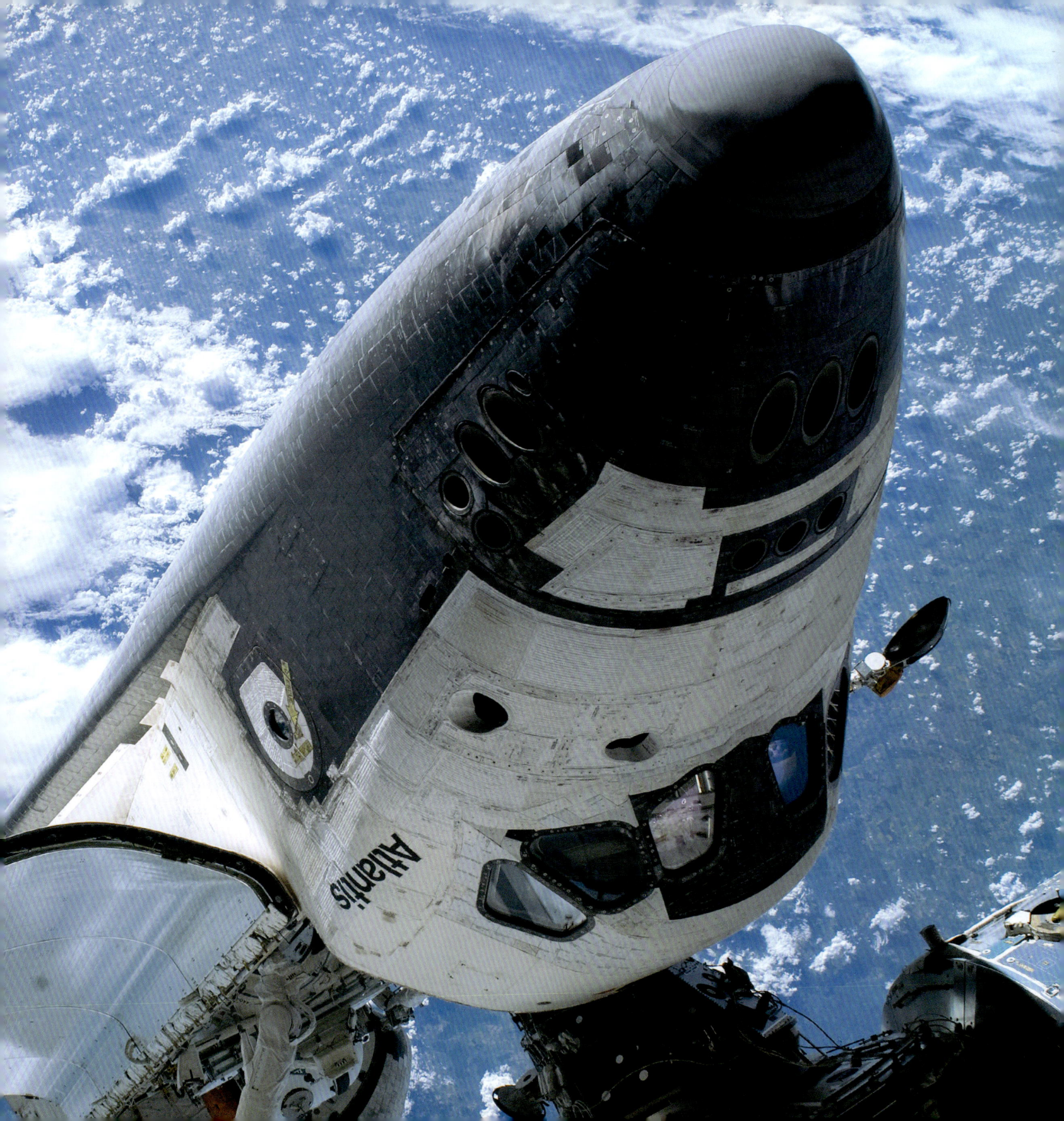

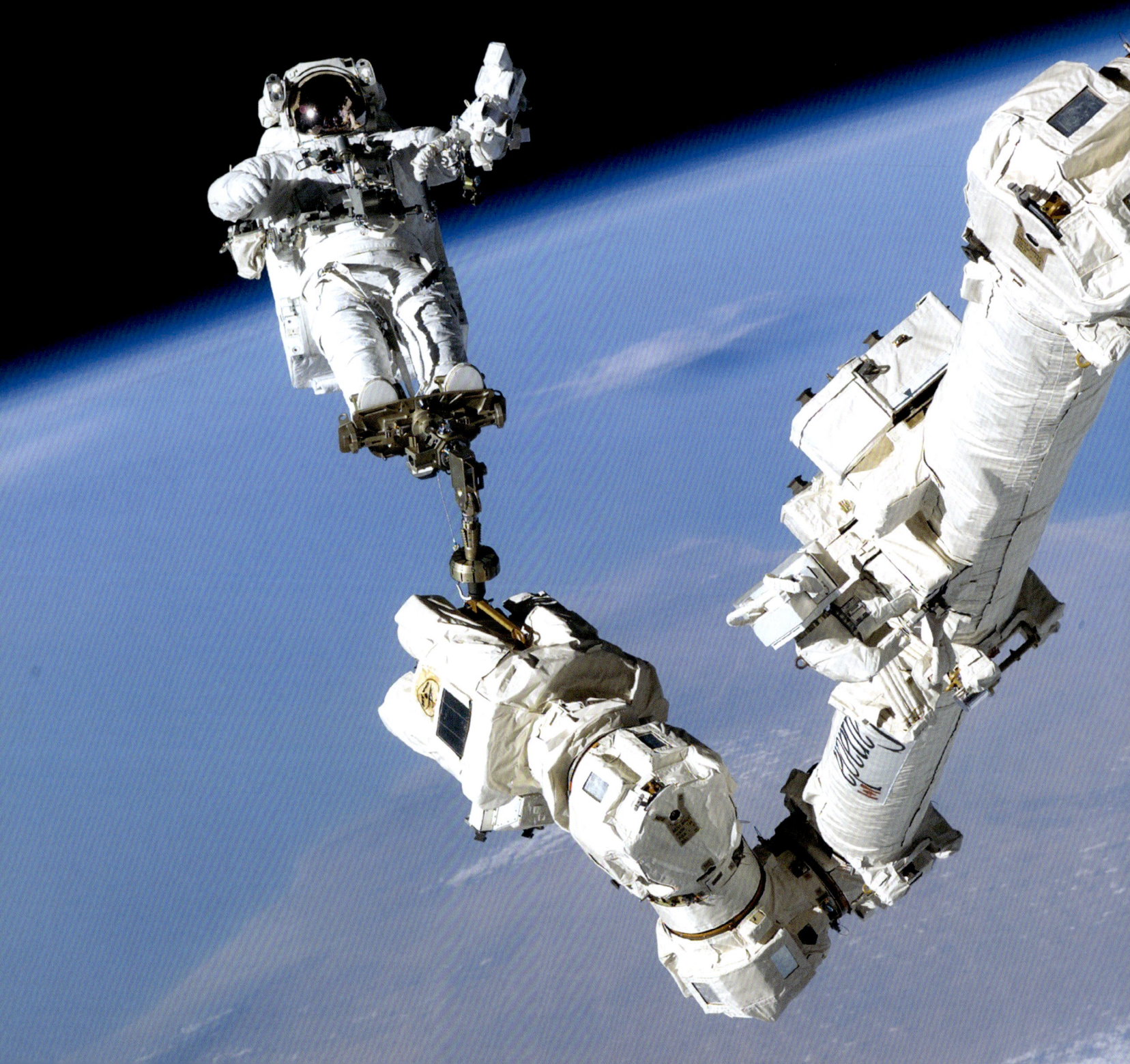

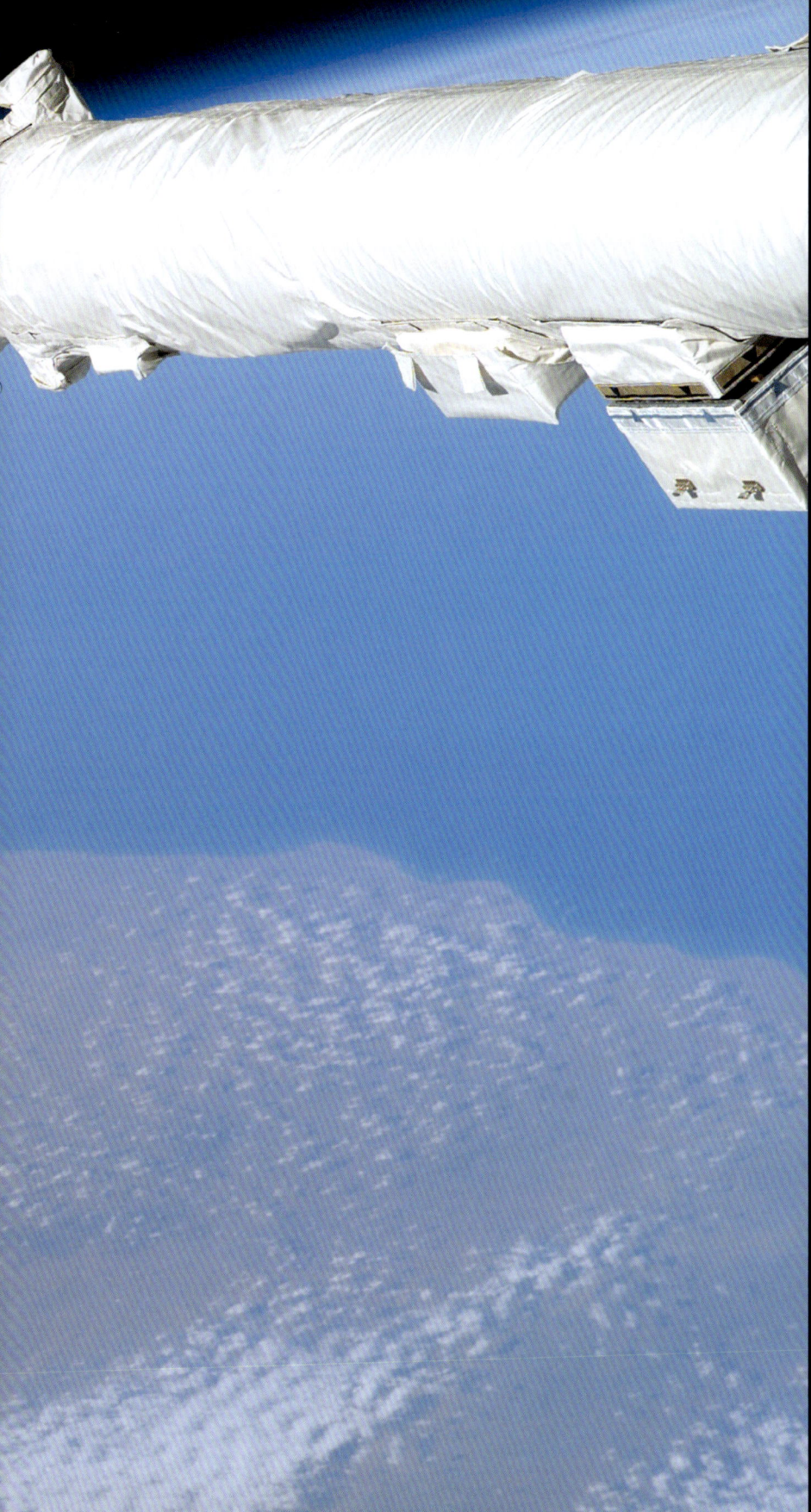

26 July to 9 August 2005

While Discovery is safely docked to the International Space Station, mission specialist Stephen Robinson hangs in space while attached to the station's Canadarm2 during the mission's third Extra Vehicular Activity (EVA).

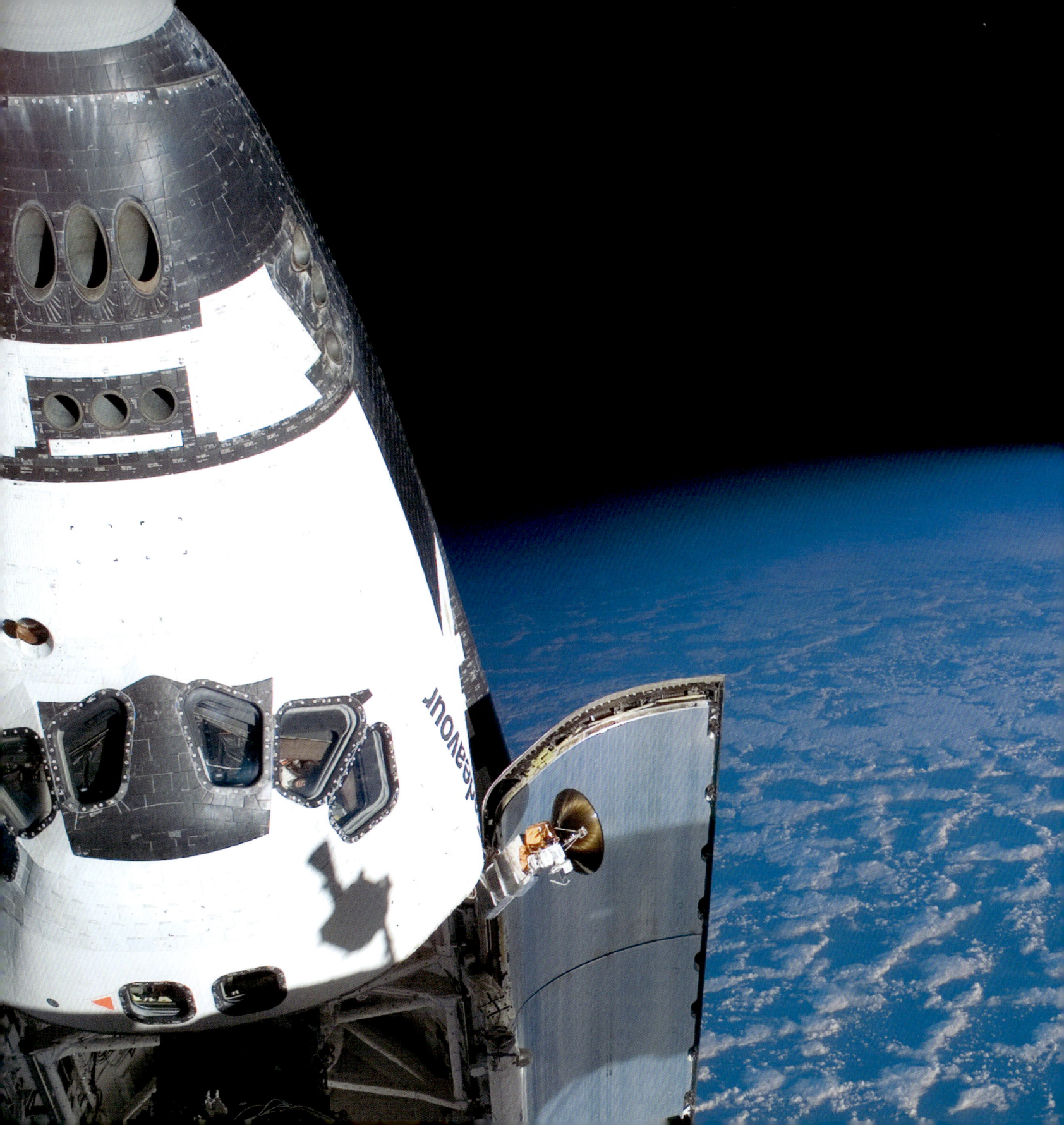

Endeavour set against the rich-blue Earth while docked with the International Space Station (ISS). The seven-person crew was assigned to continue construction of the ISS, including the installation of a third starboard truss segment. STS-118 was the 22nd shuttle mission to the space station.

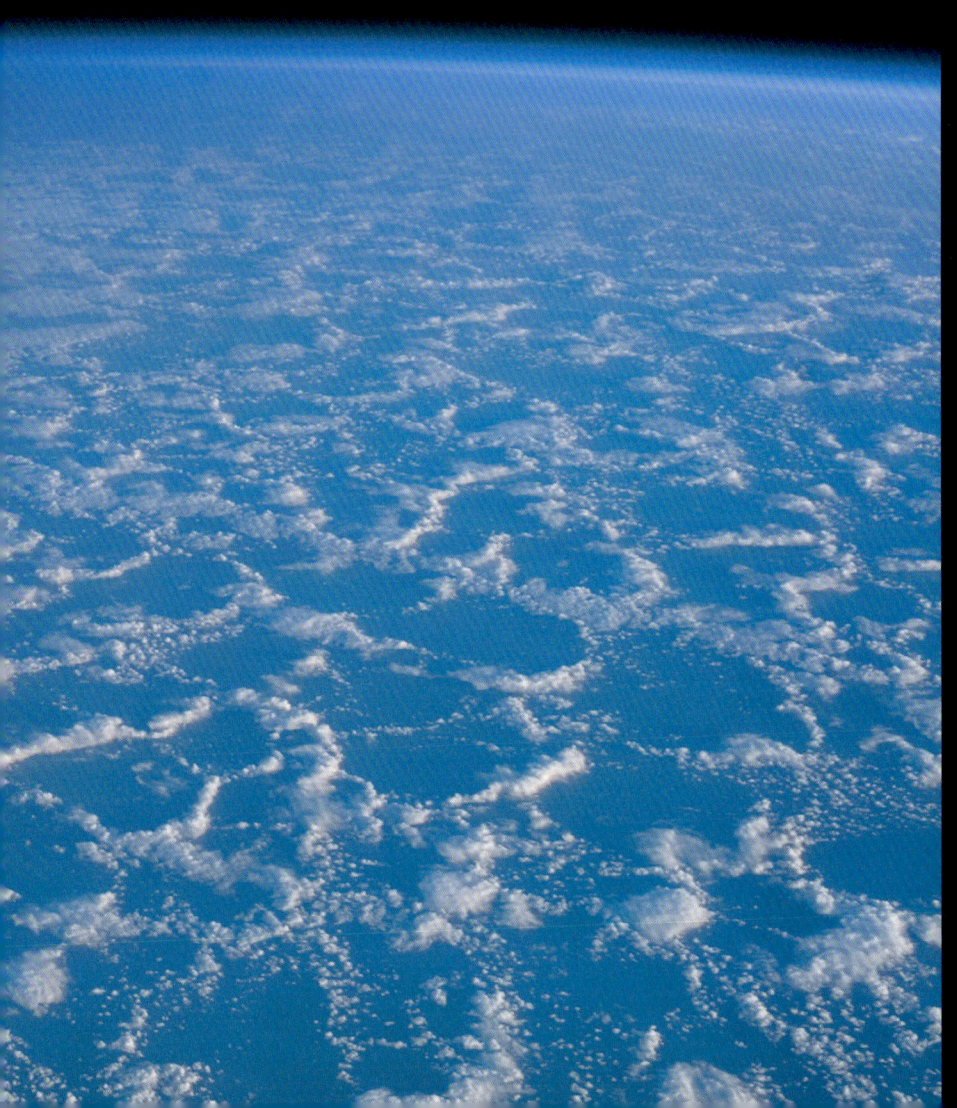

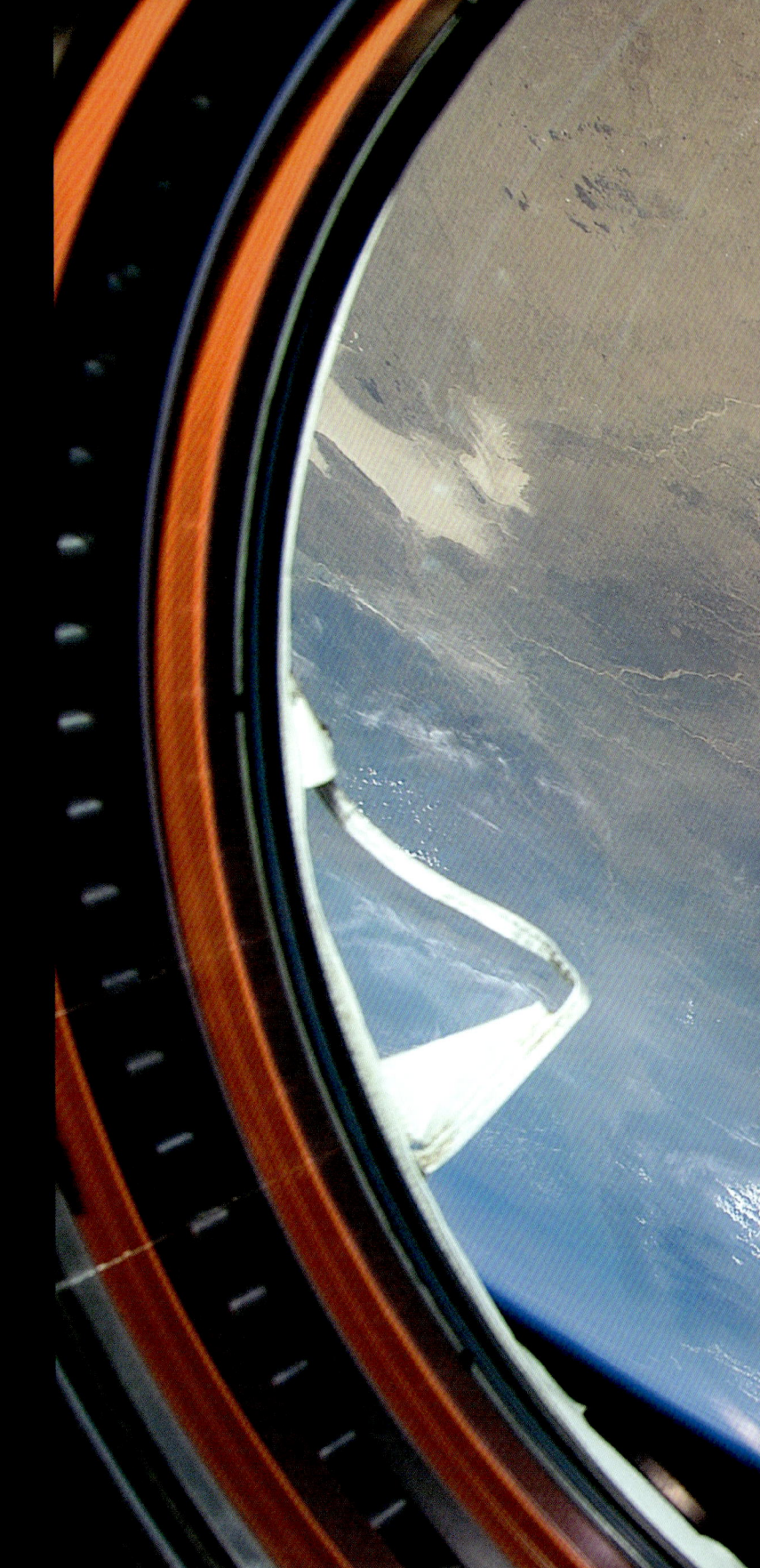

STS-102
8–21 March 2001

Watching the world go by: Space shuttle Discovery and Earth as seen from the International Space Station (ISS). The mission objectives were mainly the delivery of new supplies to the ISS, as well as replacing the Expedition 1 space station crew with the Expedition 2 crew after nearly five months aboard the ISS.

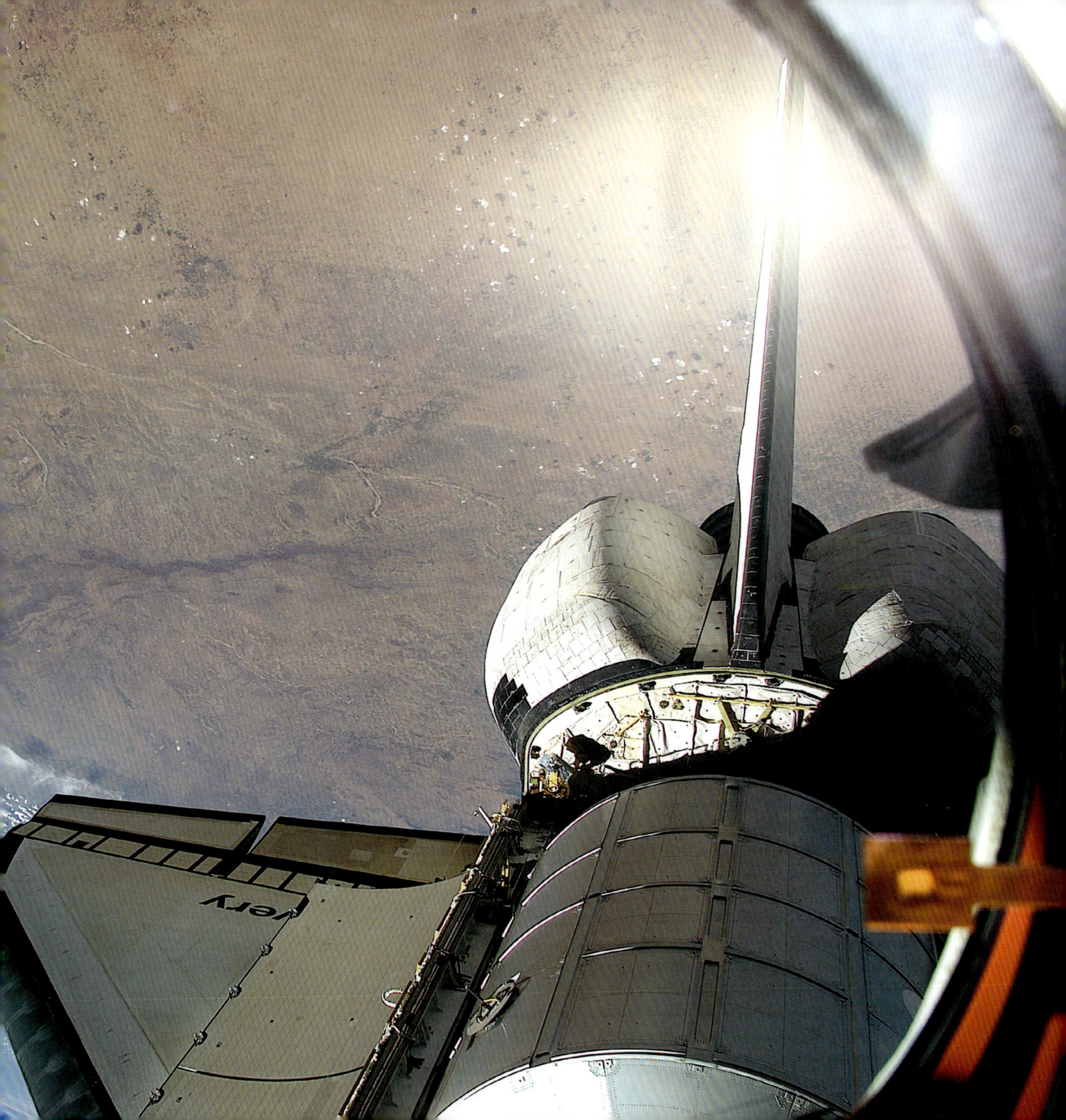

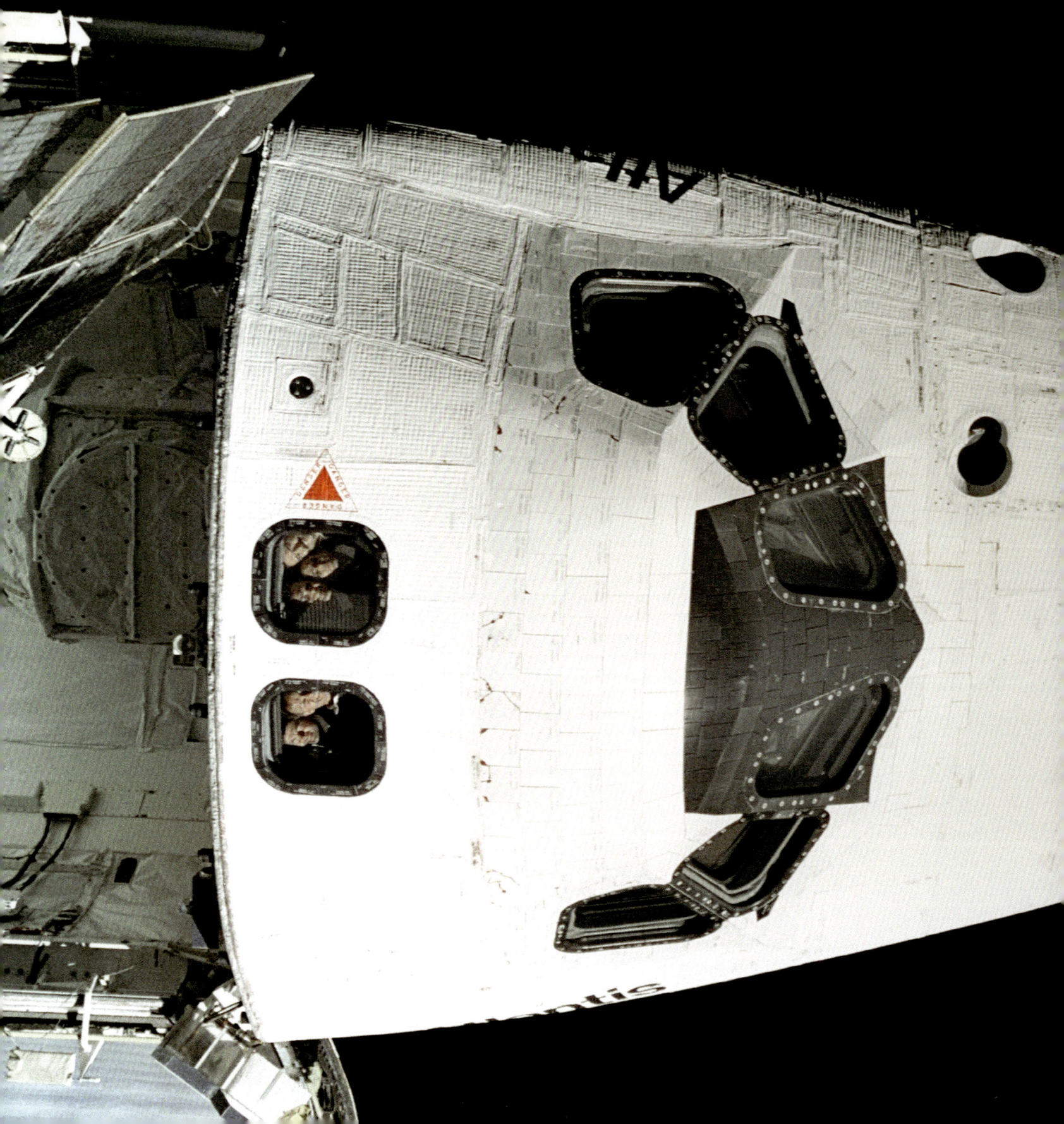

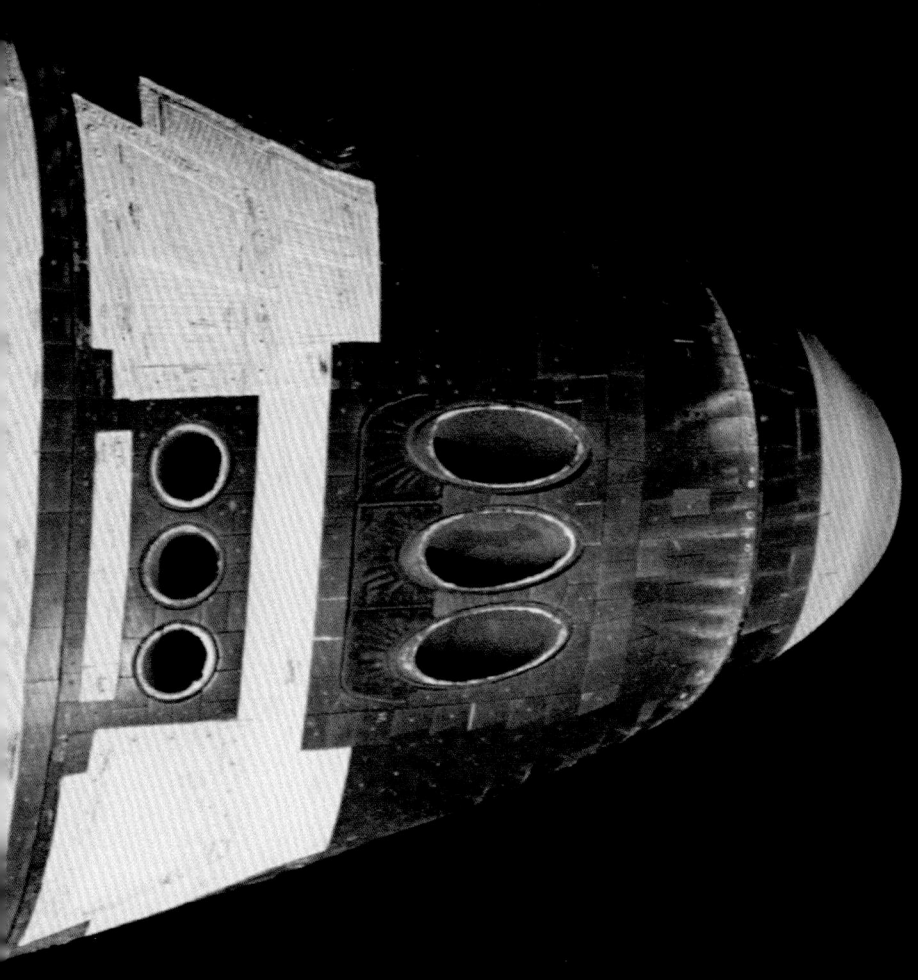

STS-74
12–20 November 1995

As Atlantis approaches the Russian Mir space station for rendezvous, crew members aboard Mir capture a fascinating image of their new guests. While the astronauts gaze out into the infinite blackness, the shuttle's robust structure provides a safe haven from the hostile and deadly vacuum of space.

STS-71
27 June to 7 July 1995

(Right) An inspiring profile view of Atlantis while docked to the Russian Mir space station. The mission marked the first time an American shuttle had docked with a Russian space station. This photograph was taken by Russian cosmonauts aboard their Soyuz spacecraft.

STS-79
16–26 September 1996

(Following spread: Left) Atlantis set against a dramatic Earth backdrop while docked with Mir during a 10-day mission. The mission conducted the first American astronaut exchange aboard the Russian space station. Shannon Lucid returned to Earth with Atlantis after serving 179 days aboard Mir and 188 total days in space. John Blaha replaced Lucid to begin a four-month stay, before returning home aboard STS-81 in January 1997.

STS-71
27 June to 7 July 1995

(Following spread: Right) Space shuttle Atlantis photographed through a fish-eye lens while docked with the Russian Mir space station. STS-71 was the 100th US piloted space flight and the first US space mission to return home with more crew members than when it departed. Atlantis delivered two cosmonauts to the Mir space station and returned to Earth with the Mir EO-18 crew, consisting of two cosmonauts and one US astronaut.

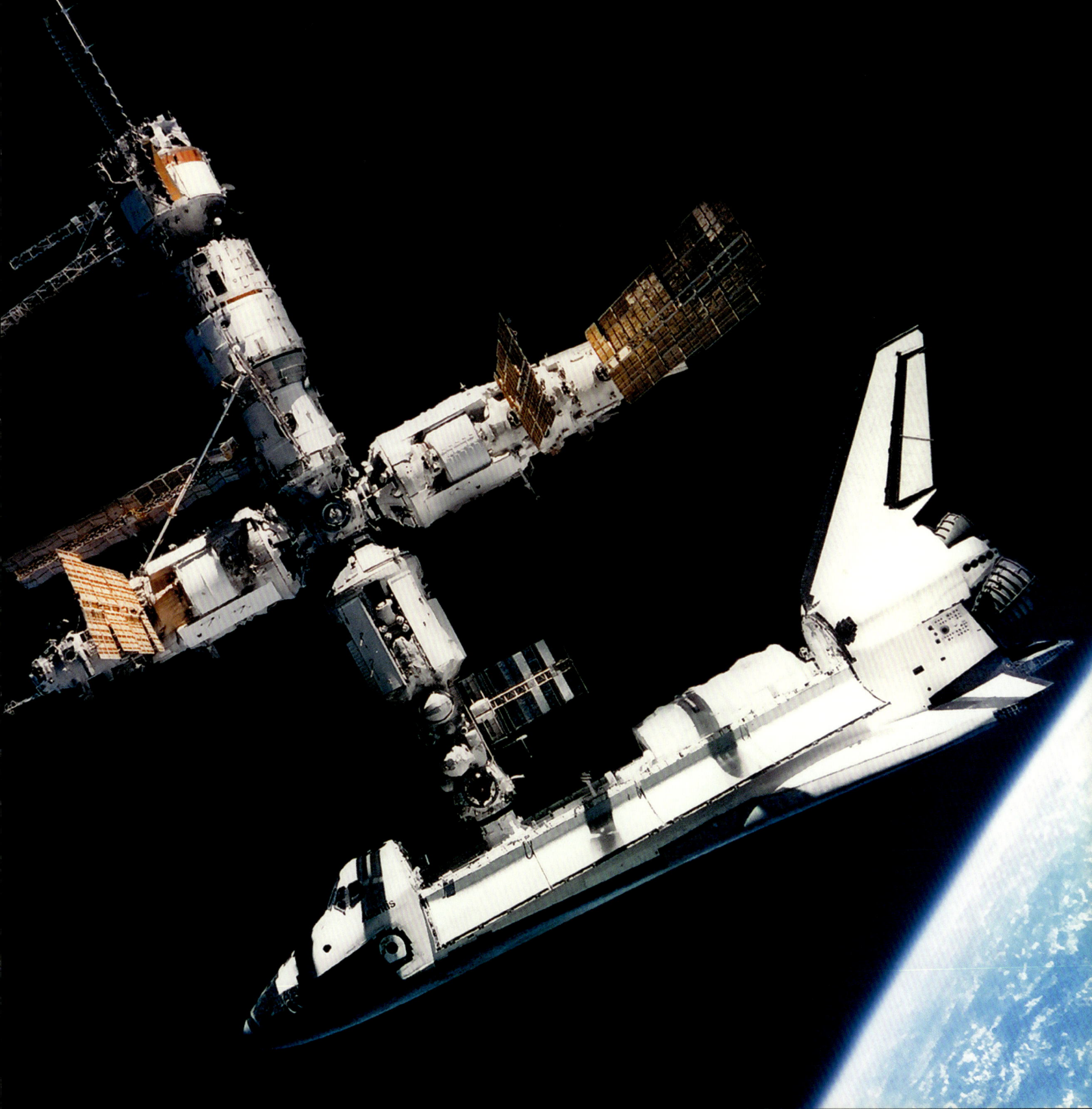

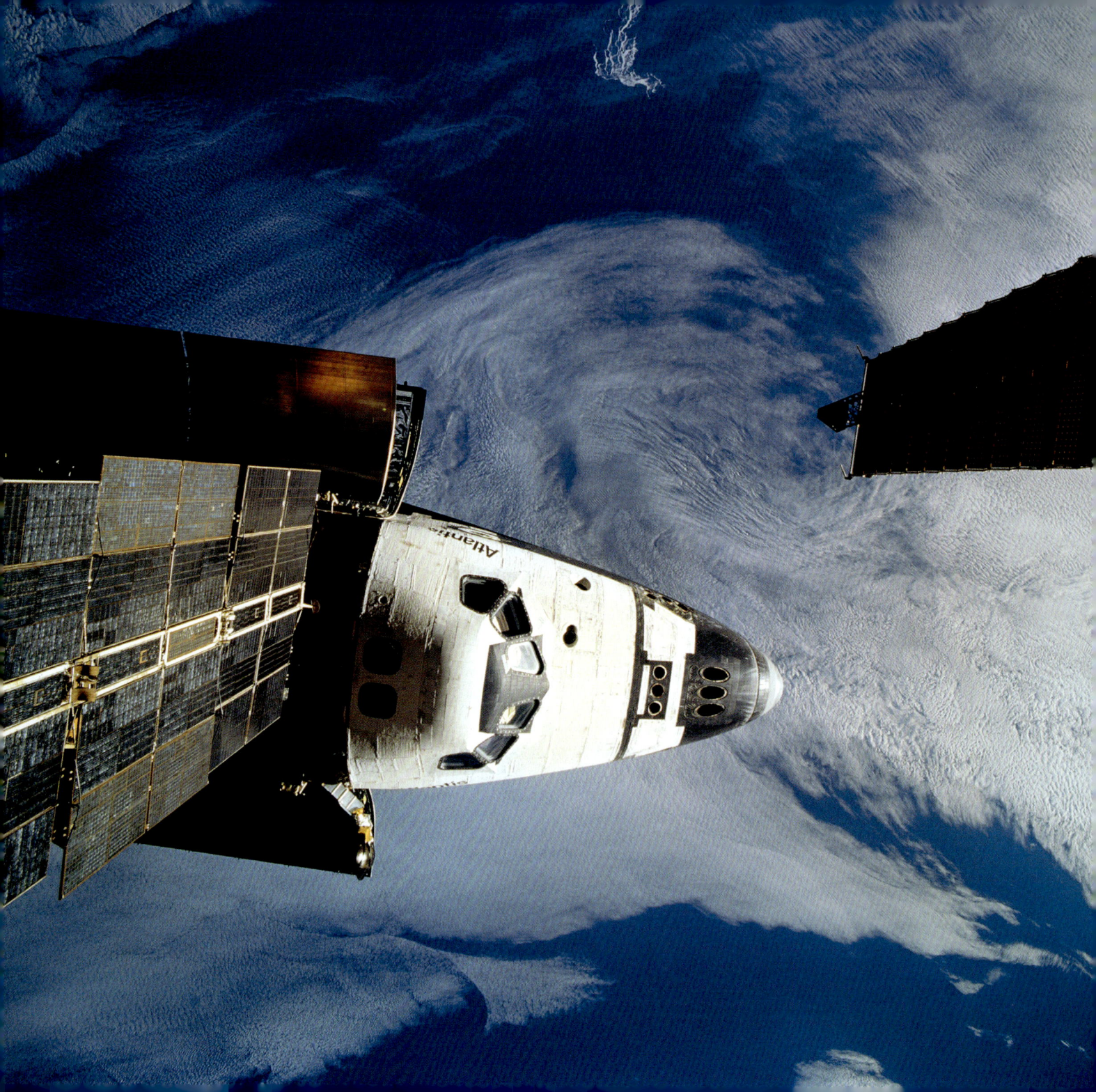

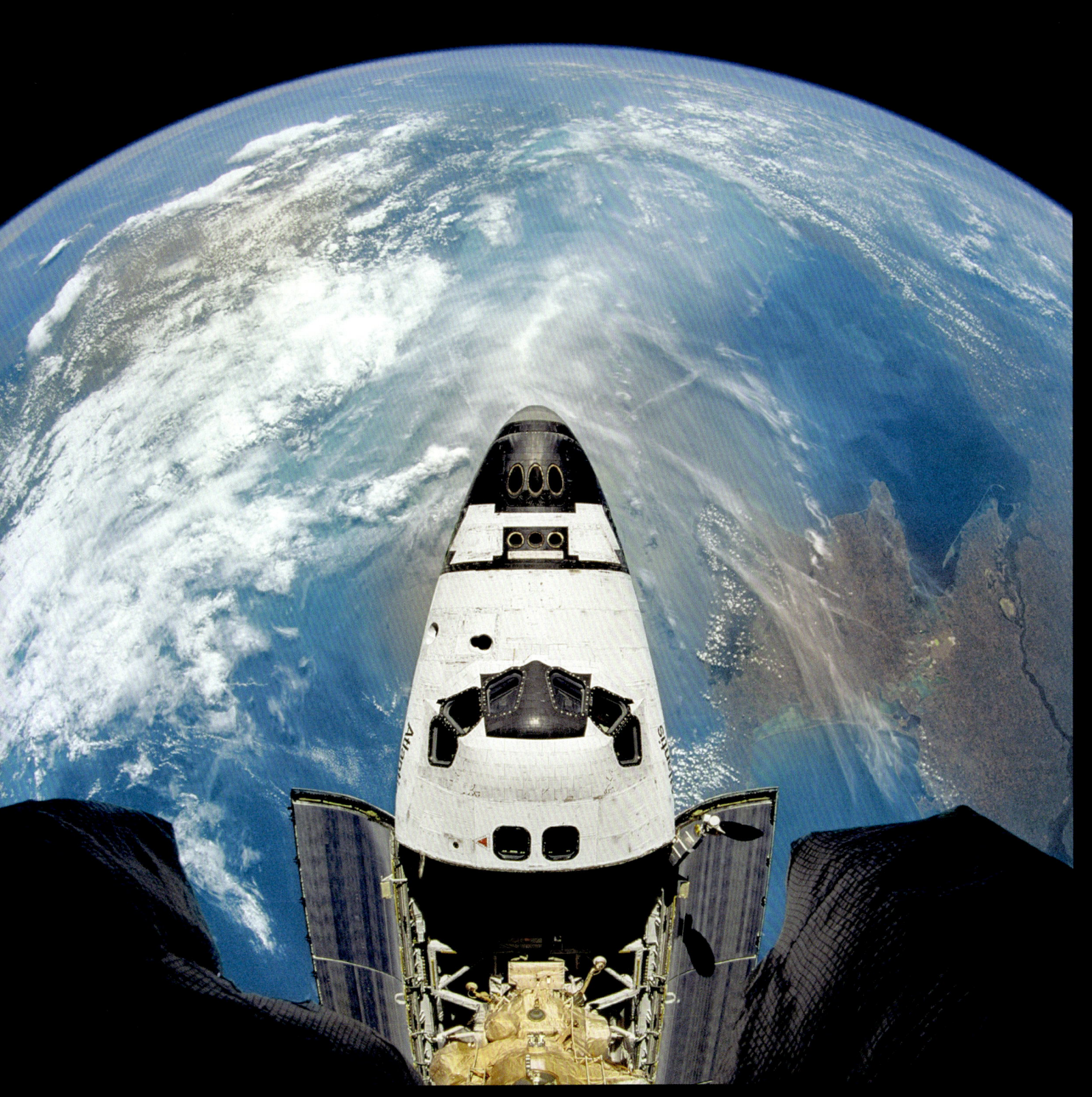

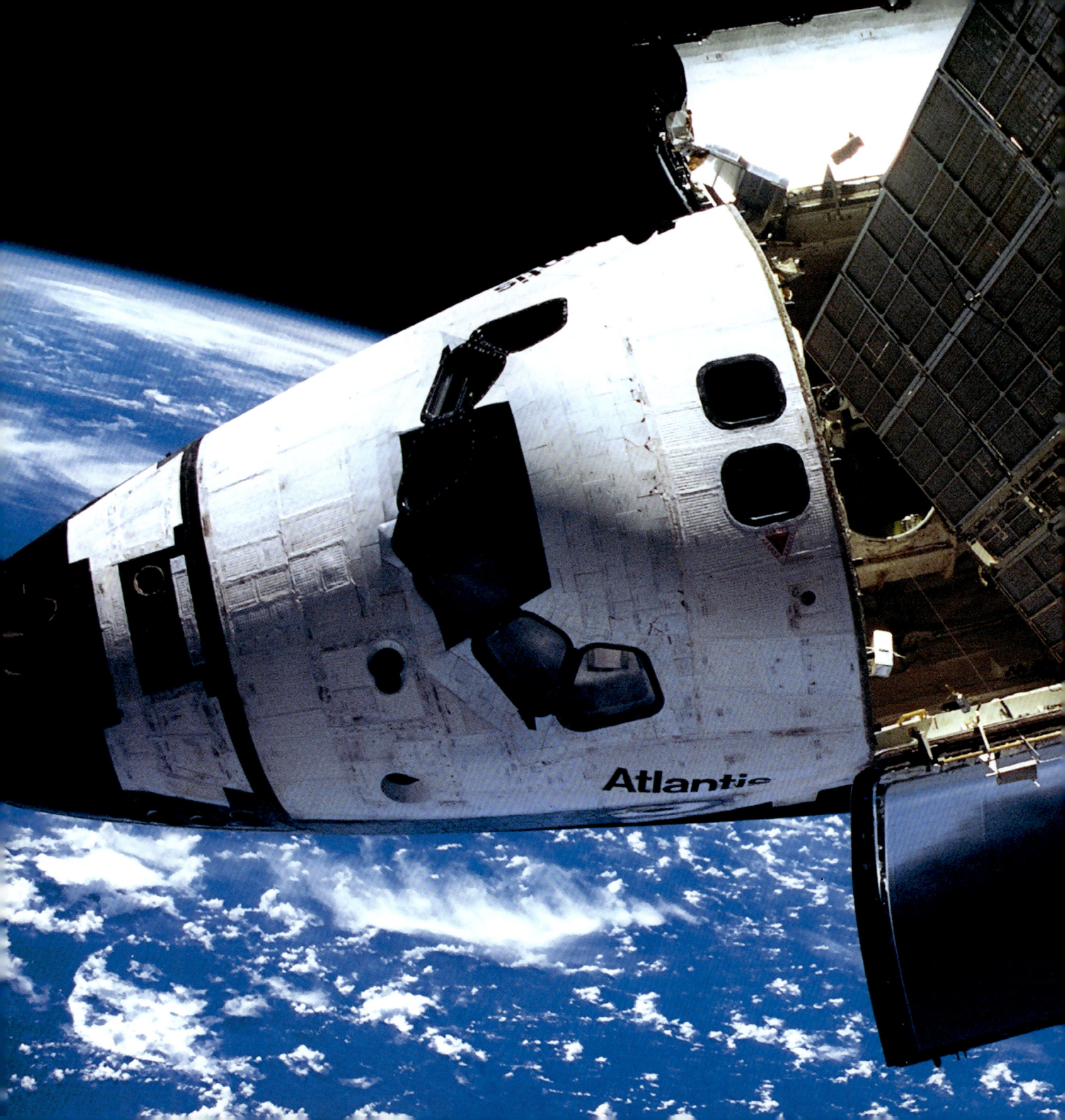

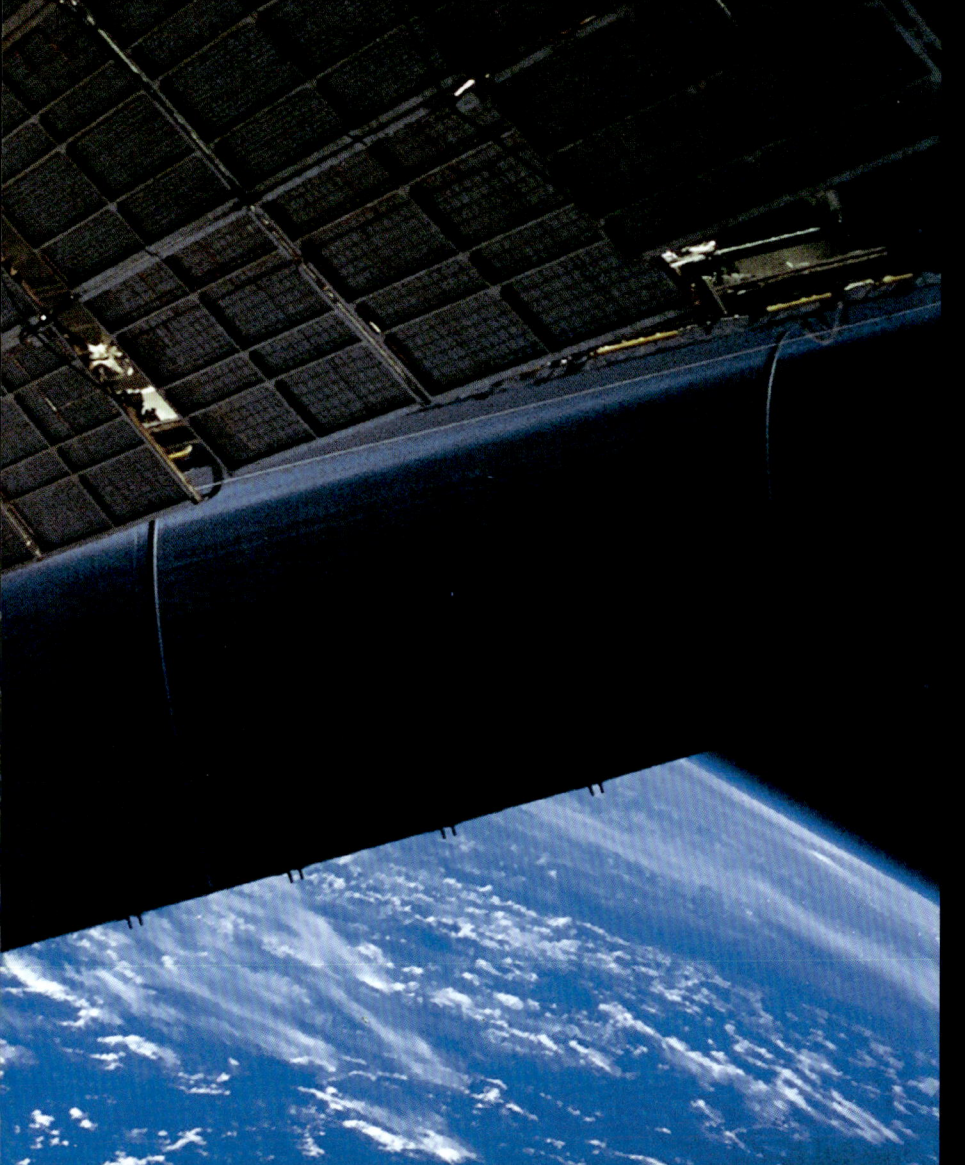

more living and working space. After the successful 13-day mission, Columbia and her gallant crew would meet their tragic fate during the dangerous Earth re-entry procedure.

STS-102
8–21 March 2001

(Following spread: Right) Space shuttle Discovery's aft section is framed by the window of Destiny, a Laboratory Module on the International Space Station (ISS). A cloudy Earth sky provides a dramatic backdrop as the shuttle rests peacefully while docked with the ISS.

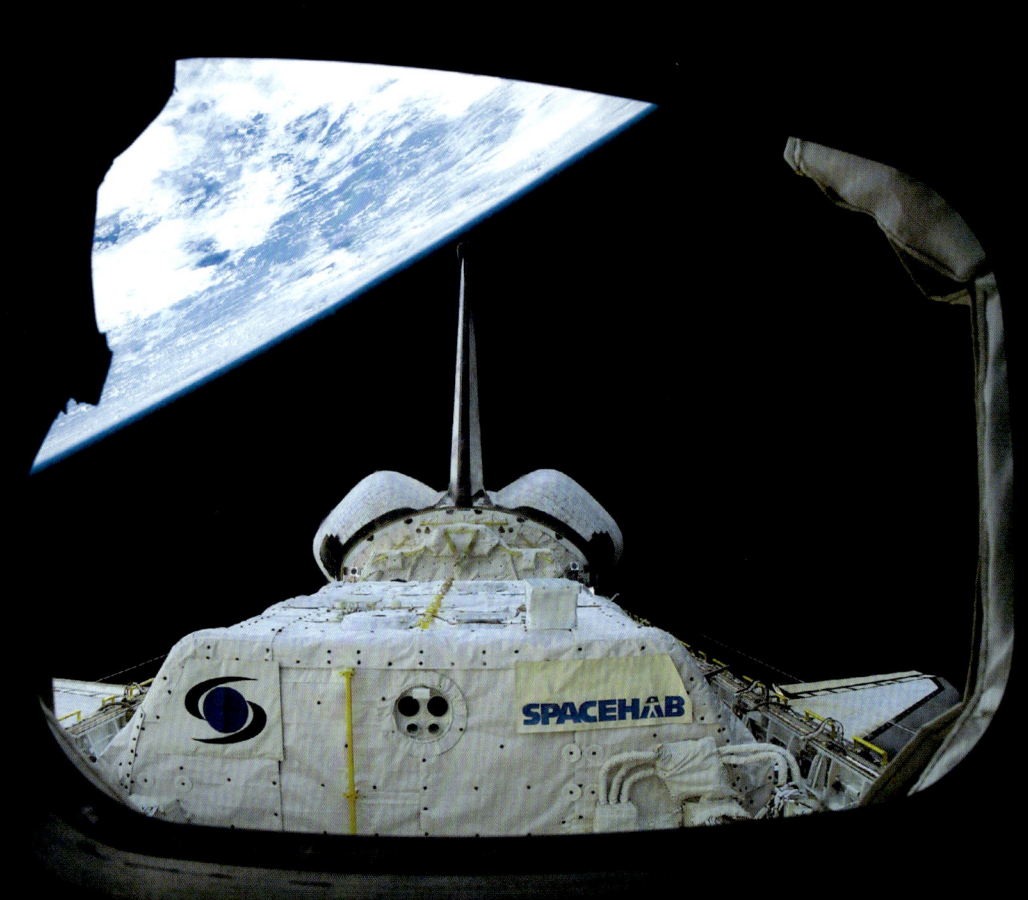

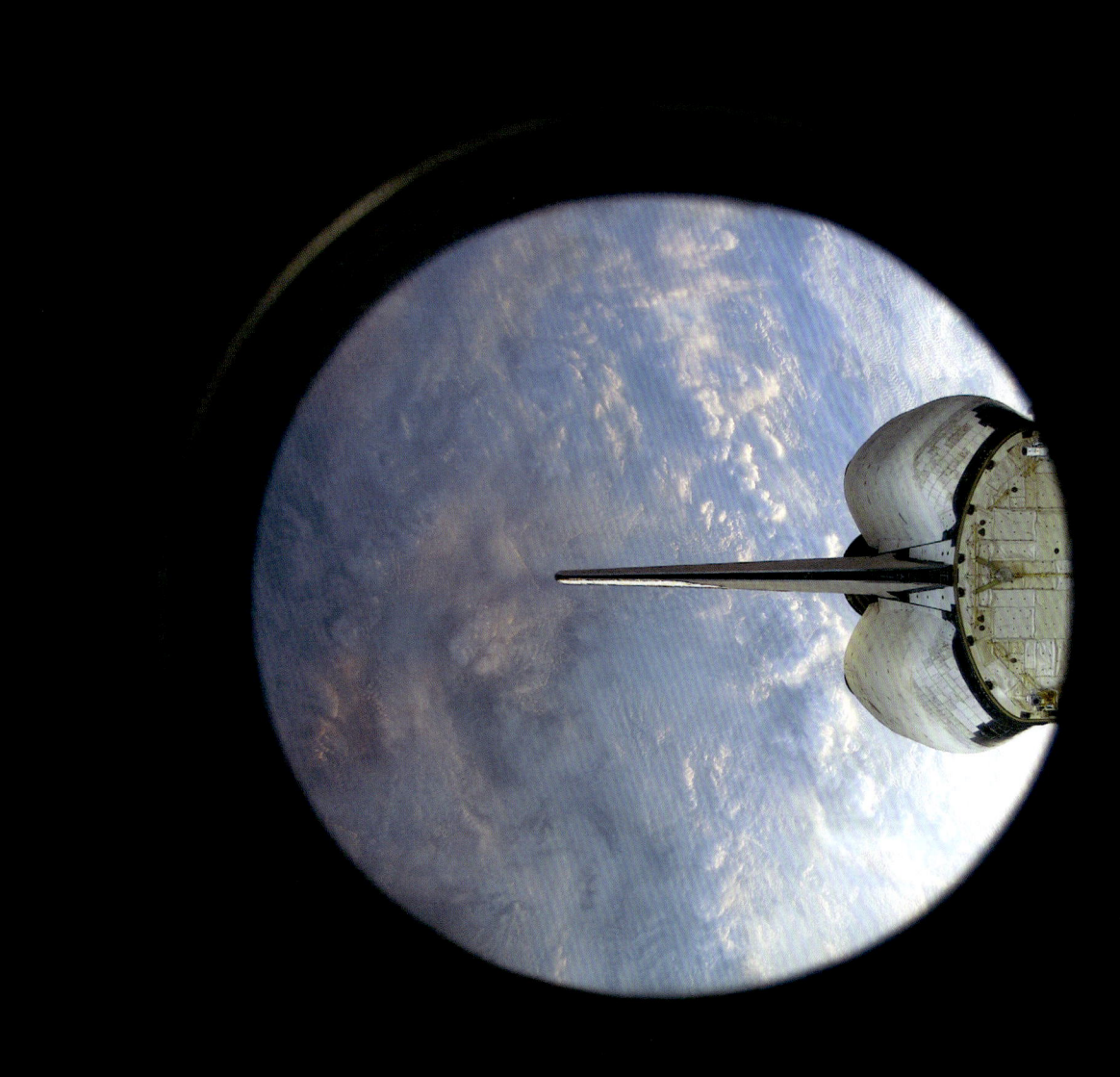

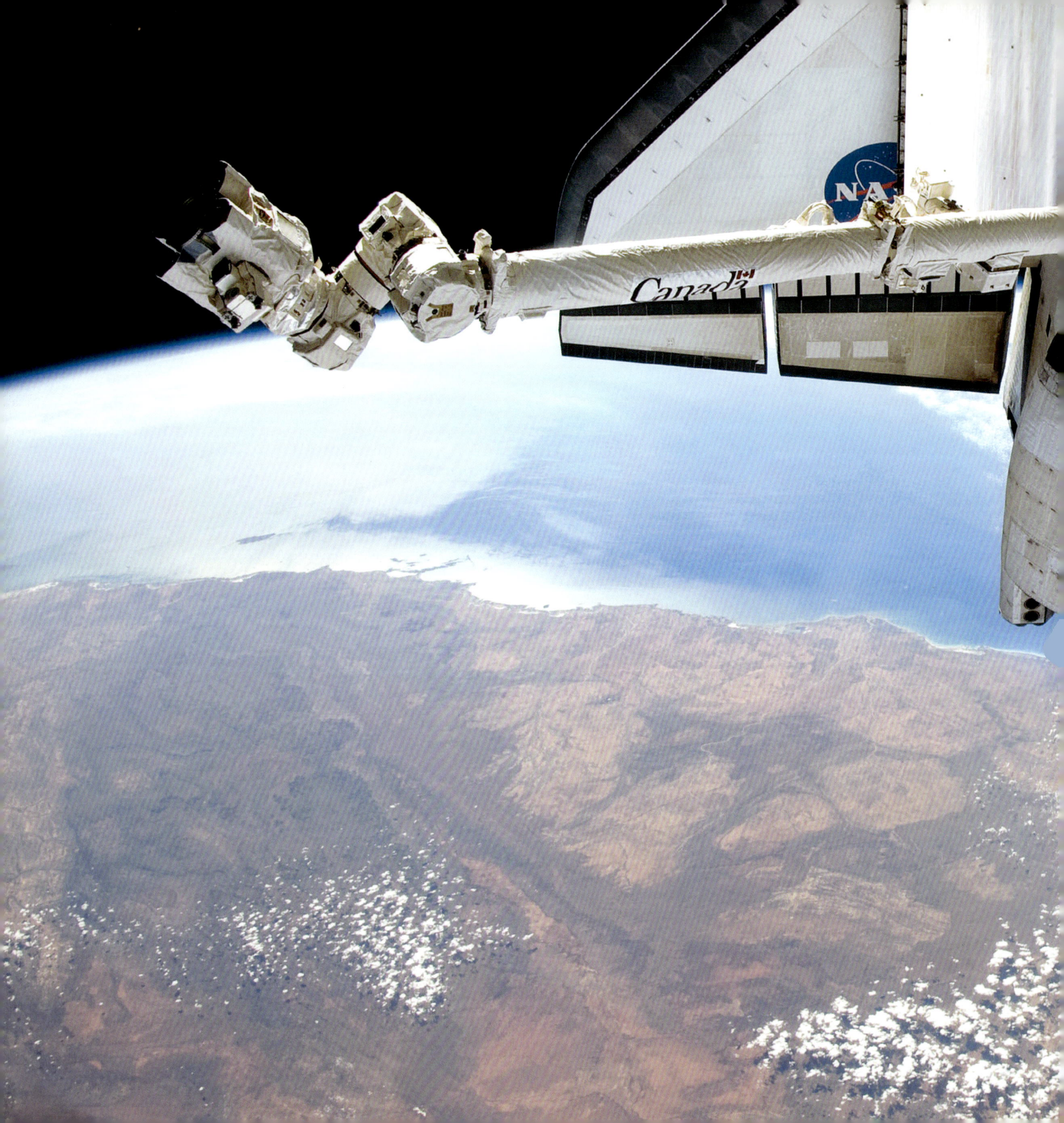

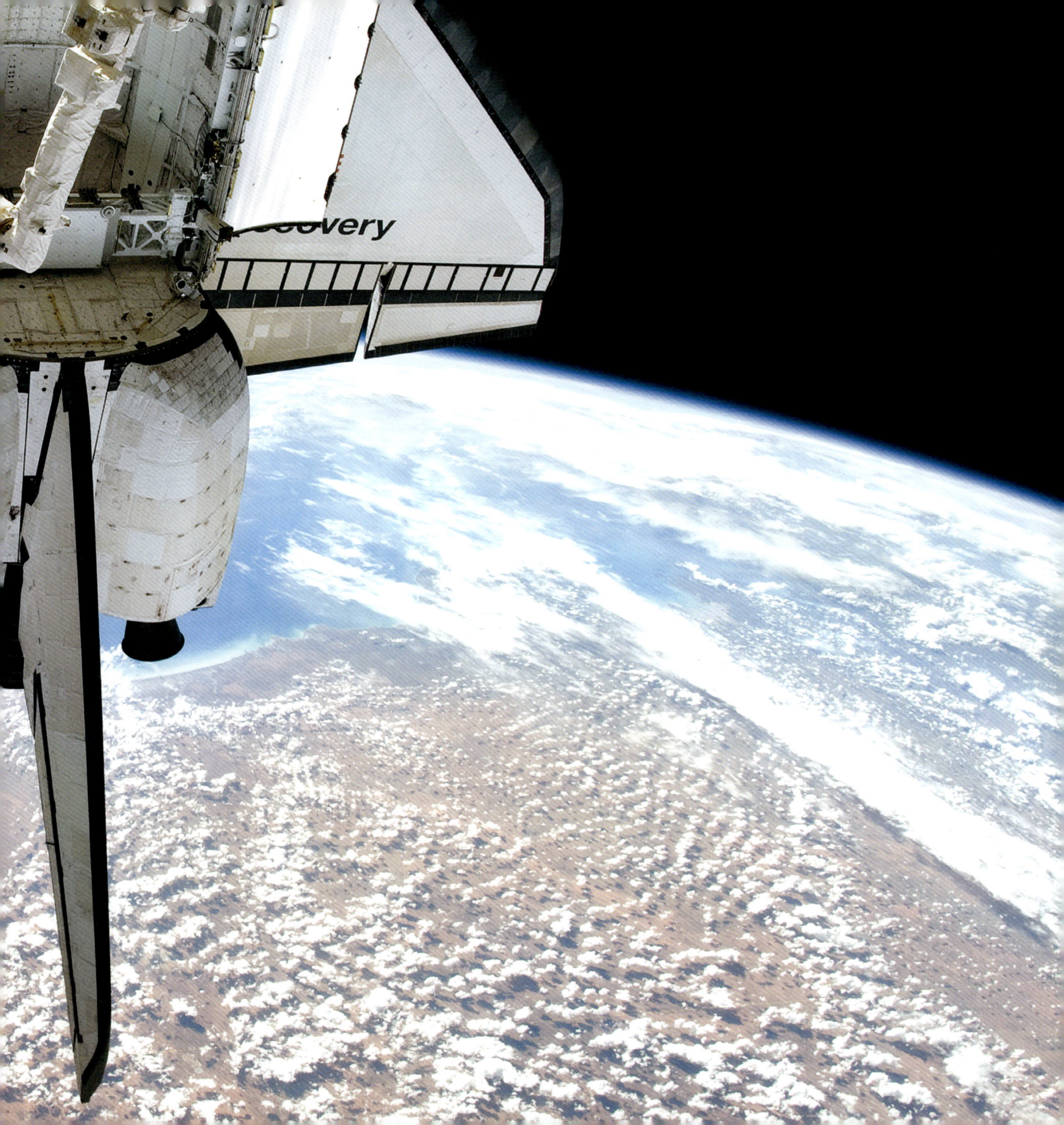

STS-131
5–20 April 2010

(Previous spread) Discovery hangs over a vibrant Earth while docked with the International Space Station (ISS). The north-west coast of Australia and the Indian Ocean can be seen in the photograph taken by a crew member aboard the ISS.

STS-109
1–12 March 2002

(Right) A beautiful and painting-like Earth horizon provides a dramatic setting in this image photographed from Columbia's aft flight deck. The shuttle's payload bay can be seen showcasing the new solar arrays for the Hubble space telescope.

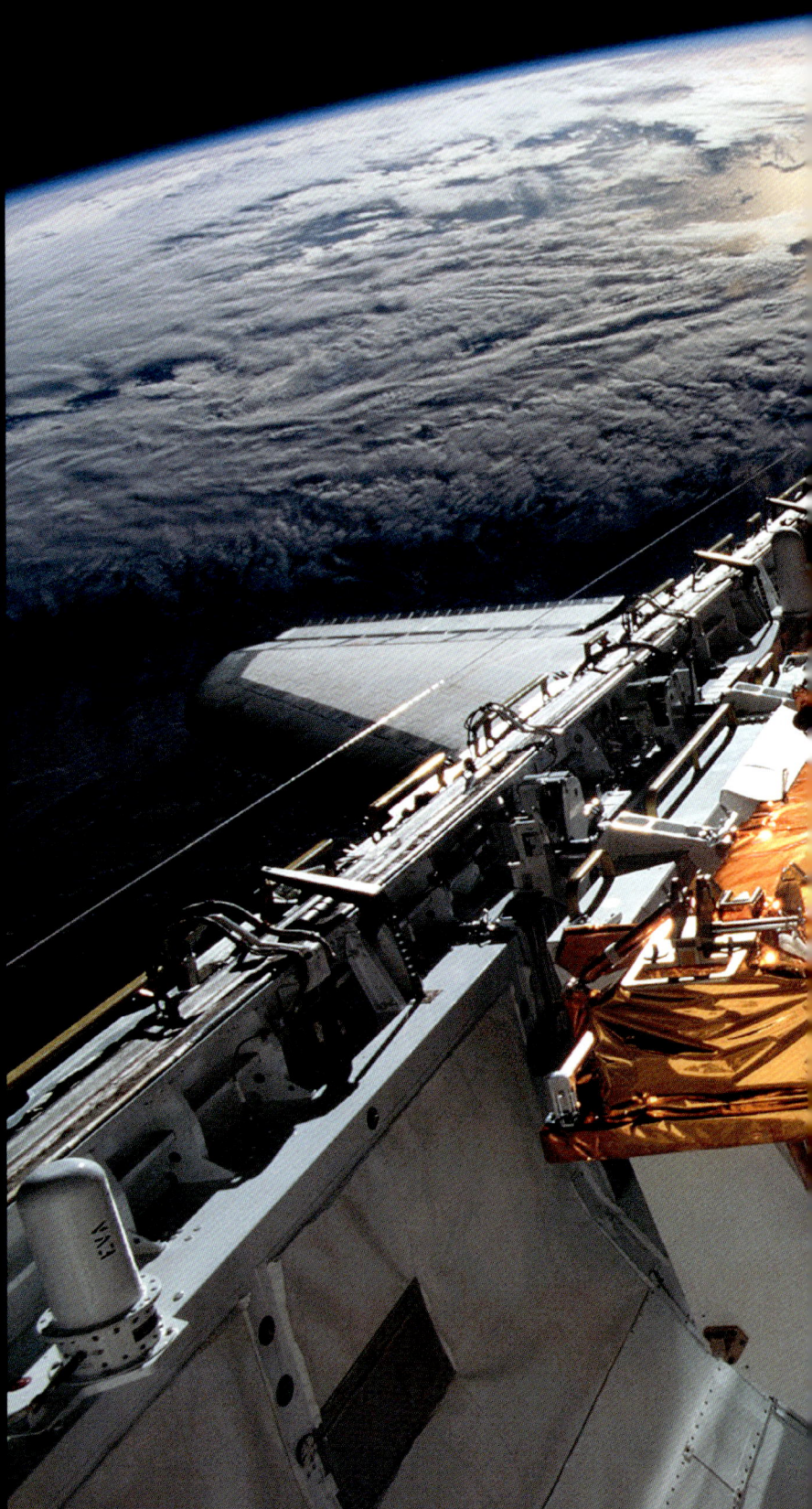

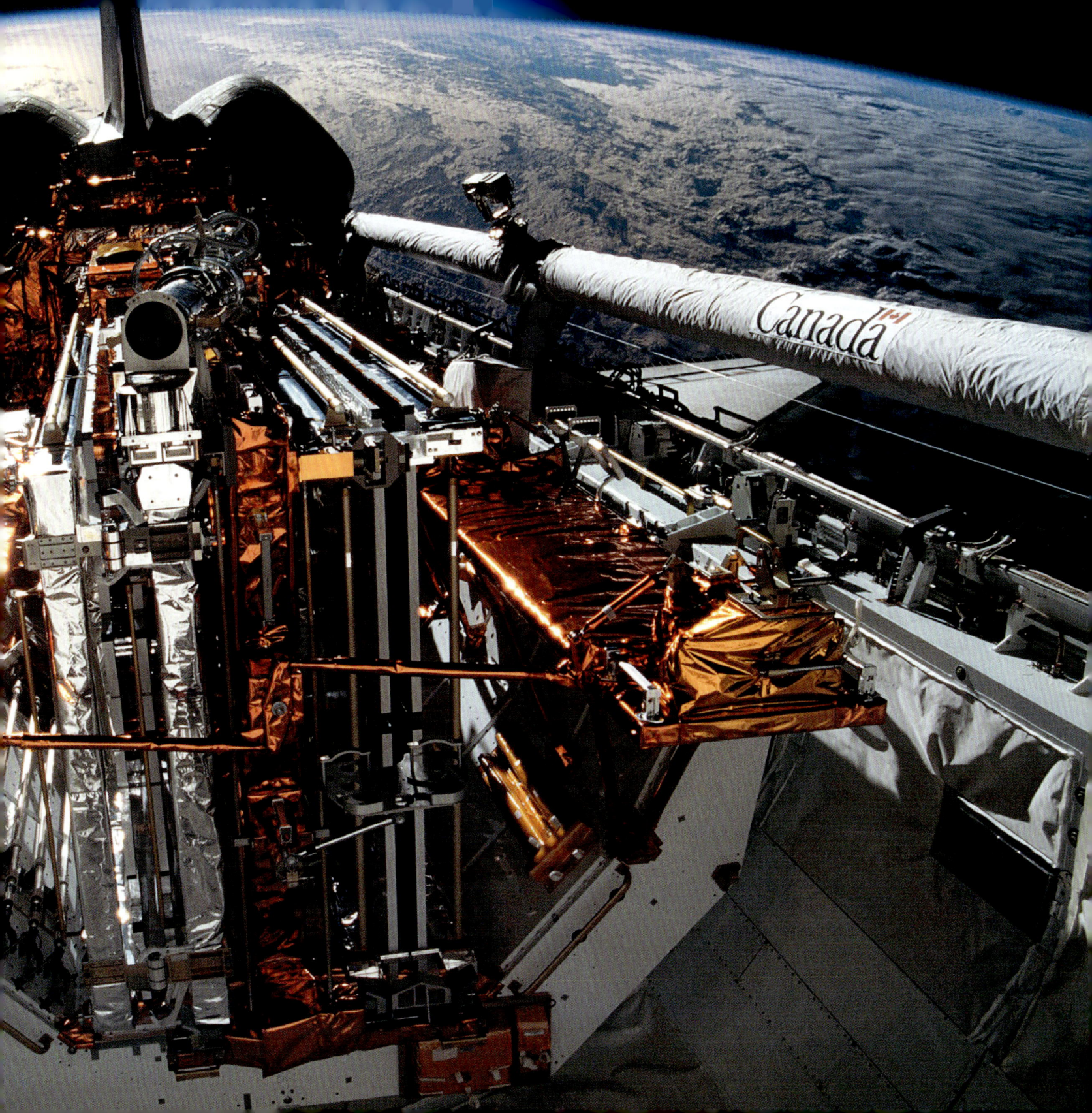

STS-95
29 October to 8 November 1998

(Right) From Discovery's aft windows, crew members get a fascinating view as the Moon creeps out from behind Earth's horizon. The mission marked a return to space for astronaut John Glenn, America's first man to orbit Earth in 1962. At age 77, this was Glenn's first space flight since his legendary Mercury mission aboard the tiny spacecraft he named Friendship 7, in which he completed three Earth orbits.

STS-49
7–16 May 1992

(Following spread: Left) The glowing Earth serves as the perfect backdrop for Thomas D. Akers and Kathryn C. Thornton (red stripes on spacesuit) as they get to work inside Endeavour's busy payload bay. Their task is to build the ASEM (Assembly of Station by EVA Methods) structure during the fourth space walk. The ASEM structure was designed to verify maintenance and assembly capabilities for extra space station freedom.

STS-61-B
26 November to 3 December 1985

(Following spread: Right) As Atlantis orbits the colourful Earth, astronaut Jerry L. Ross can be seen anchored to the Remote Manipulator System robotic arm during construction of the ACCESS device (Assembly Concept for Construction of Erectable Space Structures).

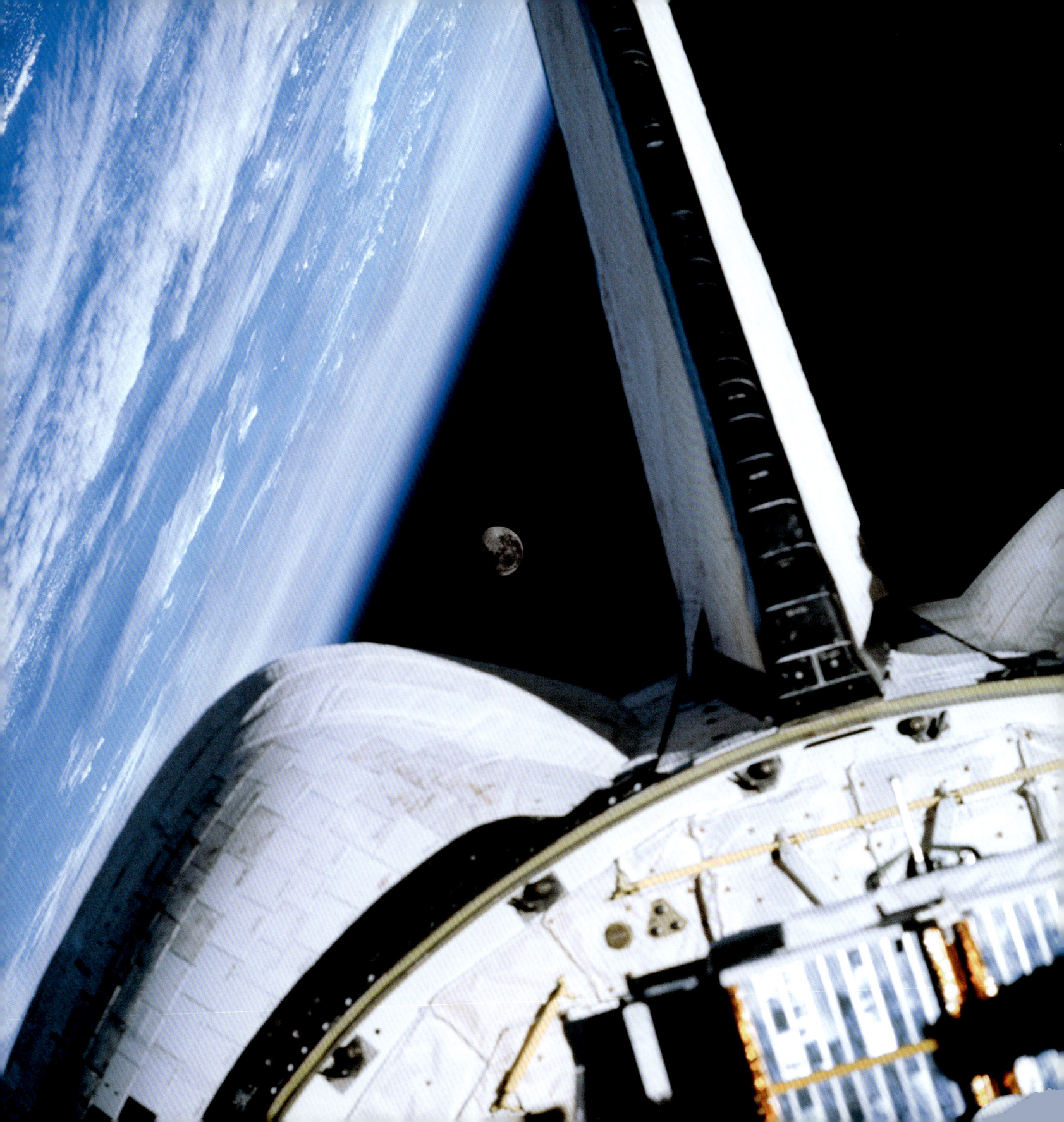

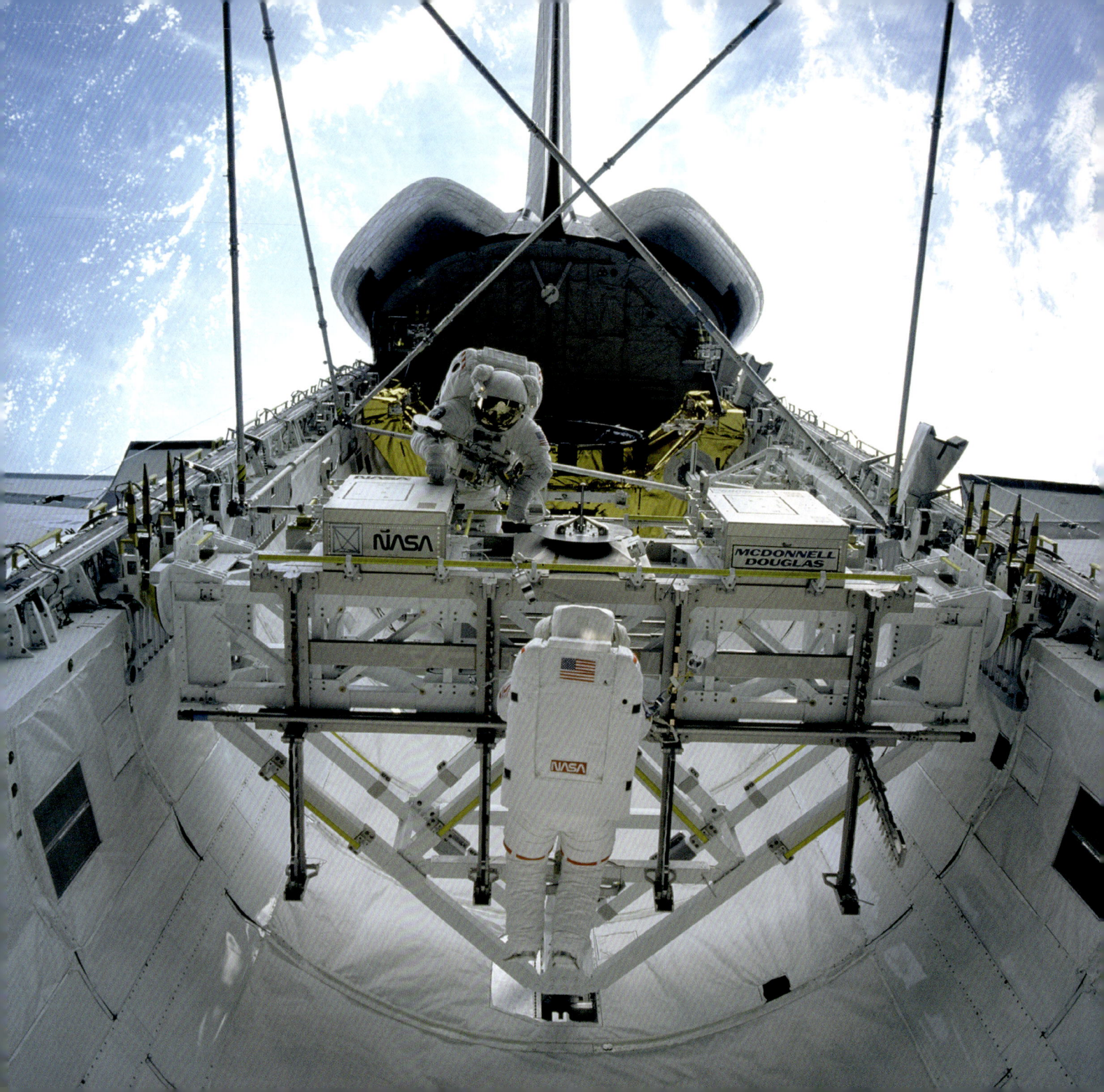

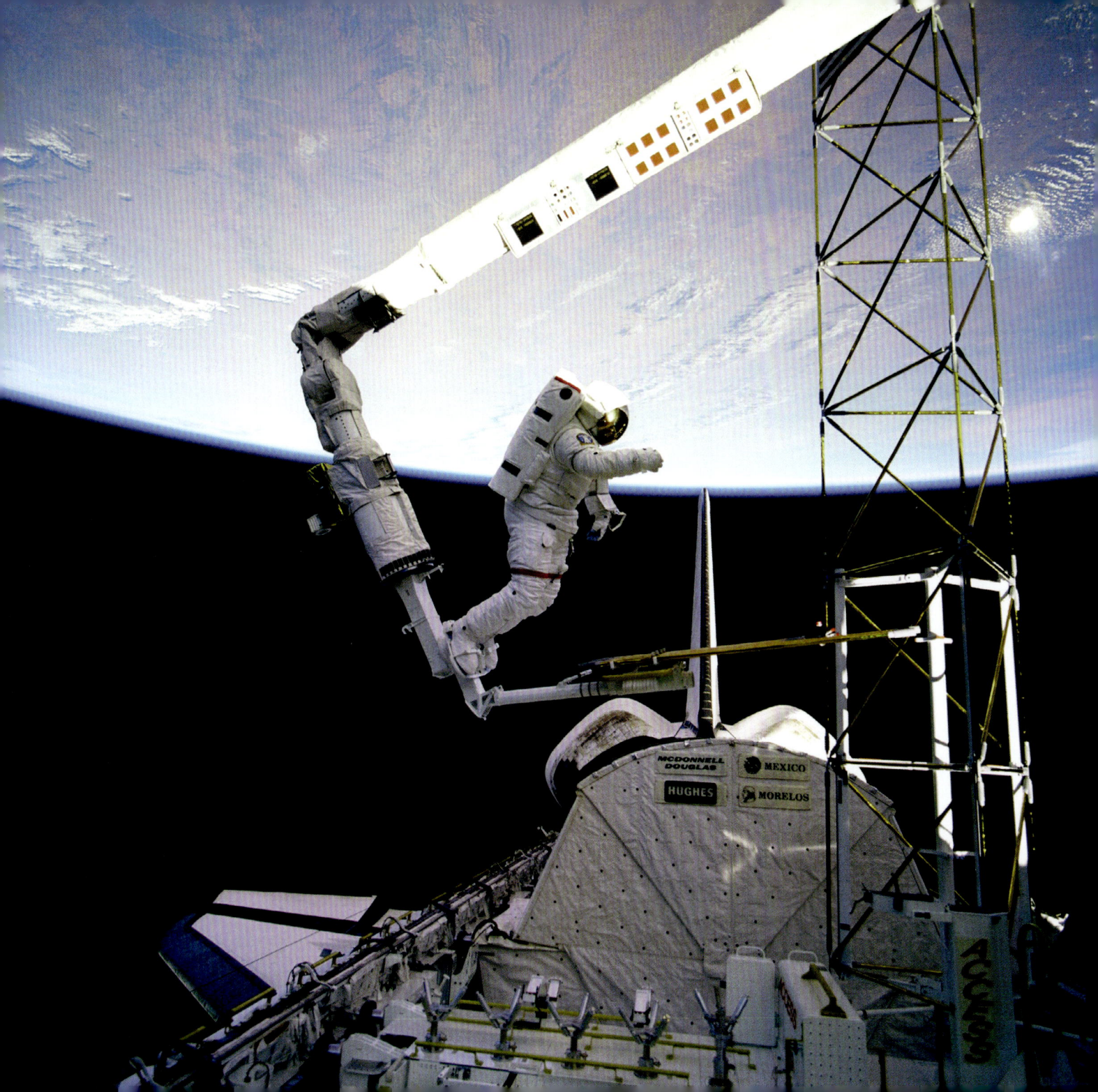

STS-131
5–20 April 2010

(Right) A long way from home. This inspiring and thought-provoking photograph from STS-131 shows astronaut Clayton Anderson with his Earth home in front of him and behind him a solar panel of the International Space Station, mankind's home in space. The payload bay of Discovery can be seen reflected in Anderson's visor with astronaut Rick Mastracchio performing his Extra Vehicular Activity (EVA) duties.

STS-61
2–13 December 1993

(Following spread: Left) Astronauts F. Story Musgrave and Jeffery A. Hoffman continue maintenance work on the Hubble Space Telescope. Musgrave is seen here anchored to Endeavour's Remote Manipulator System robotic arm, preparing to install new protective covers on the telescope's magnetometers.

STS-64
9–20 September 1994

(Following spread: Right) Astronaut Mark C. Lee tests out the new SAFER system (Simplified Aid For EVA Rescue). SAFER is a small propulsive jet pack which gives astronauts the ability to return safely to their spacecraft in the event of becoming untethered. This photograph was taken from Discovery's payload bay with the cloud-covered Earth 130 nautical miles below.

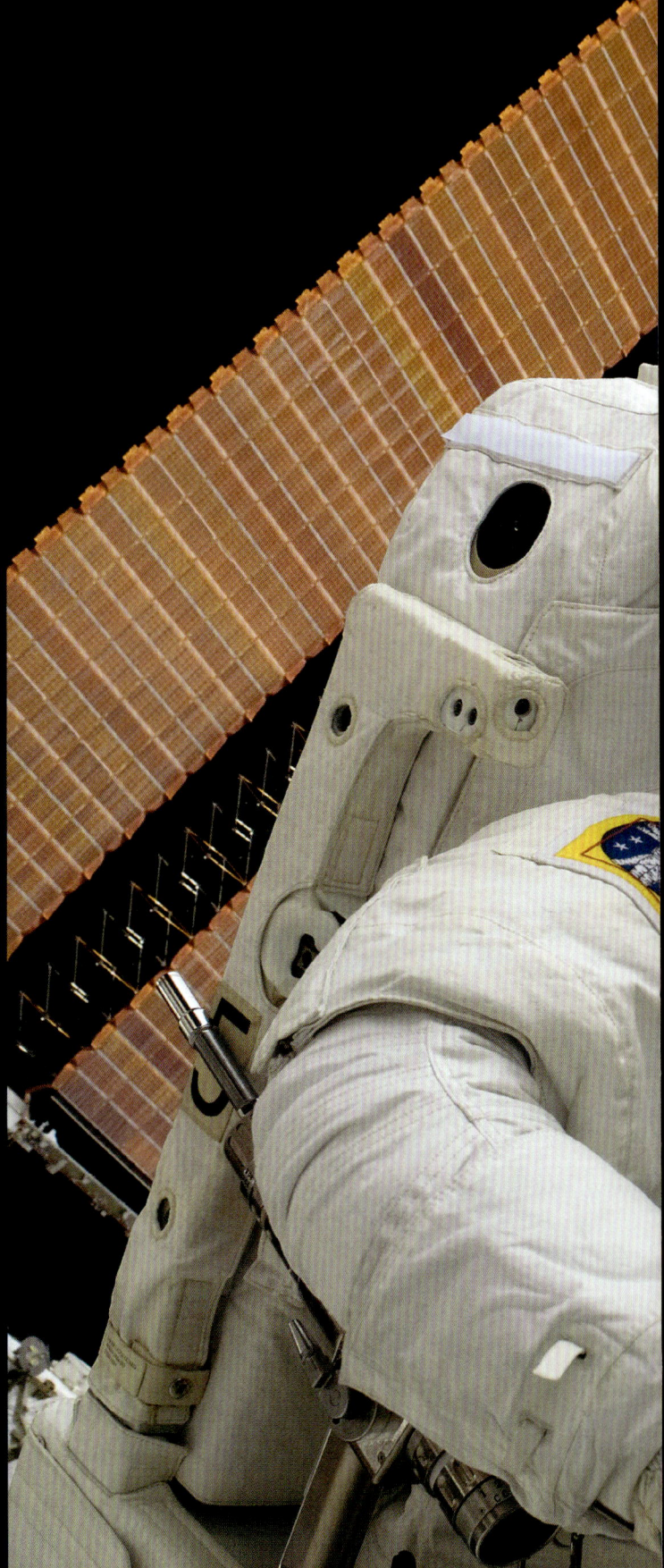

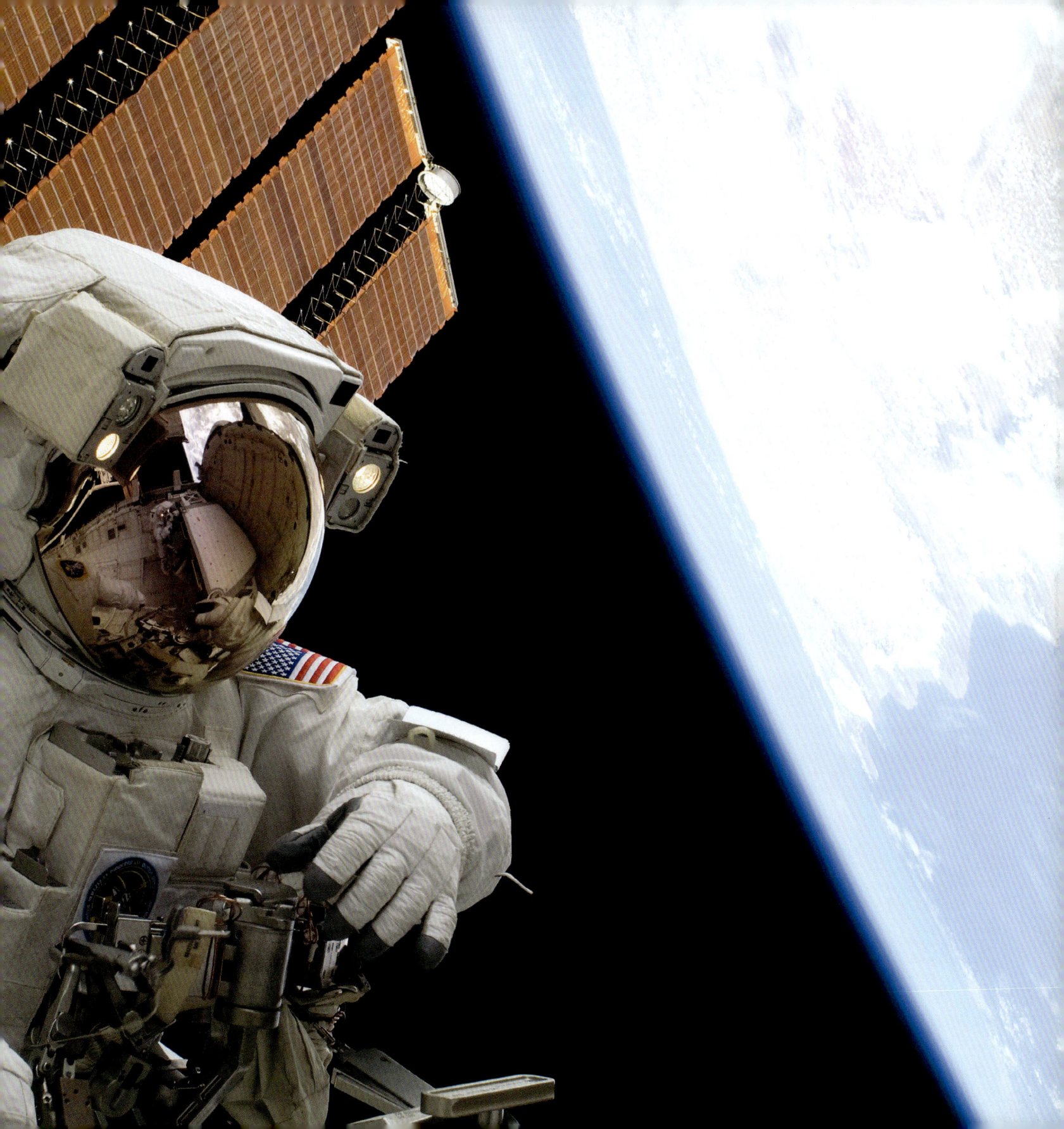

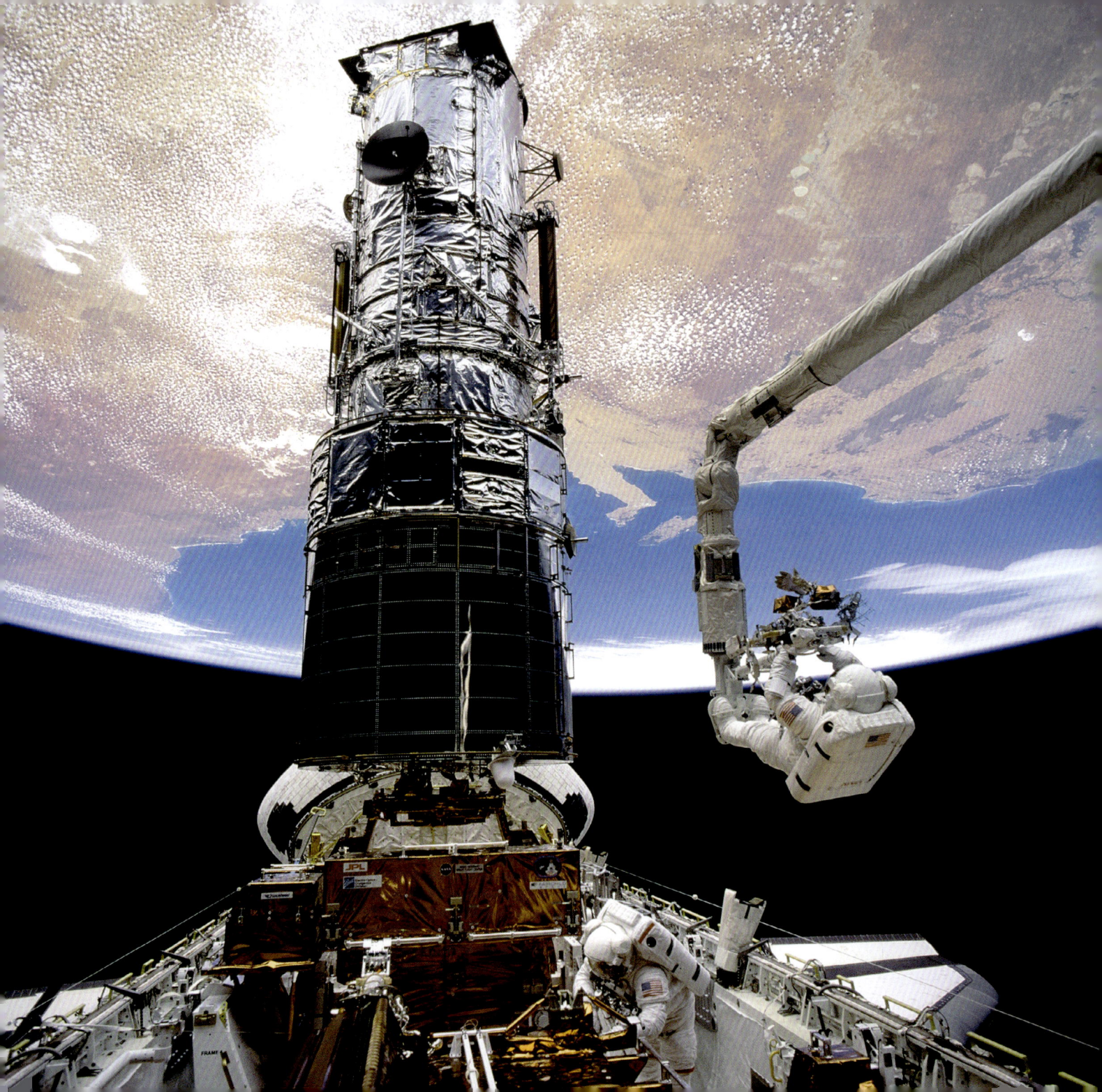

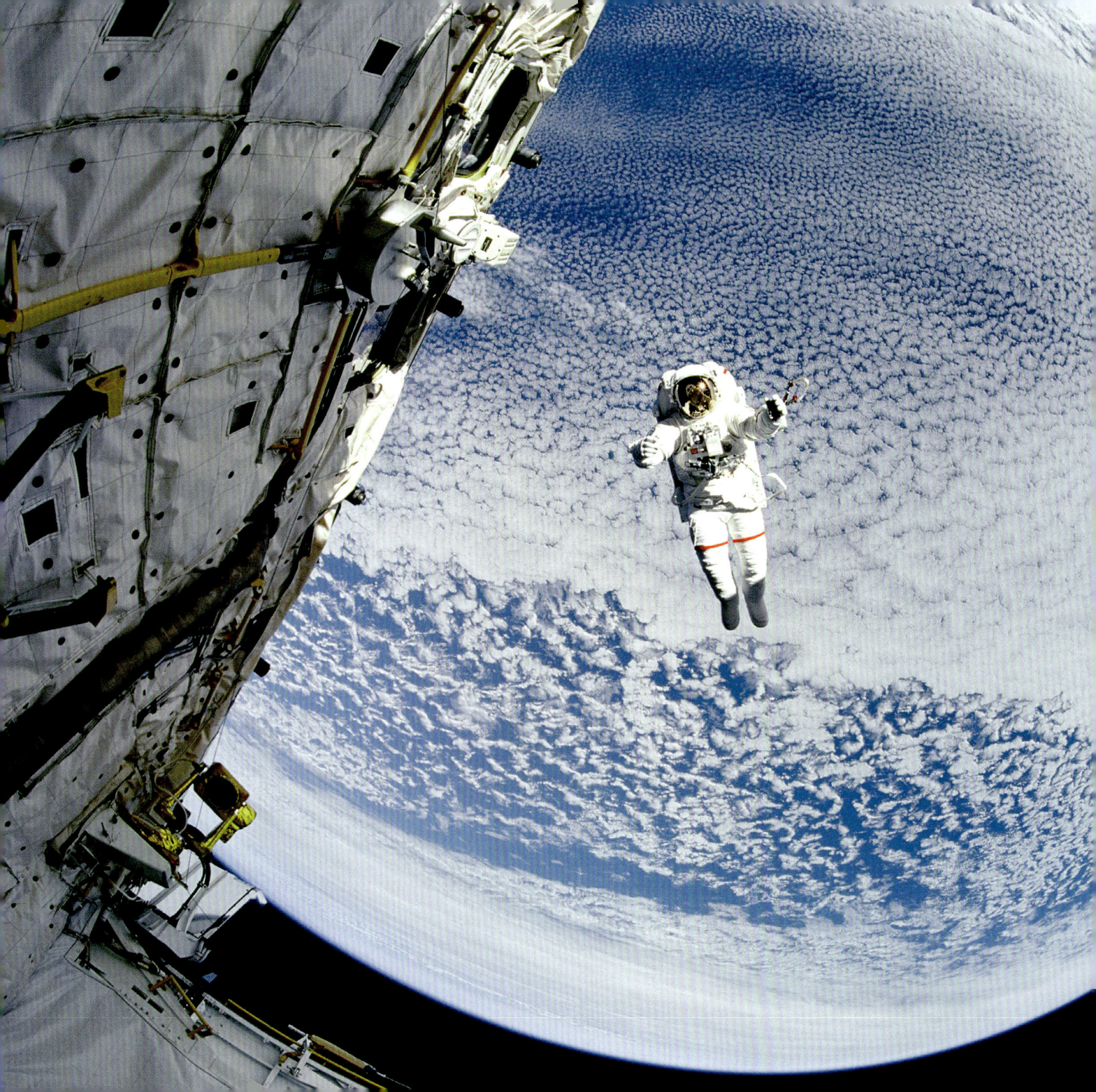

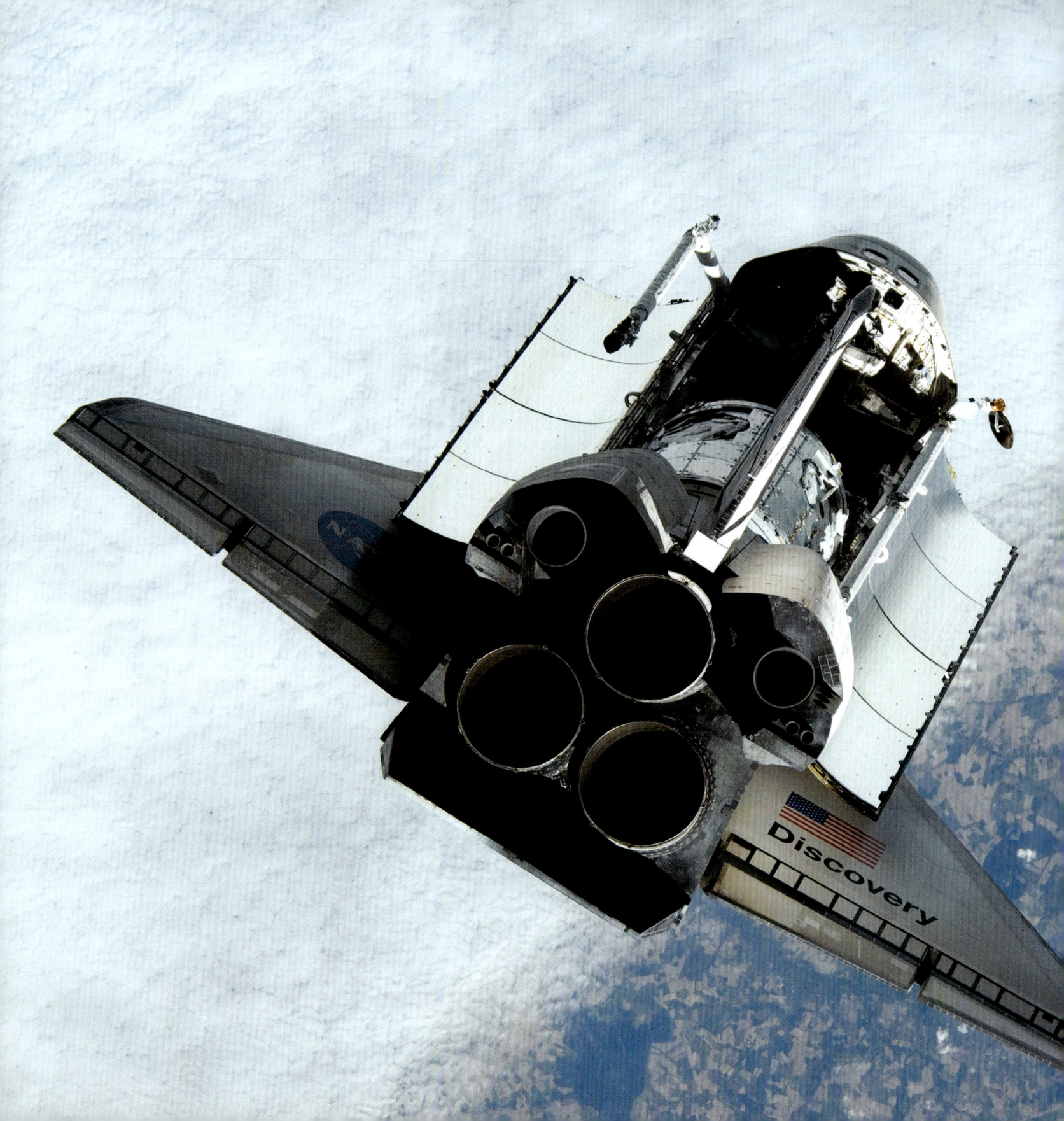

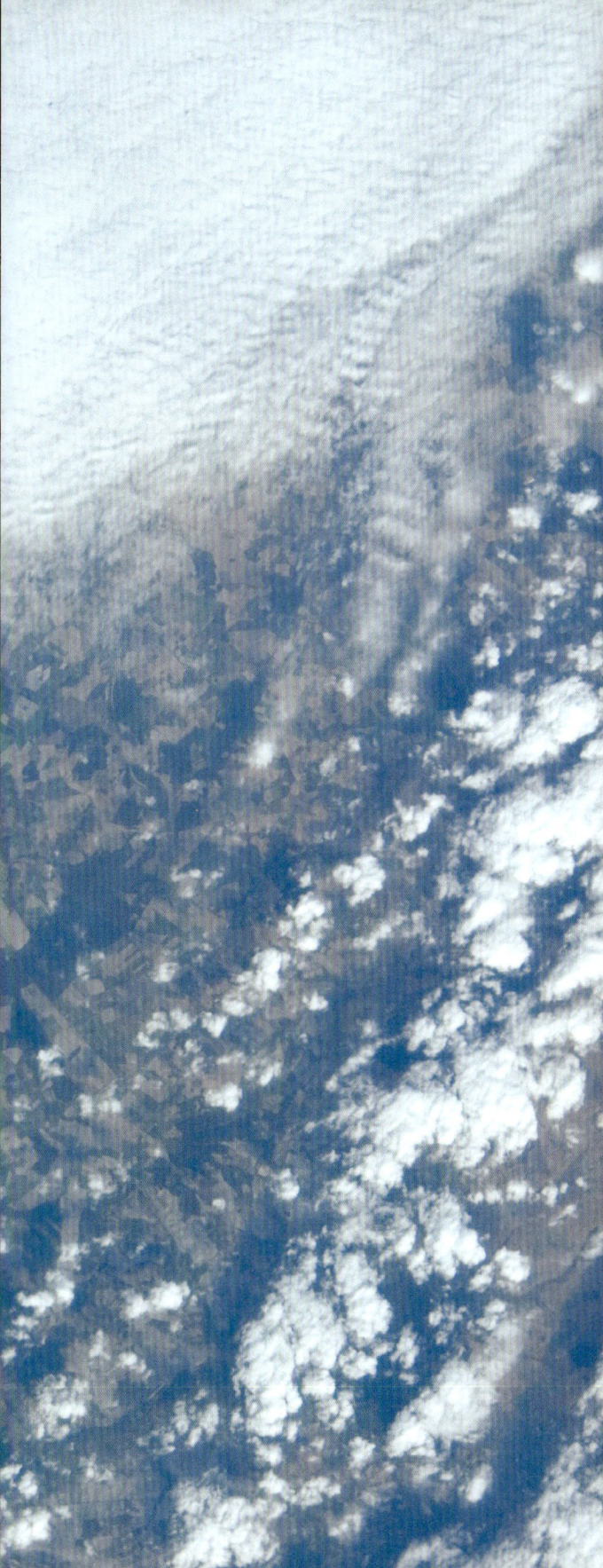

STS-120
23 October to 7 November 2007

(Left) Space shuttle Discovery flying high above the cloudy Earth at nearly 17,500 mph. This is the view as seen by crew members aboard the International Space Station.

STS-108
5–17 December 2001

(Following spread) The vertical stabilizer of space shuttle Endeavour can be seen in these two fascinating images pointing out some remarkable Earth landmarks: Queensland, Australia, and the Great Barrier Reef *(left image)* and the Emi Koussi Volcano in the Tibesti Mountains of Chad in Saharan Africa *(right image)*.

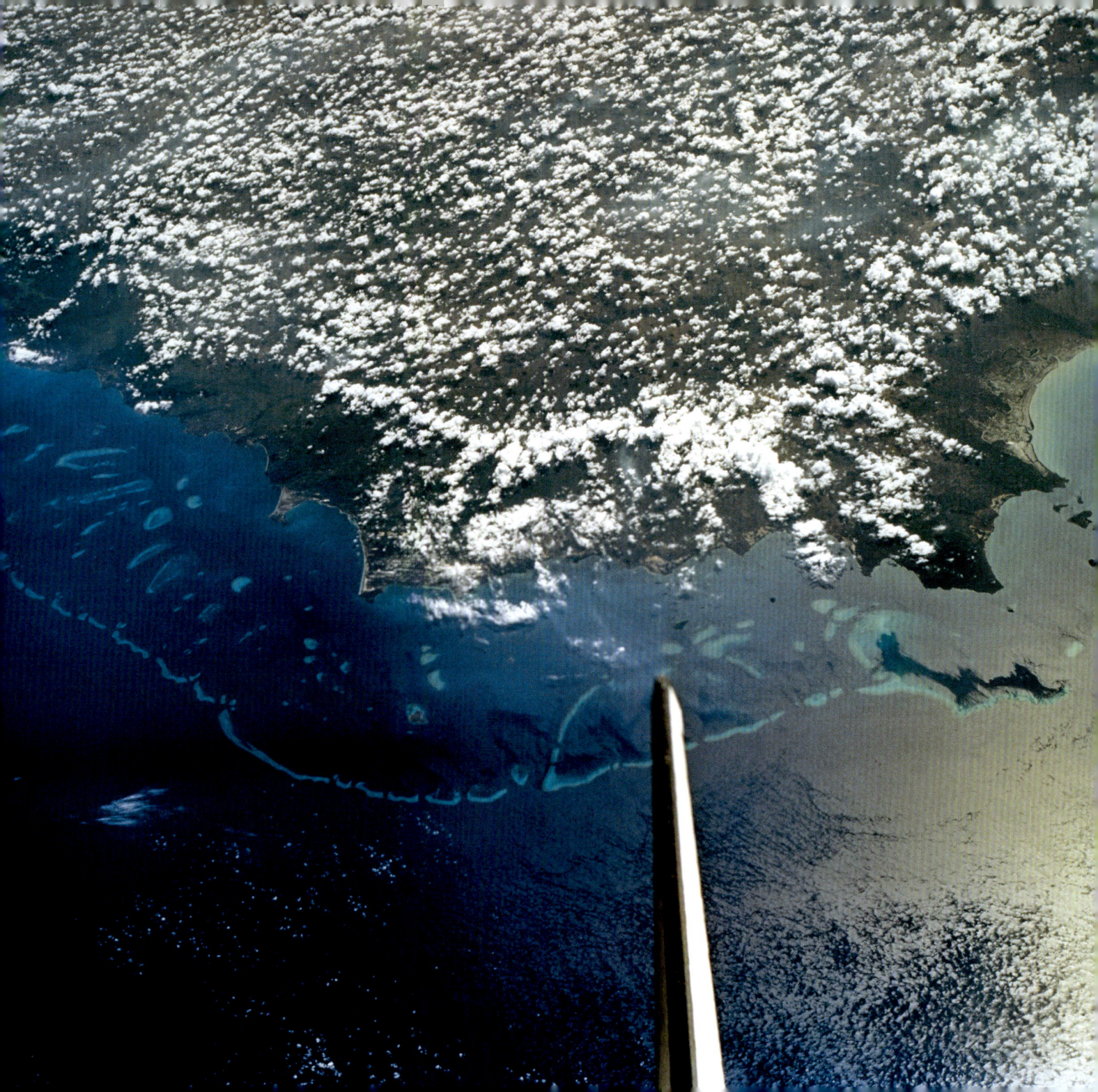

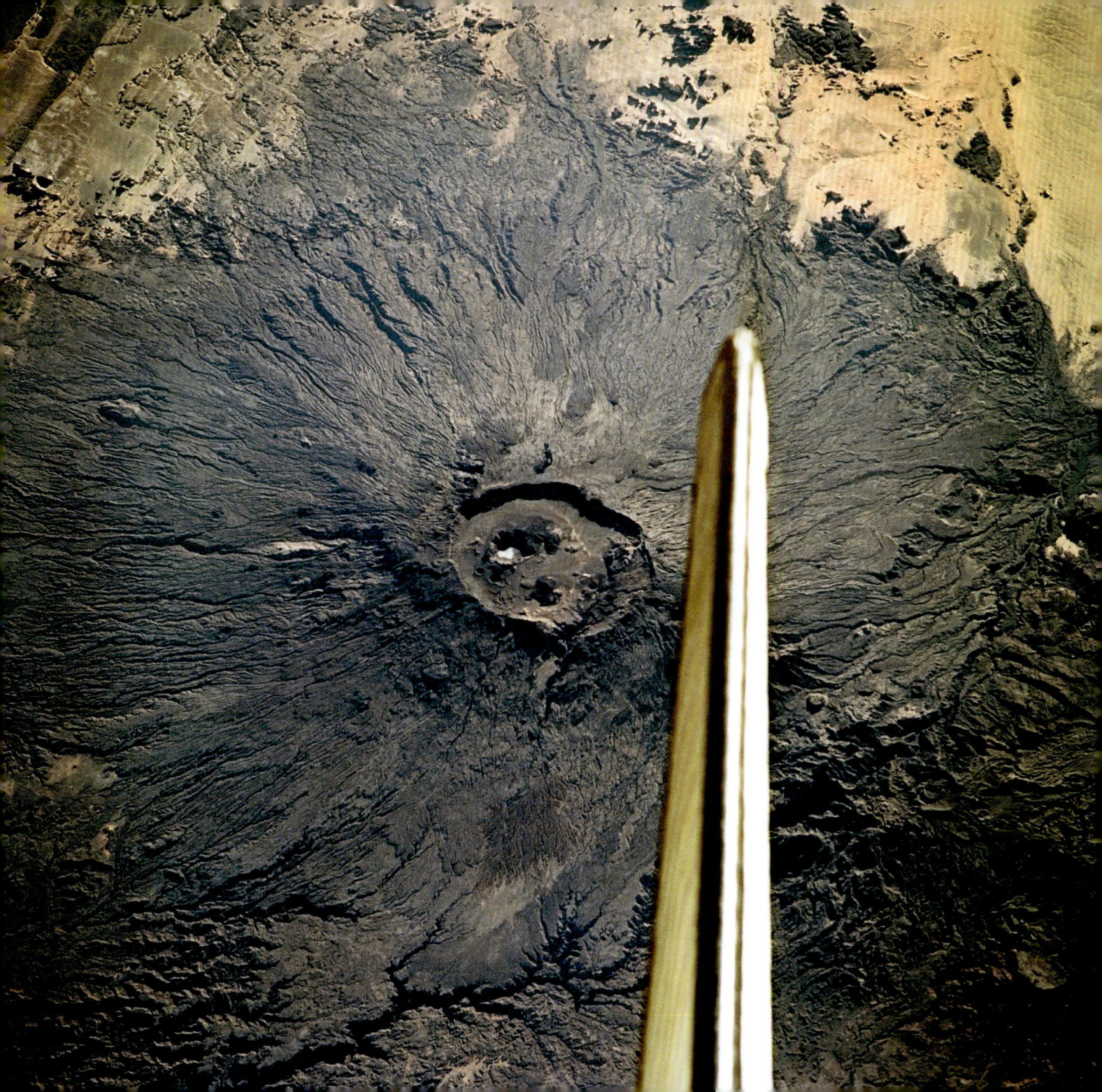

Space shuttle Endeavour deploys the STARSHINE 2 (Student Tracked Atmospheric Research Satellite for Heuristic International Networking Experiment), a small satellite which will be tracked by thousands of school students from 660 different schools in 26 countries. The students will collect information from the satellite's eight-month journey orbiting Earth, to help study and calculate the density of our upper atmosphere.

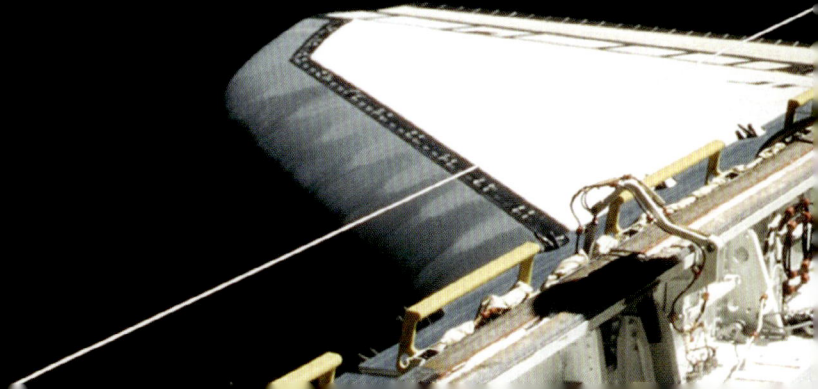

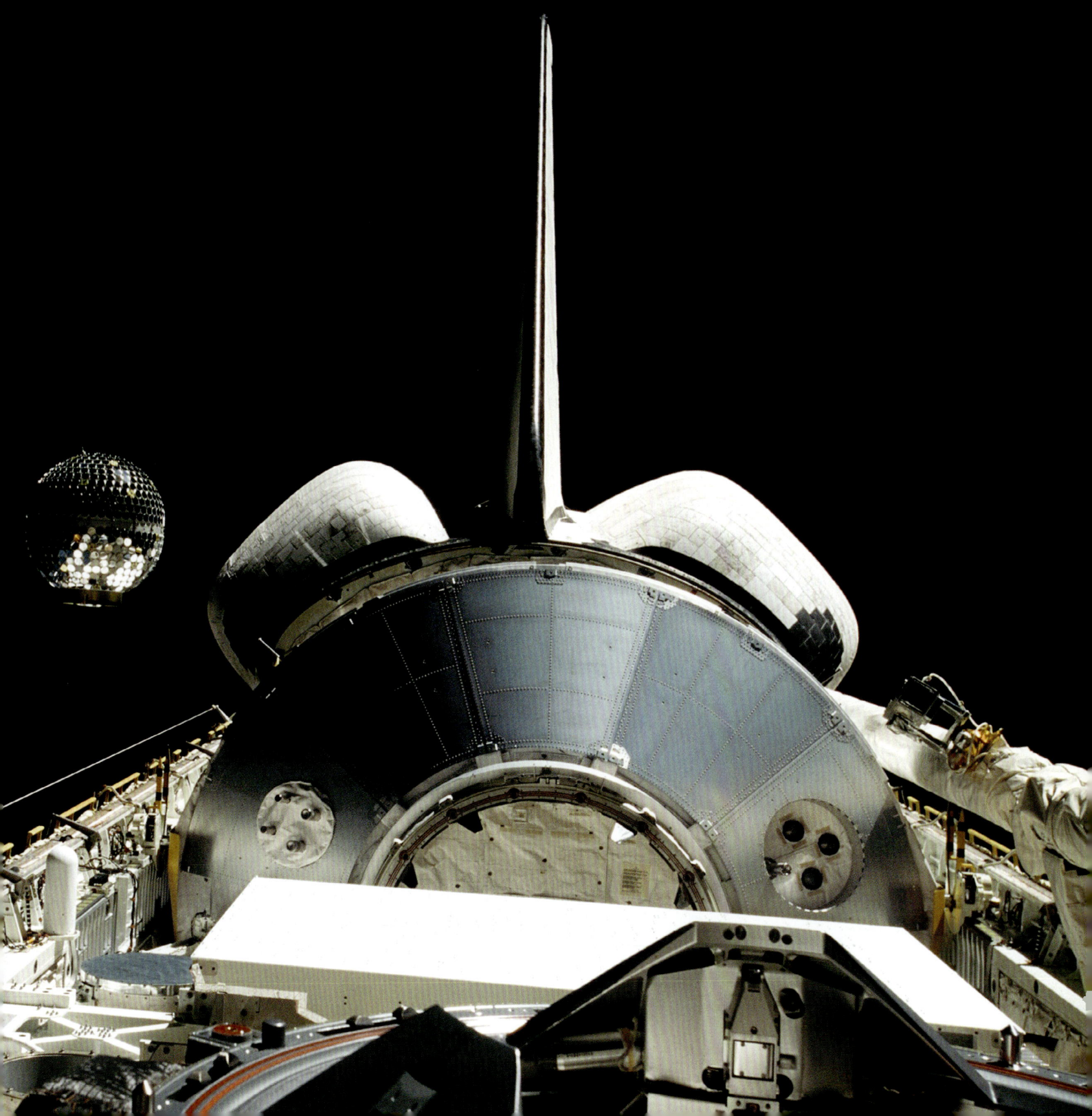

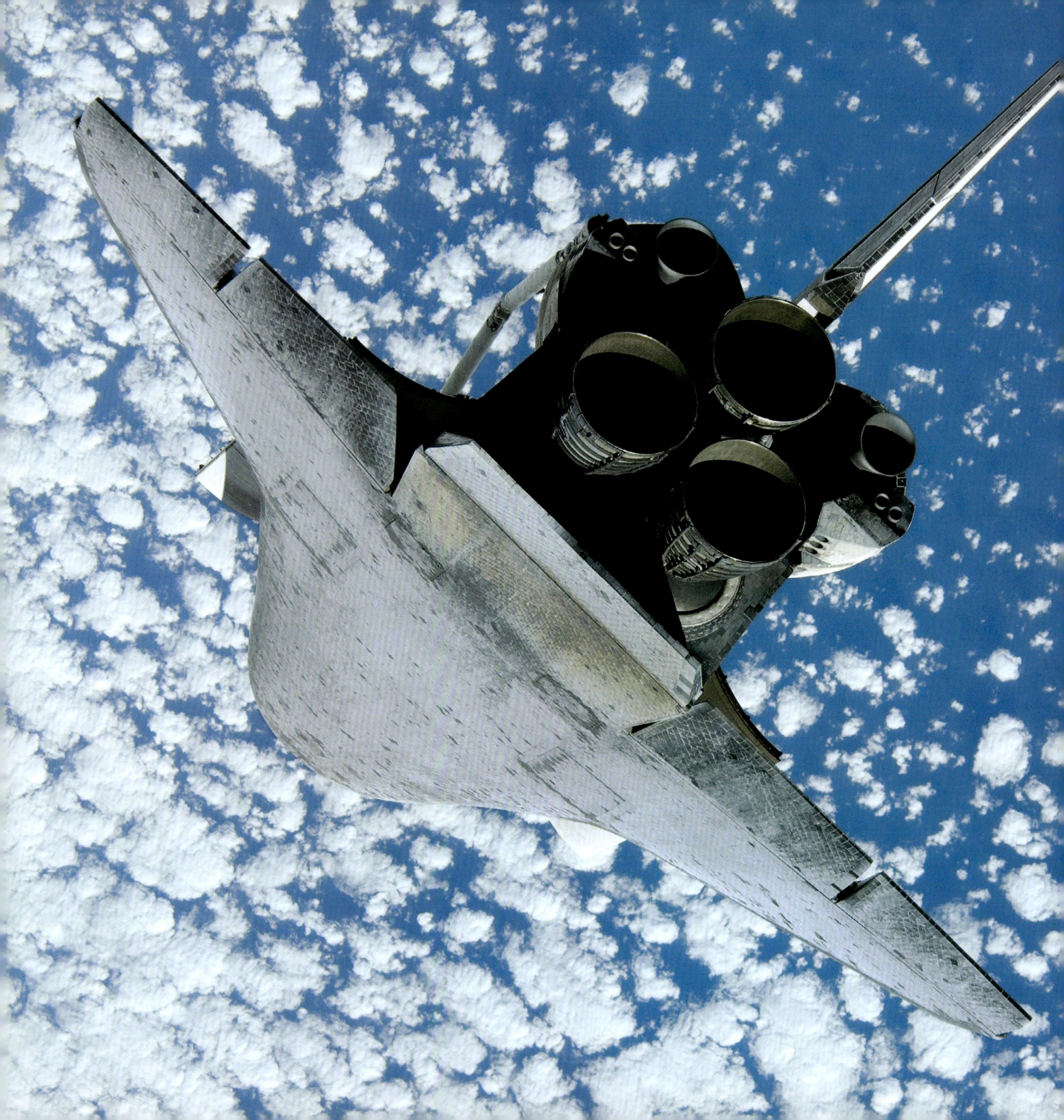

STS-120
23 October to 7 November 2007

(Left) Space shuttle Discovery performs rendezvous techniques with the International Space Station. Photographs were taken from many different angles to provide a detailed review of the shuttle's condition before re-entry to Earth's atmosphere.

STS-96
27 May to 6 June 1999

(Following spread: Left) The first STARSHINE satellite was deployed from Discovery's payload bay in 1999. The STARSHINE satellites closely resembled a disco ball with hundreds of small mirrors that reflect sunlight in order to study solar activity.

STS-57
21 June to 1 July 1993

(Following spread: Right) With the most amazing view while at work, astronaut G. David Low manoeuvres fellow astronaut Peter J. K. Wisoff while attached to Endeavour's Remote Manipulator System robotic arm. This task was rehearsed to simulate and test the handling of large objects in space, in preparation for servicing the Hubble Space Telescope.

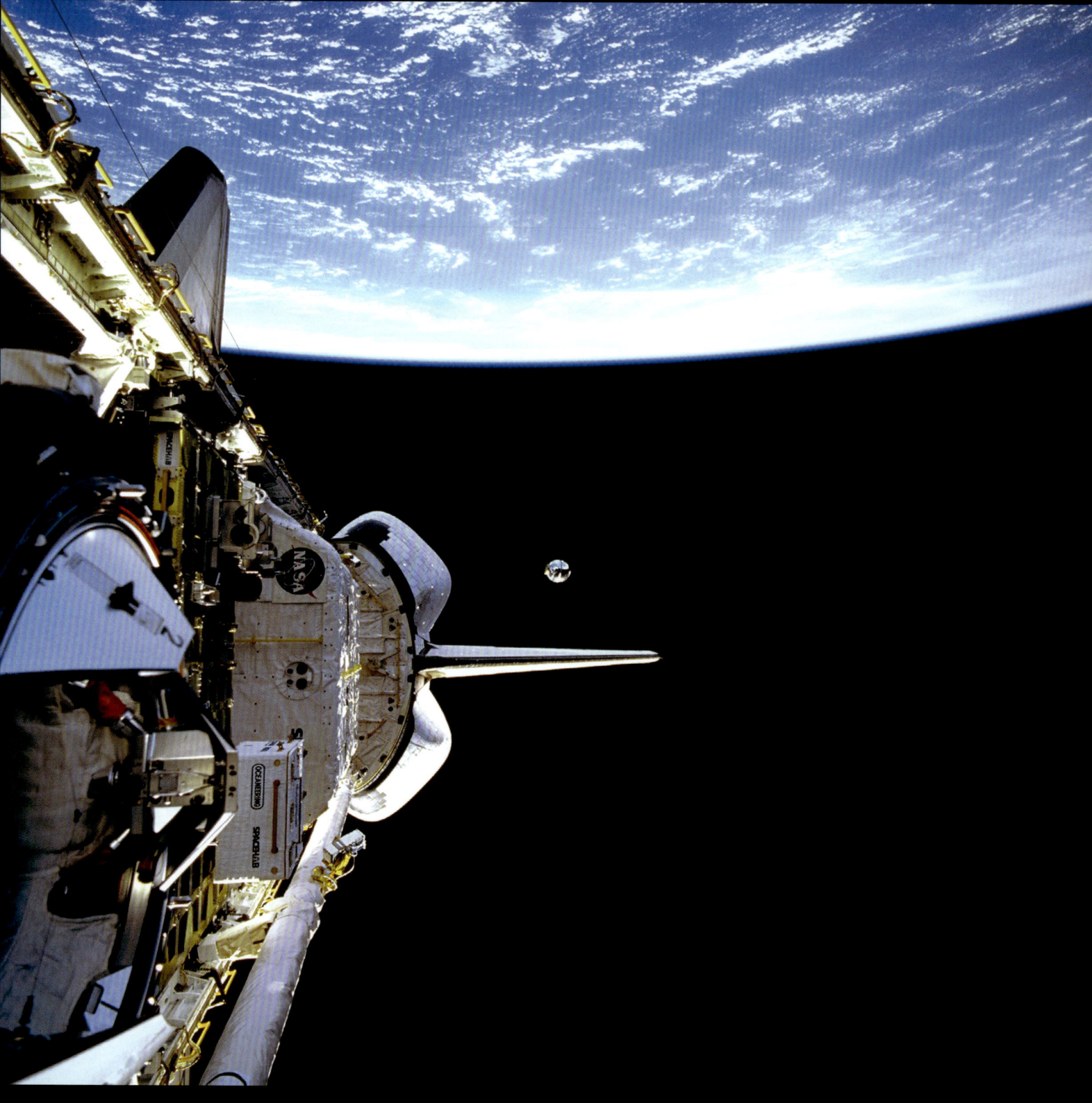

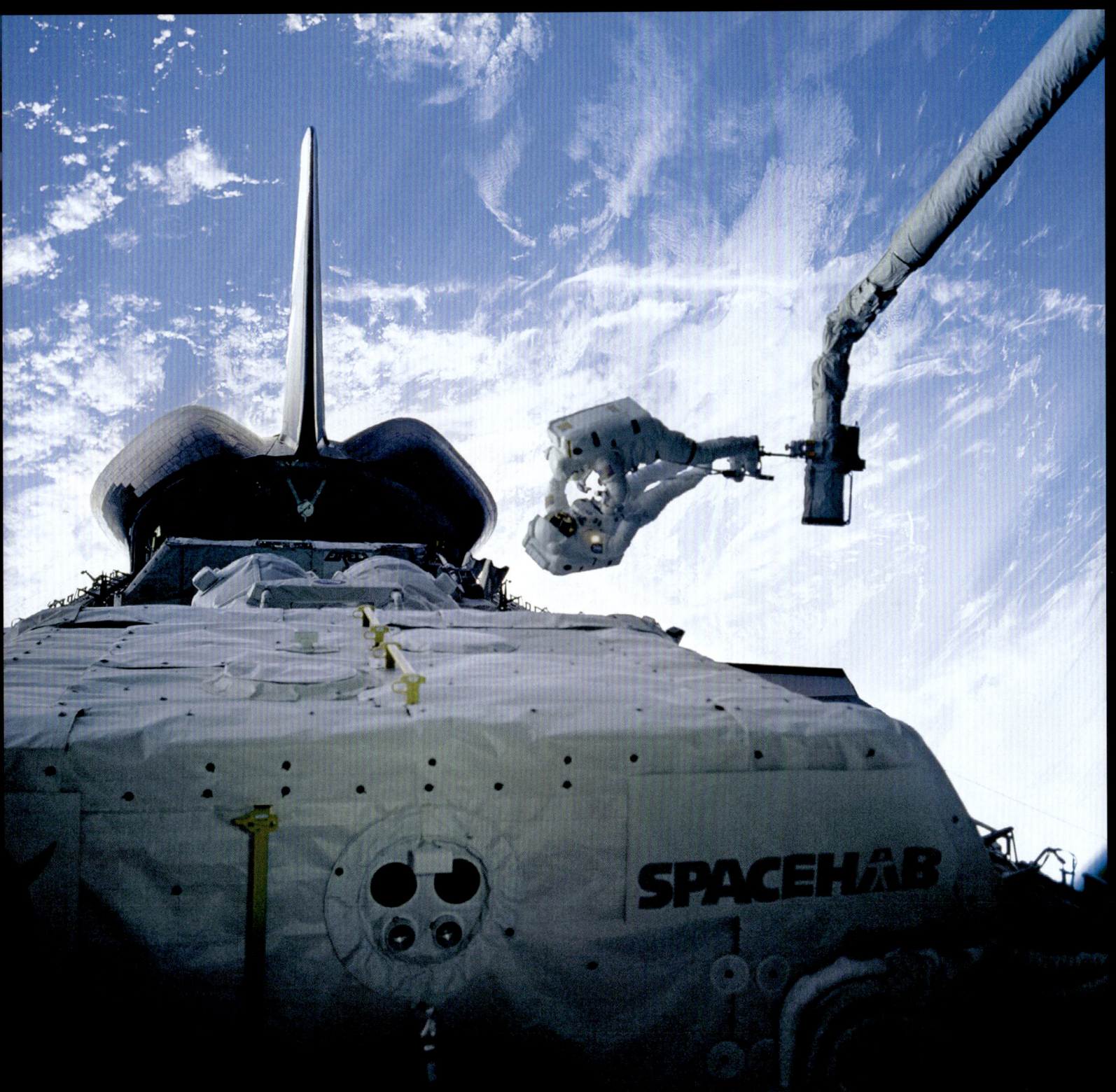

Astronaut Stephen K. Robinson inspects the shuttle's thermal protection system and removes gap fillers from Discovery's heat shield during the mission's final session of Extra Vehicular Activity (EVA), in preparation for Earth re-entry.

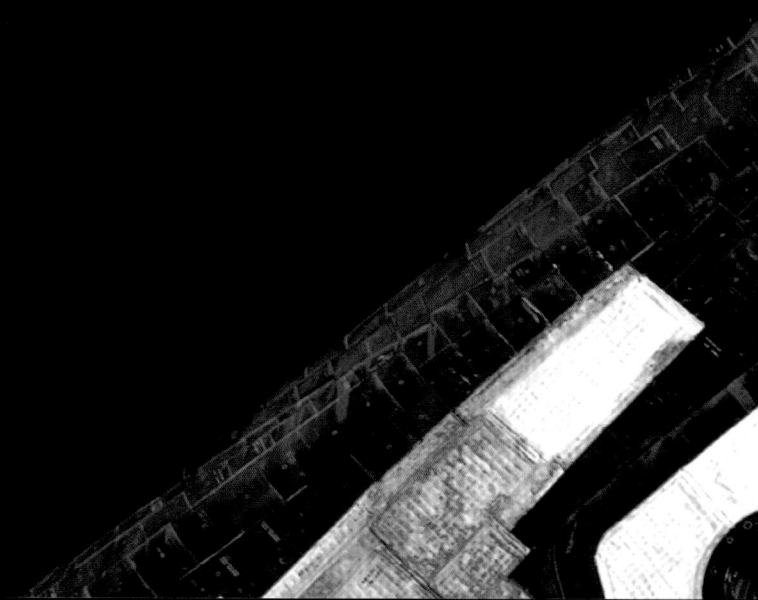

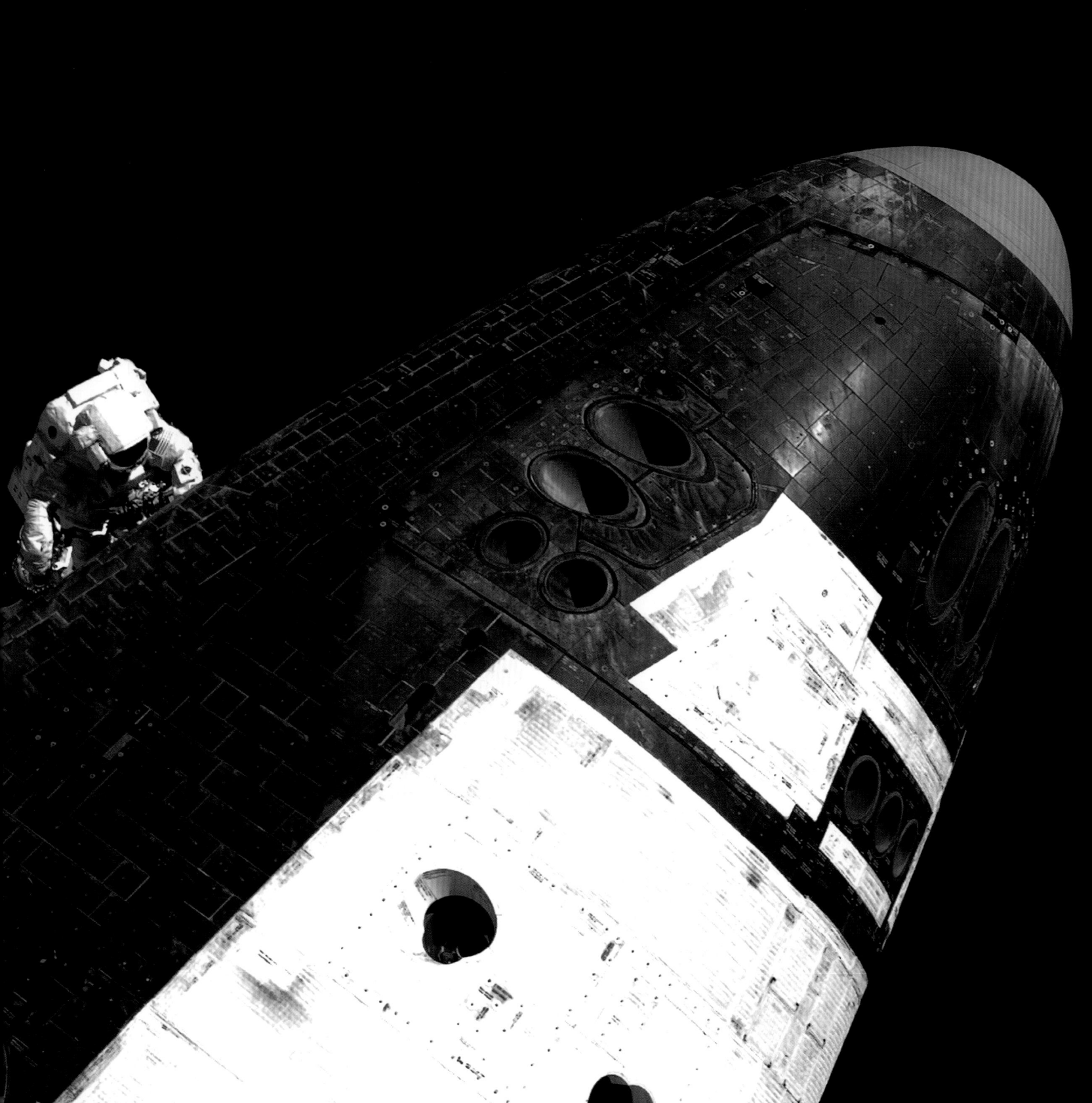

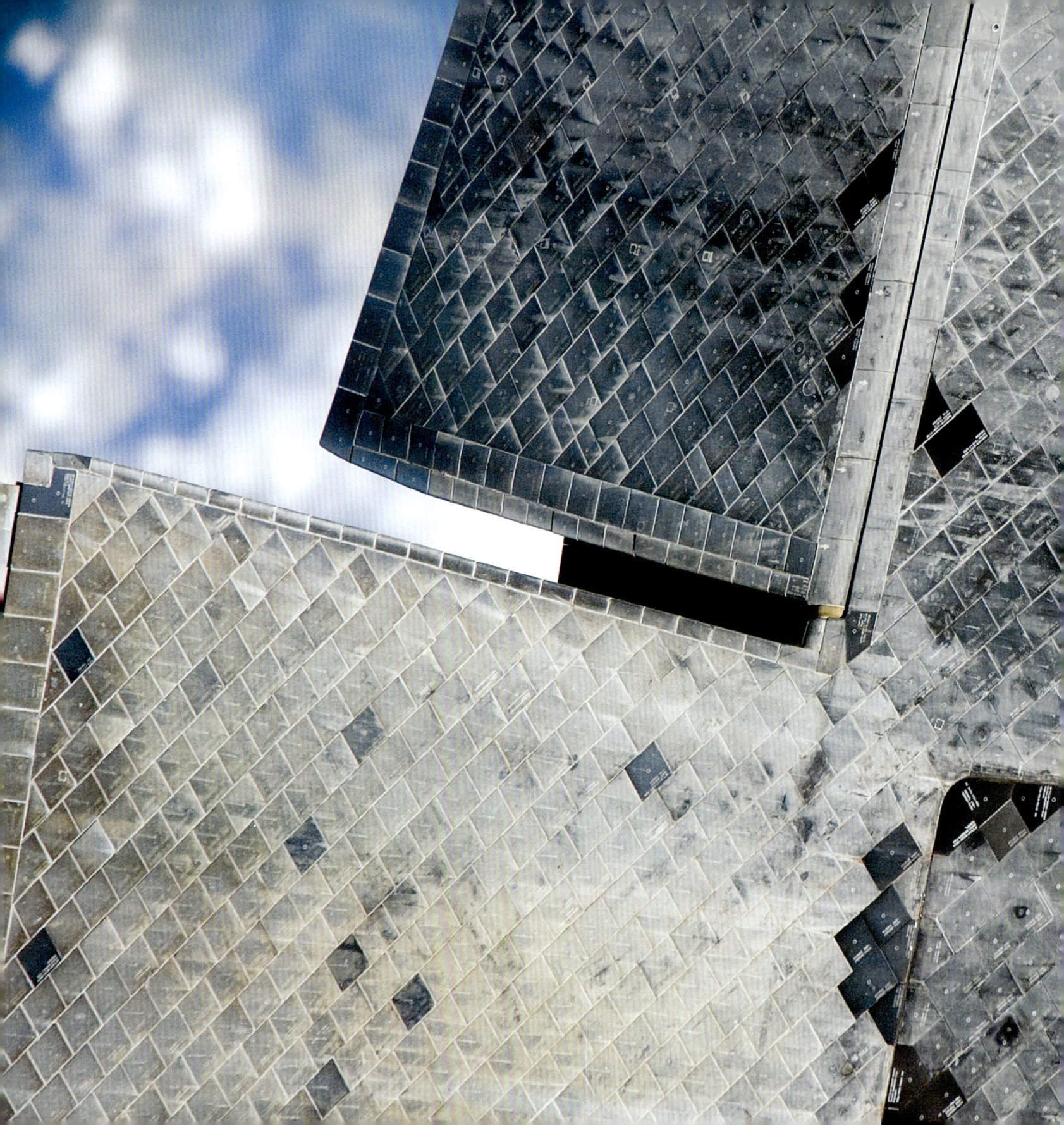

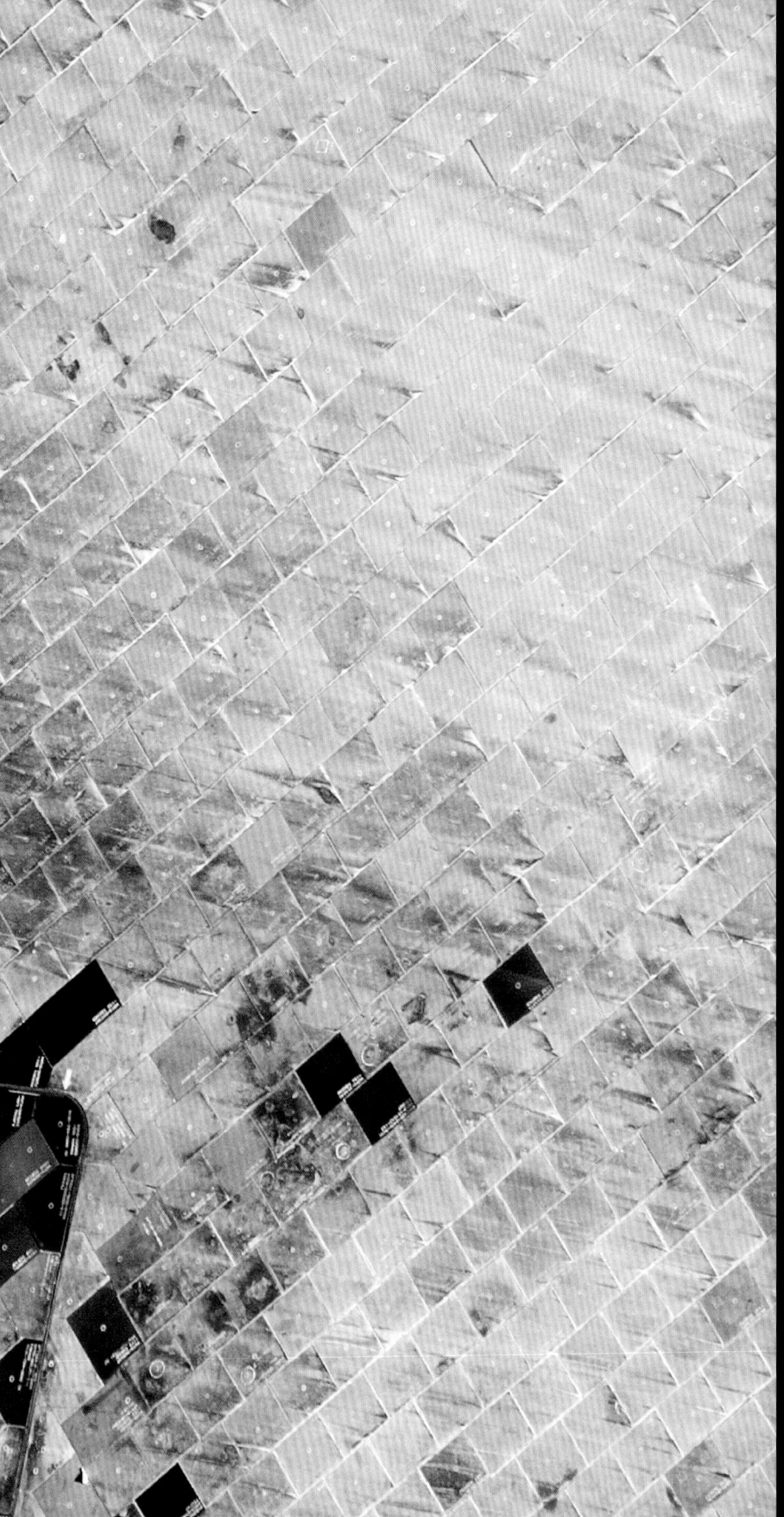

Discovery's underside thermal protection system. The shuttle is covered with protective tiles designed to withstand extremely high temperatures during the dangerous Earth re-entry procedure. More than 24,000 tiles are used to cover the shuttle's surface. Without these tiles the space shuttle would burn up and be destroyed on re-entry.

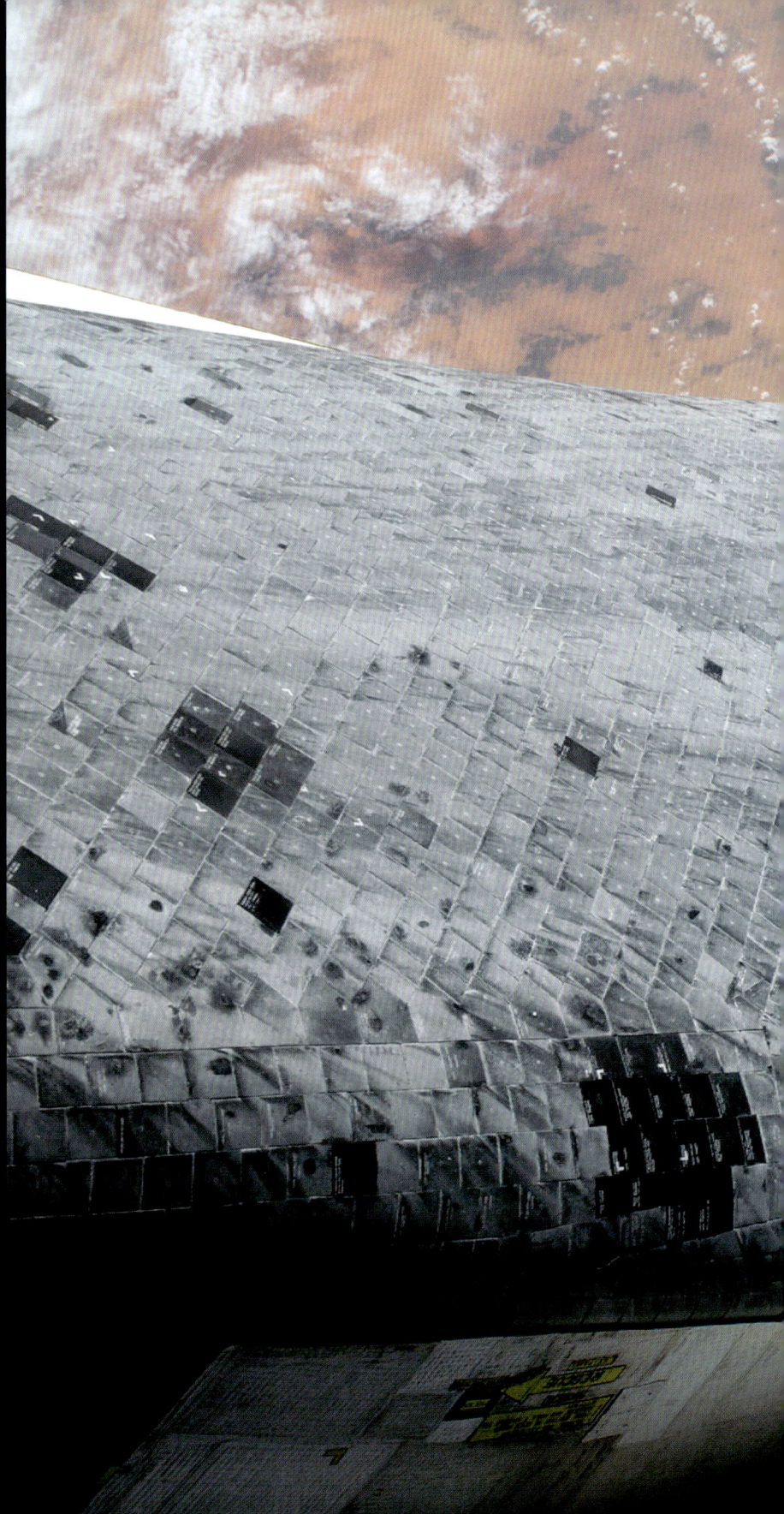

STS-114
26 July to 9 August 2005

A view from a space walk. Astronaut Stephen K. Robinson gets a detailed look at Discovery's underside thermal protection tiles. STS-114 marks the space shuttle's return to flight after the Columbia disaster of 2003, which claimed the lives of all seven crew members. Future missions saw more extensive procedures implemented, requiring thorough inspections of the shuttle's heat shield before re-entry could be approved.

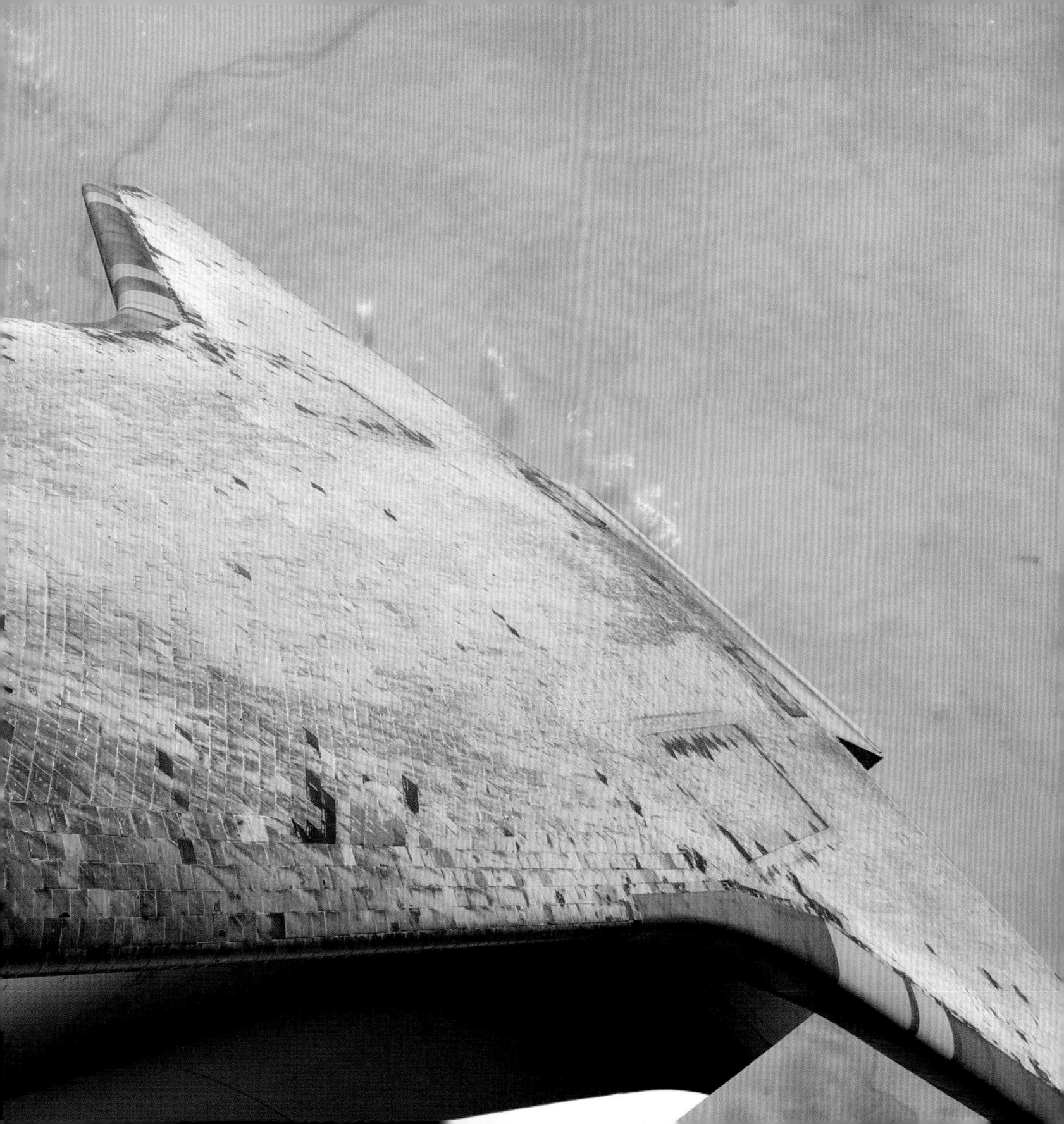

16 January to 1 February 2003

Ilan Ramon, the first Israeli astronaut and payload specialist of the ill-fated STS-107 mission, looks out in awe upon a lustrous Earth horizon. The mission ended in tragedy when Columbia disintegrated during Earth re-entry. The gallant crew never returned home.

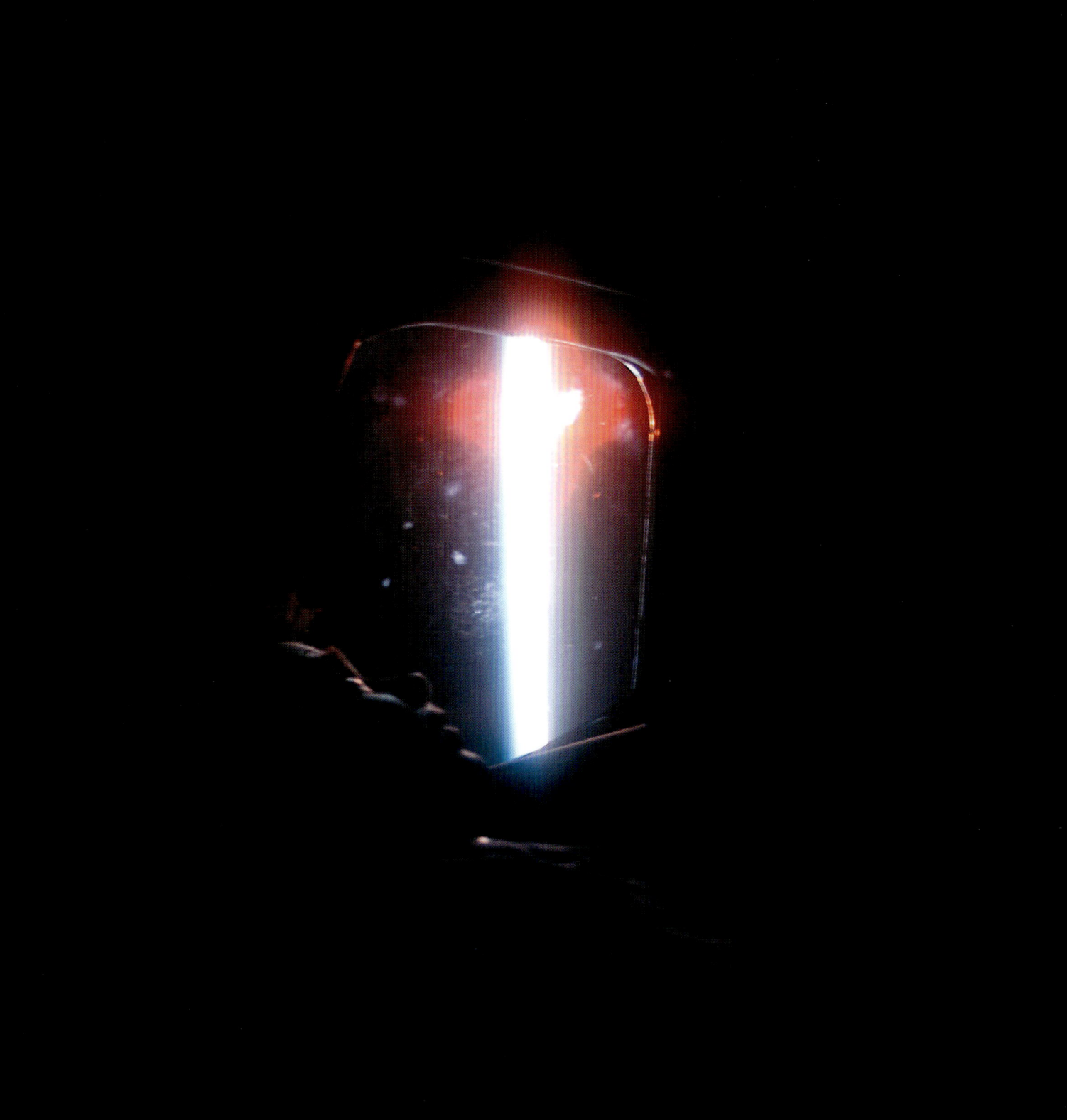

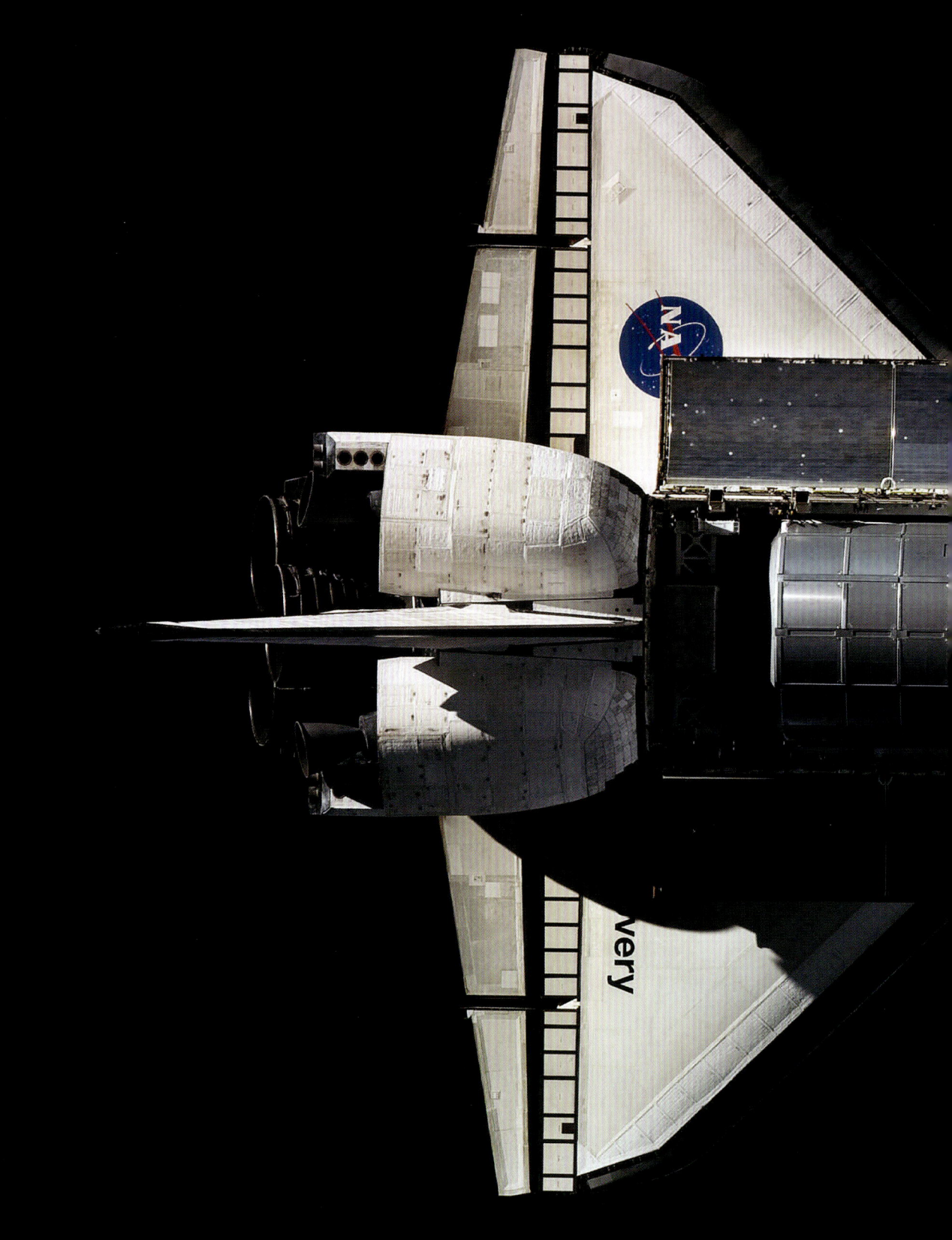

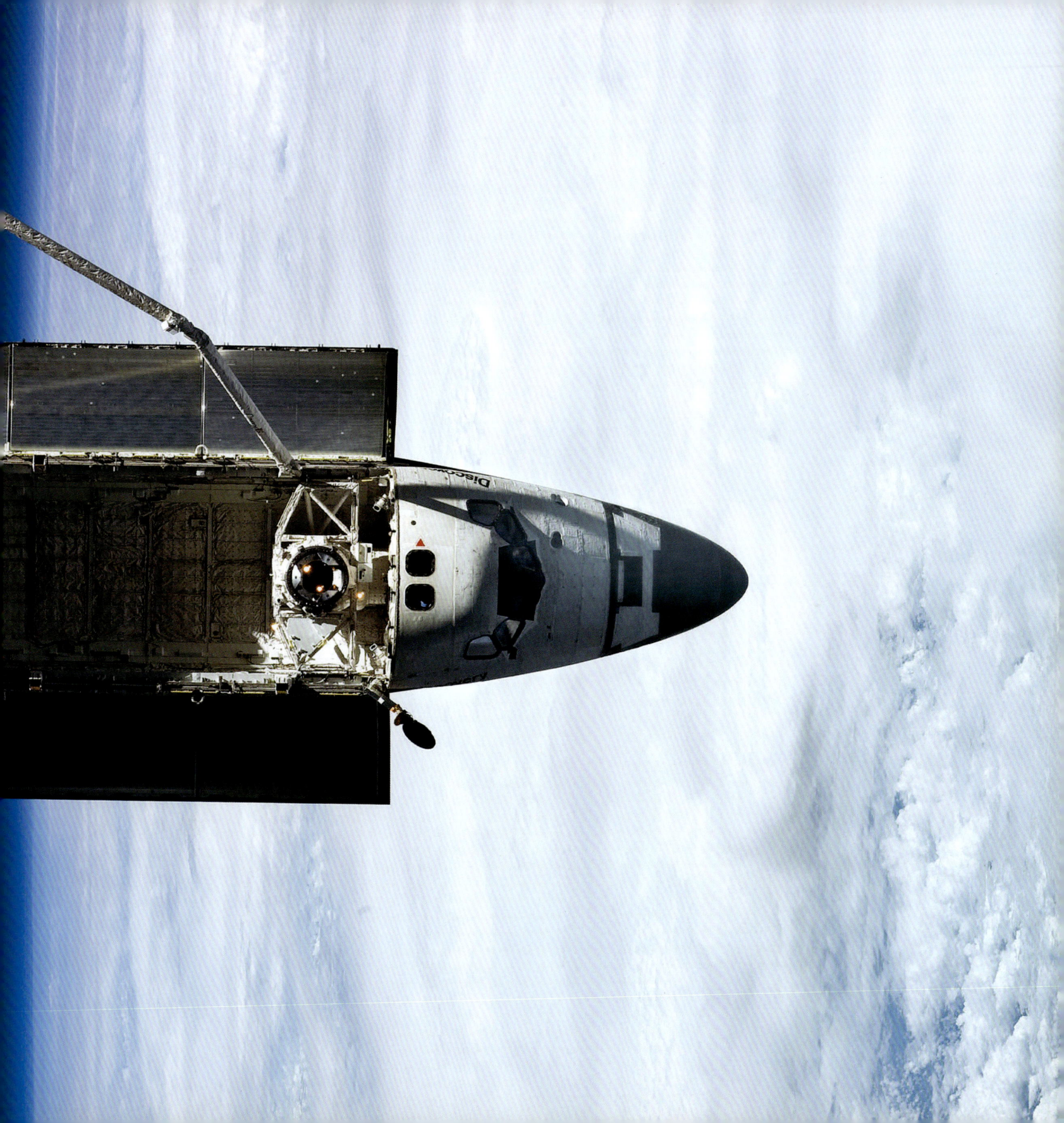

(Previous spread) What a remarkable sight! An inspiring view of Discovery saying its farewells just after undocking with the International Space Station. This image beautifully shows the contrast between the radiant, glowing Earth and the dark emptiness of infinite space.

STS-131
5–20 April 2010

(Right) As Discovery separates from the International Space Station, the crew begin preparations for their journey home after being docked in space for 10 days.

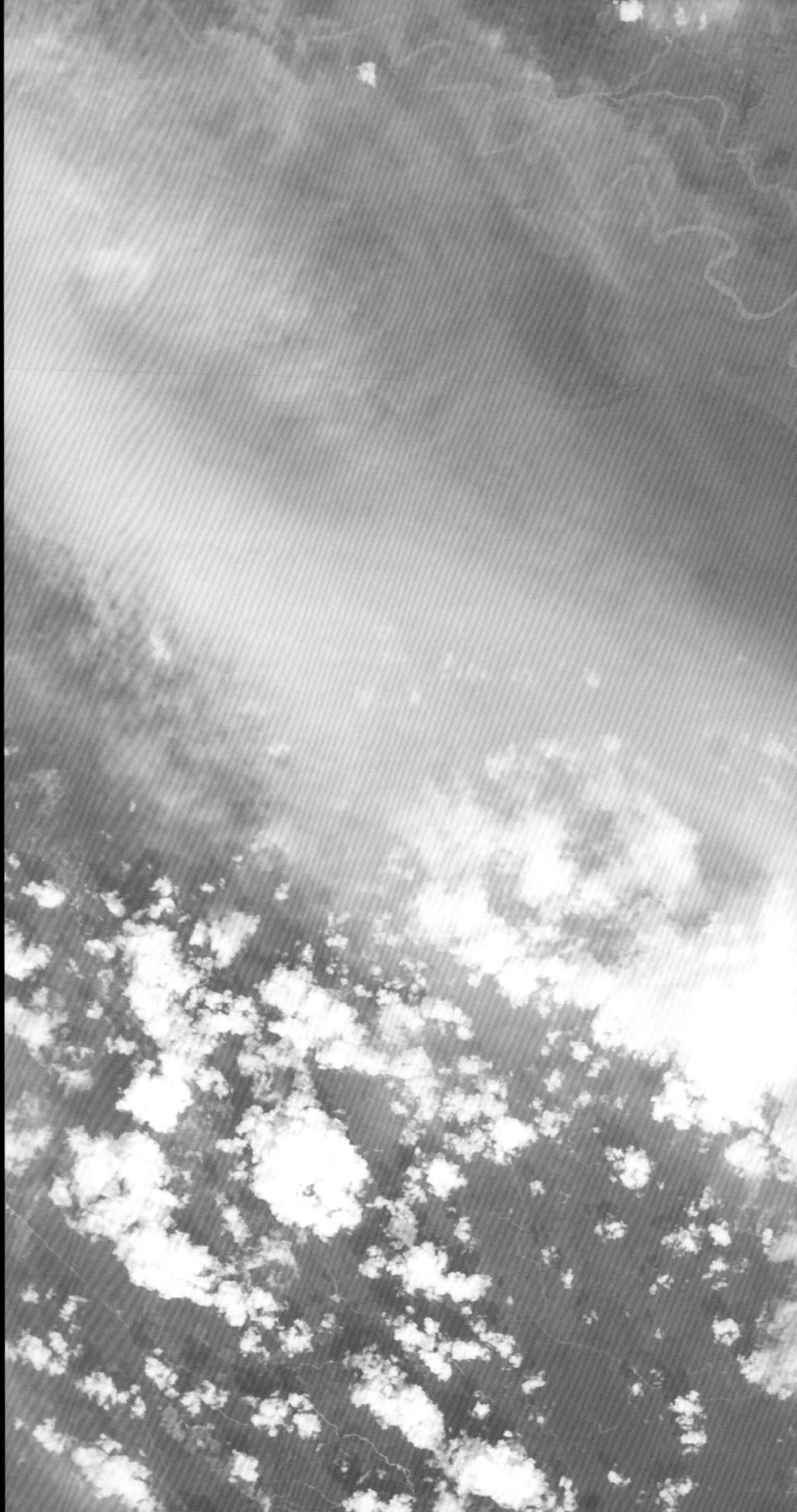

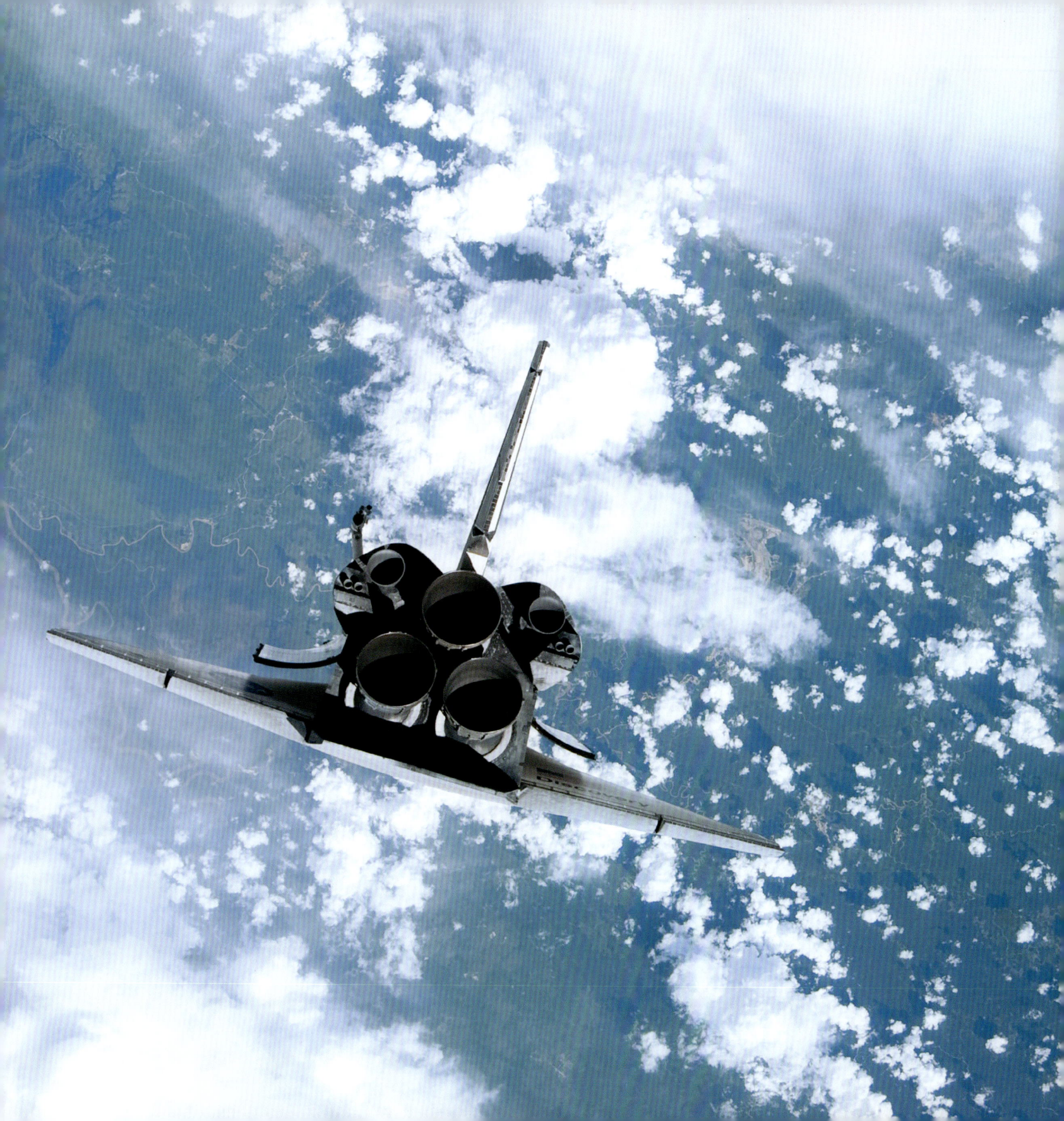

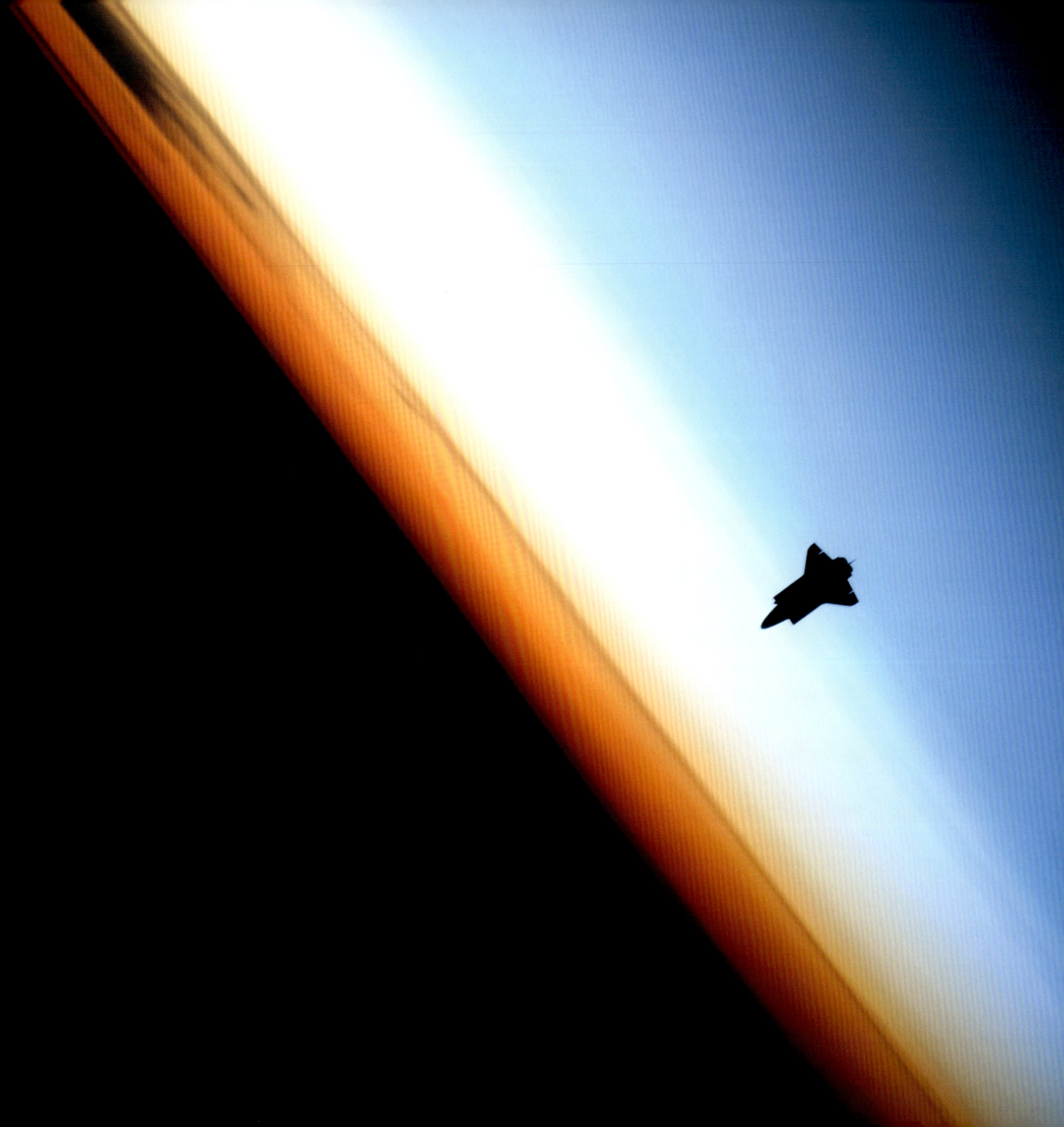

A crew member aboard the International Space Station captures this fascinating and unique exposure of Endeavour, silhouetted by the vibrant brilliance of Earth's atmosphere. The orange layer is called the troposphere, and is where Earth's weather originates from and is contained. The white layer is the stratosphere and the blue layer is the mesosphere.

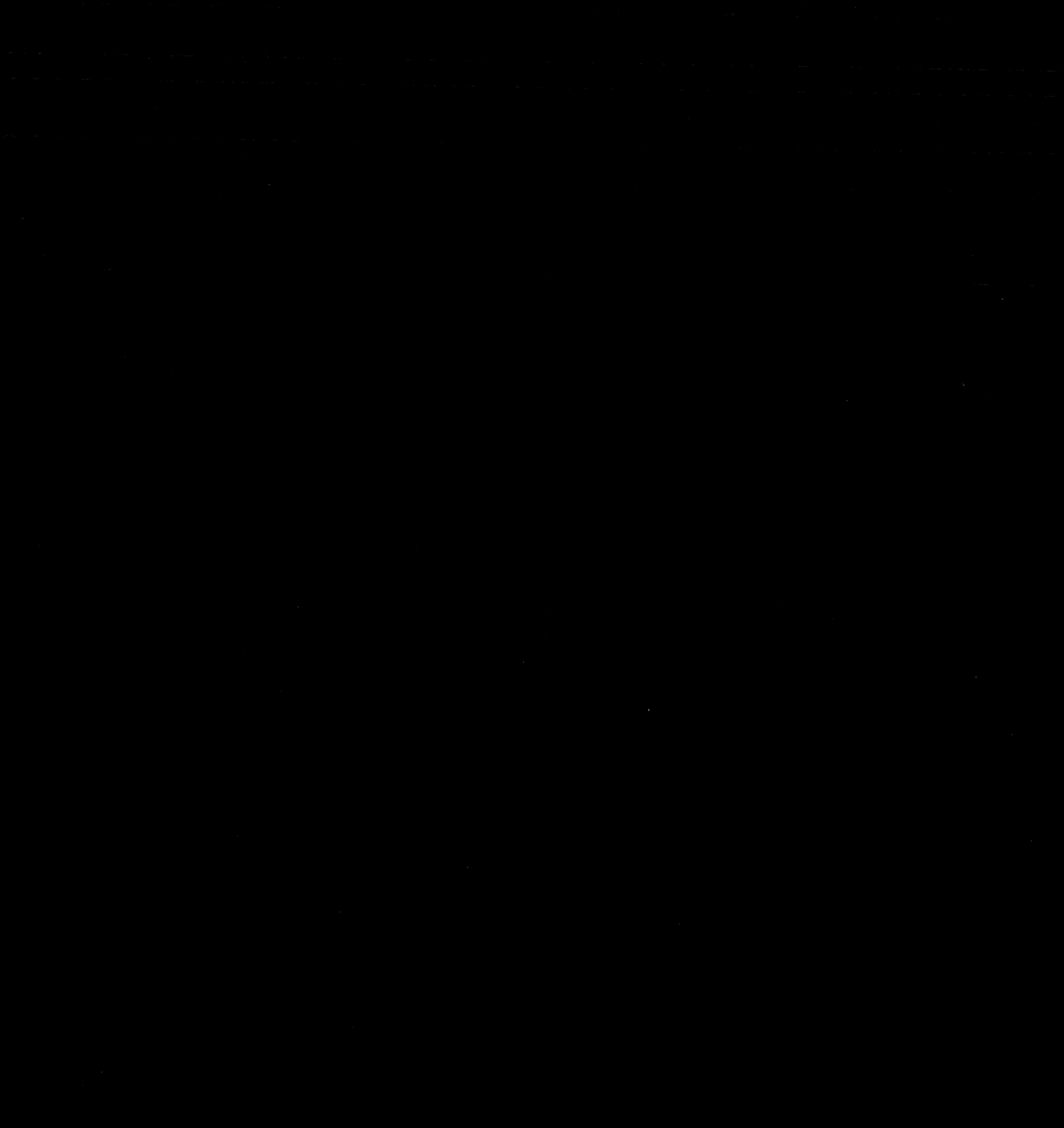

00:00:04

LANDING

After a two-day mission by astronauts Joe H. Engle and Richard H. Truly, Columbia glides on its final approach before touchdown at Edwards Air Force Base in the California desert. After this second shuttle flight, mission durations would start to become longer and, from STS-5 onwards, additional crew members were assigned to flights in order to carry out various mission objectives.

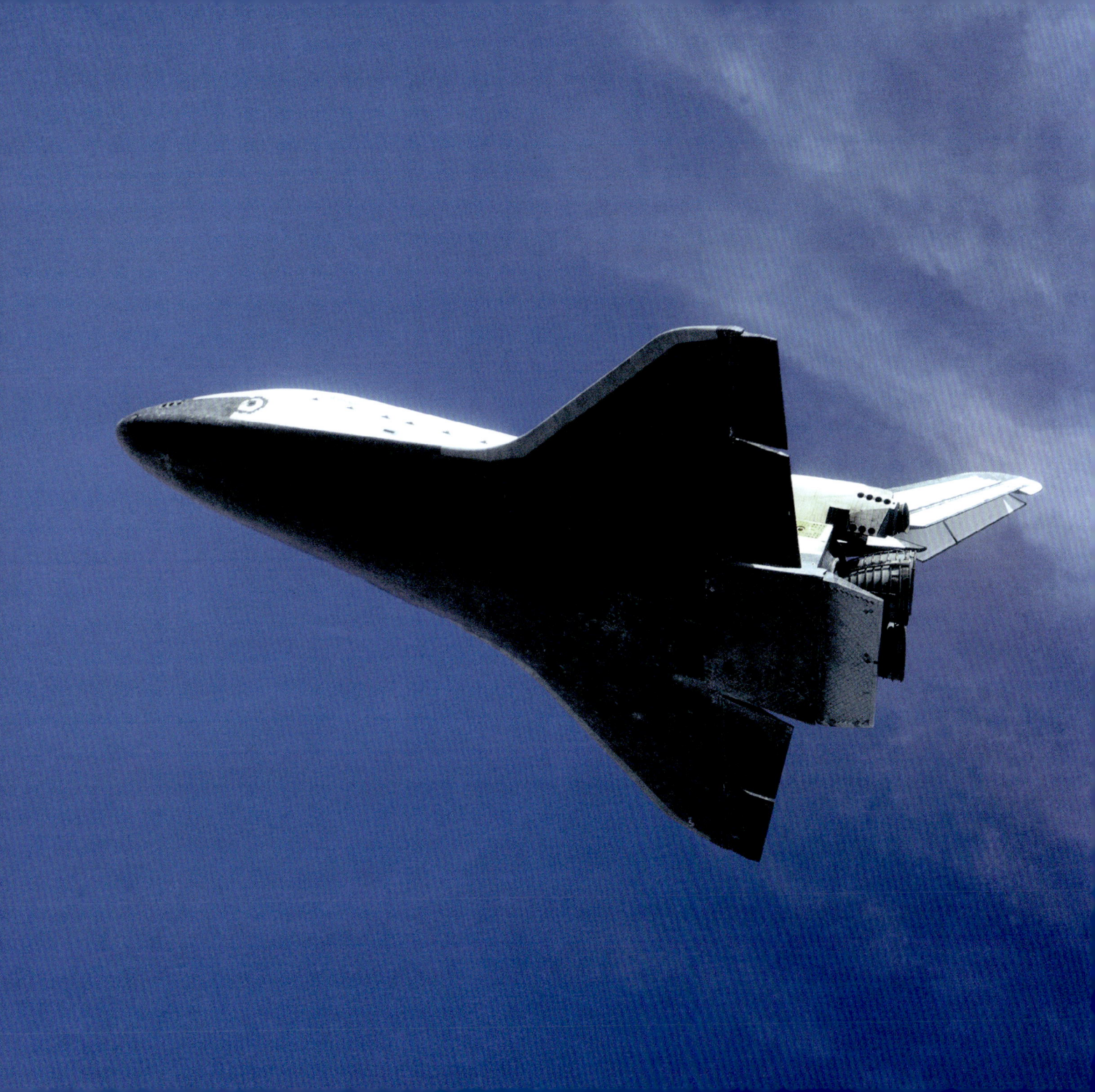

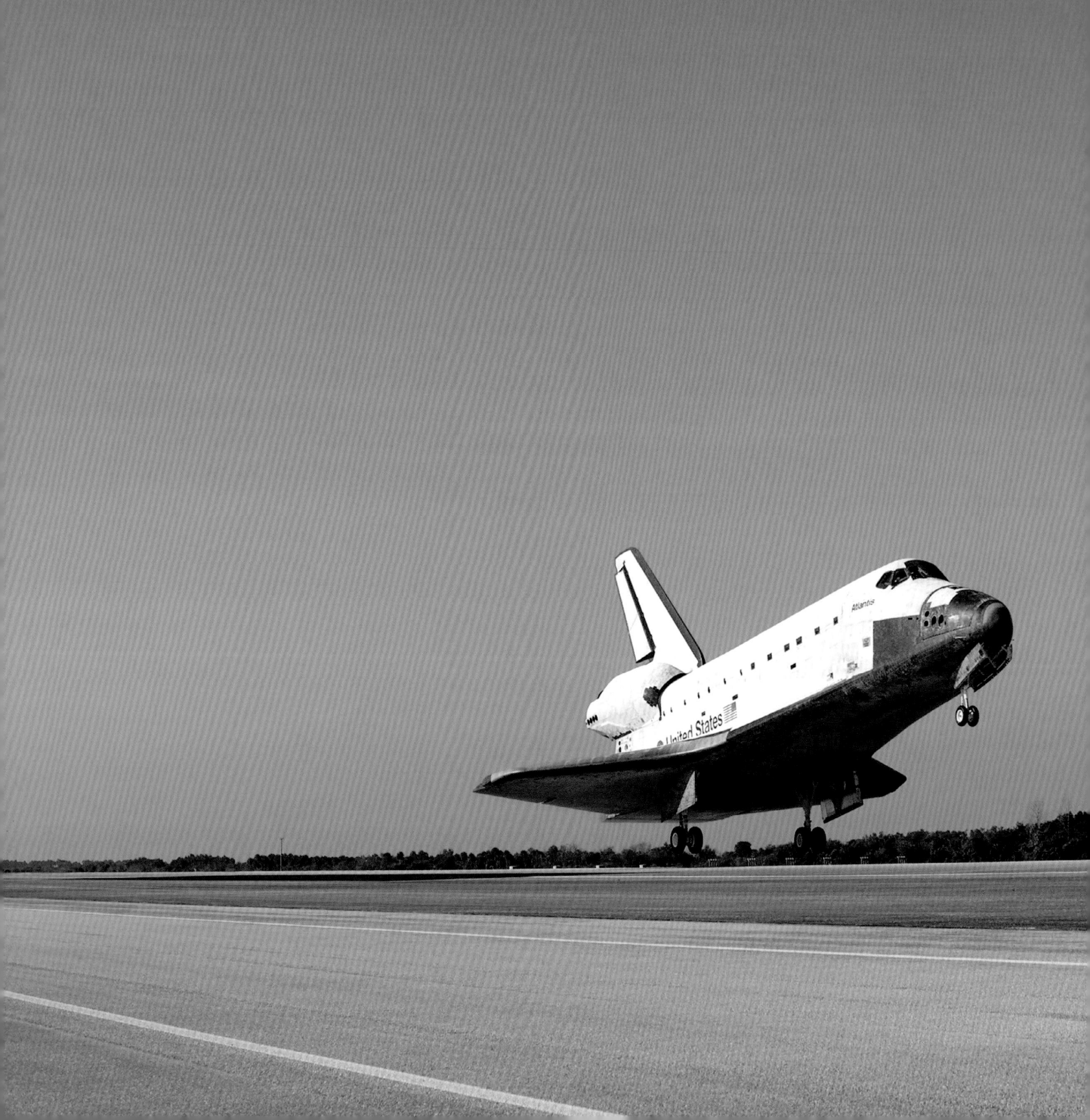

STS-132
14–26 May 2010

(Left) Atlantis nears touchdown on Runway 33 at the Kennedy Space Center landing facility, after completing a 12-day mission to the International Space Station. STS-132 was due to be the last flight for Atlantis, but in early 2011 authorization was given for the shuttle to return to space one final time.

STS-133
24 February to 9 March 2011

(Following spread) Space shuttle Discovery touches down for the final time on Runway 15 at Kennedy Space Center, upon completion of its last ever mission. This was the 35th shuttle mission to the International Space Station.

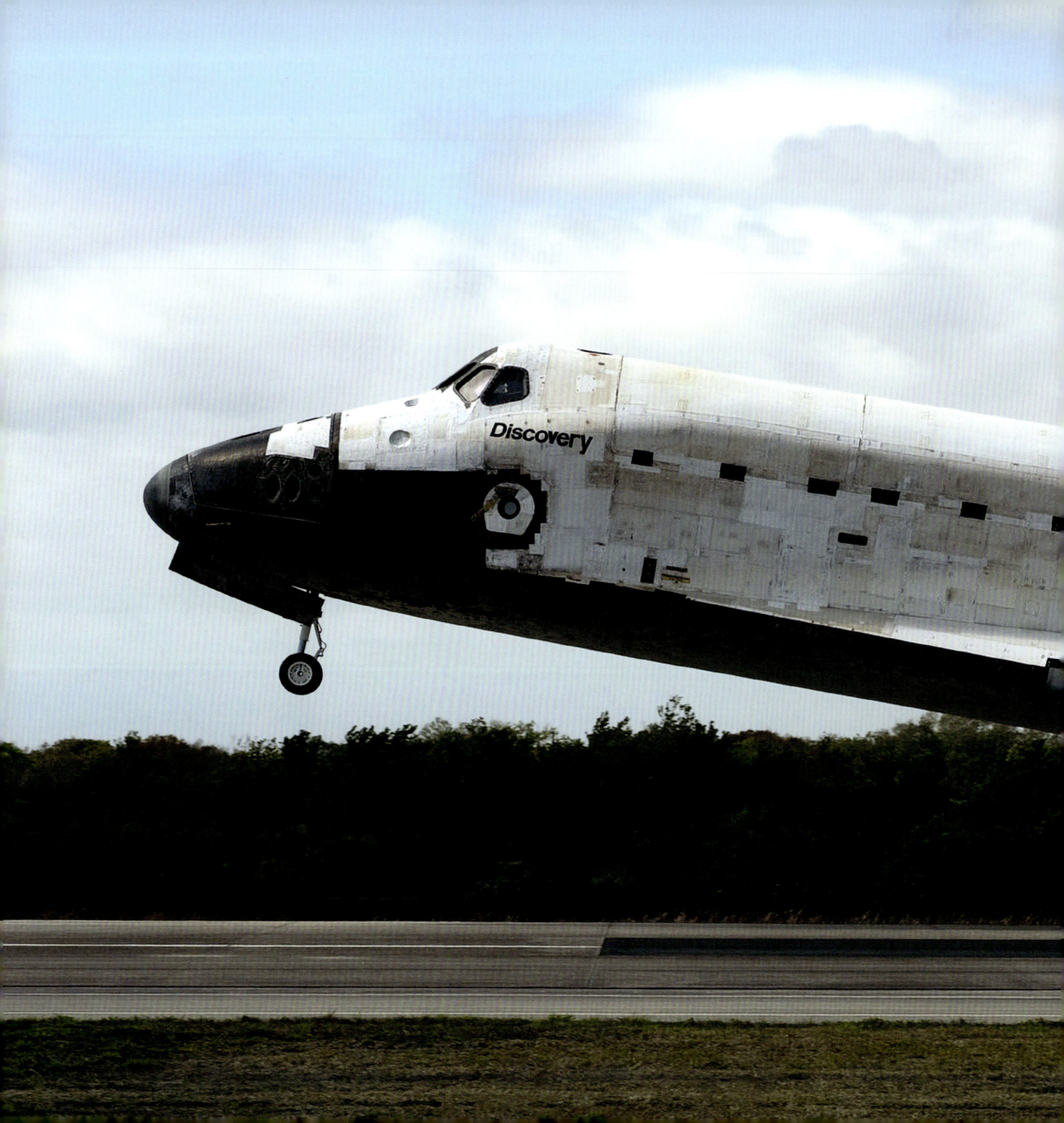

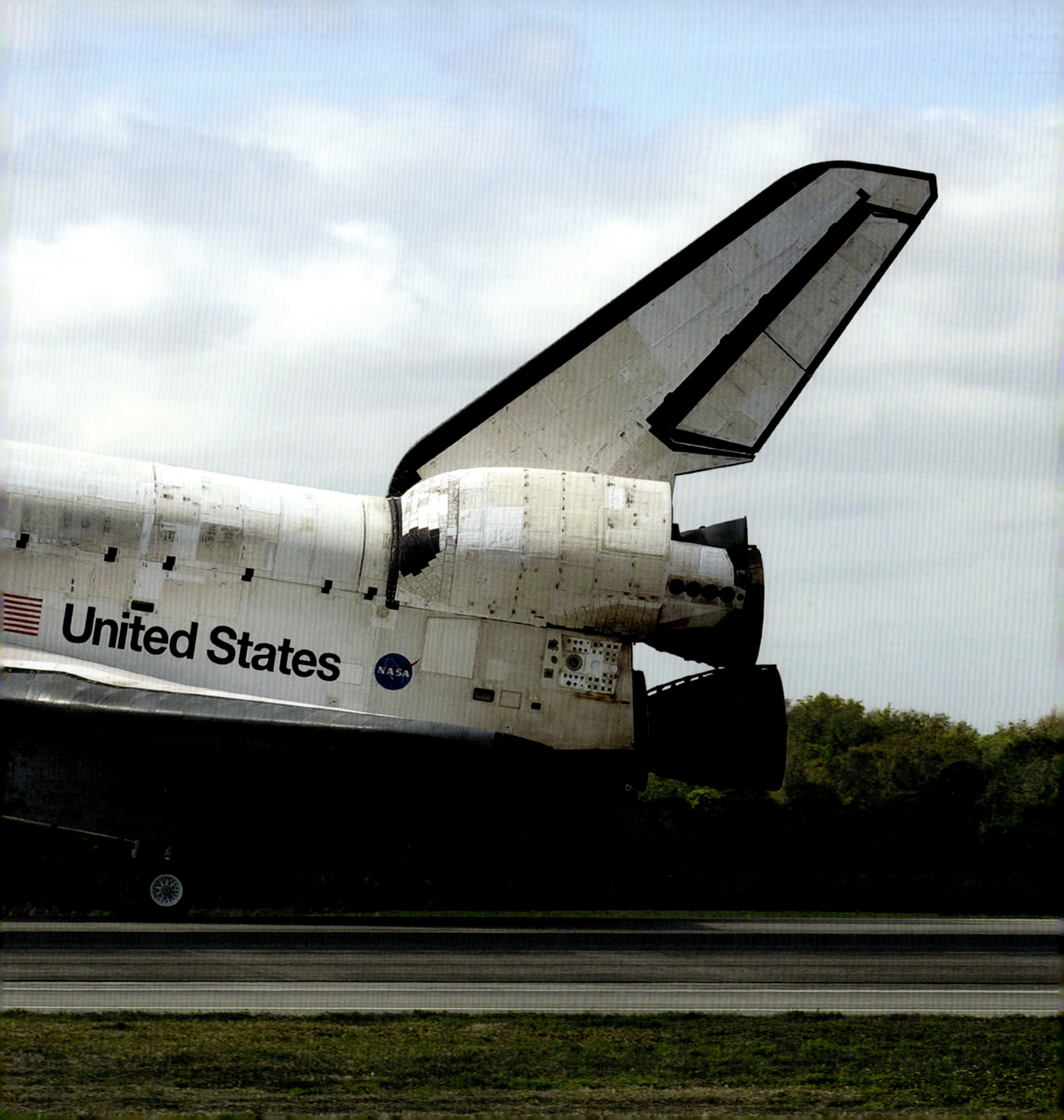

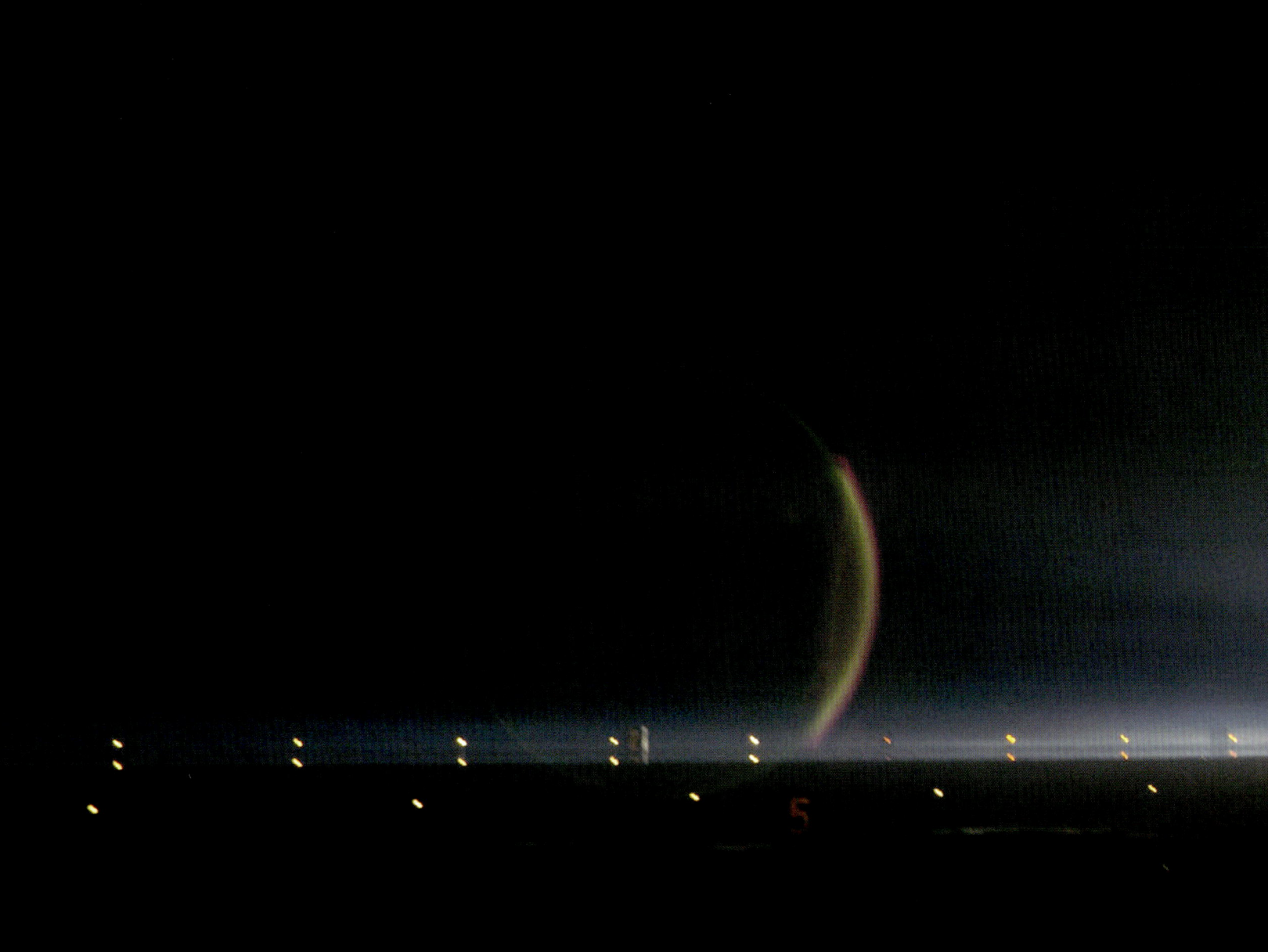

STS-135

8–21 July 2011

After 30 glorious years among the stars and 135 missions launched in total, the familiar and comforting lights at Kennedy Space Center guide Atlantis and her brave crew home for the 33rd and final time. It was the 36th shuttle voyage to the International Space Station and the concluding flight of the exceptional space shuttle program.

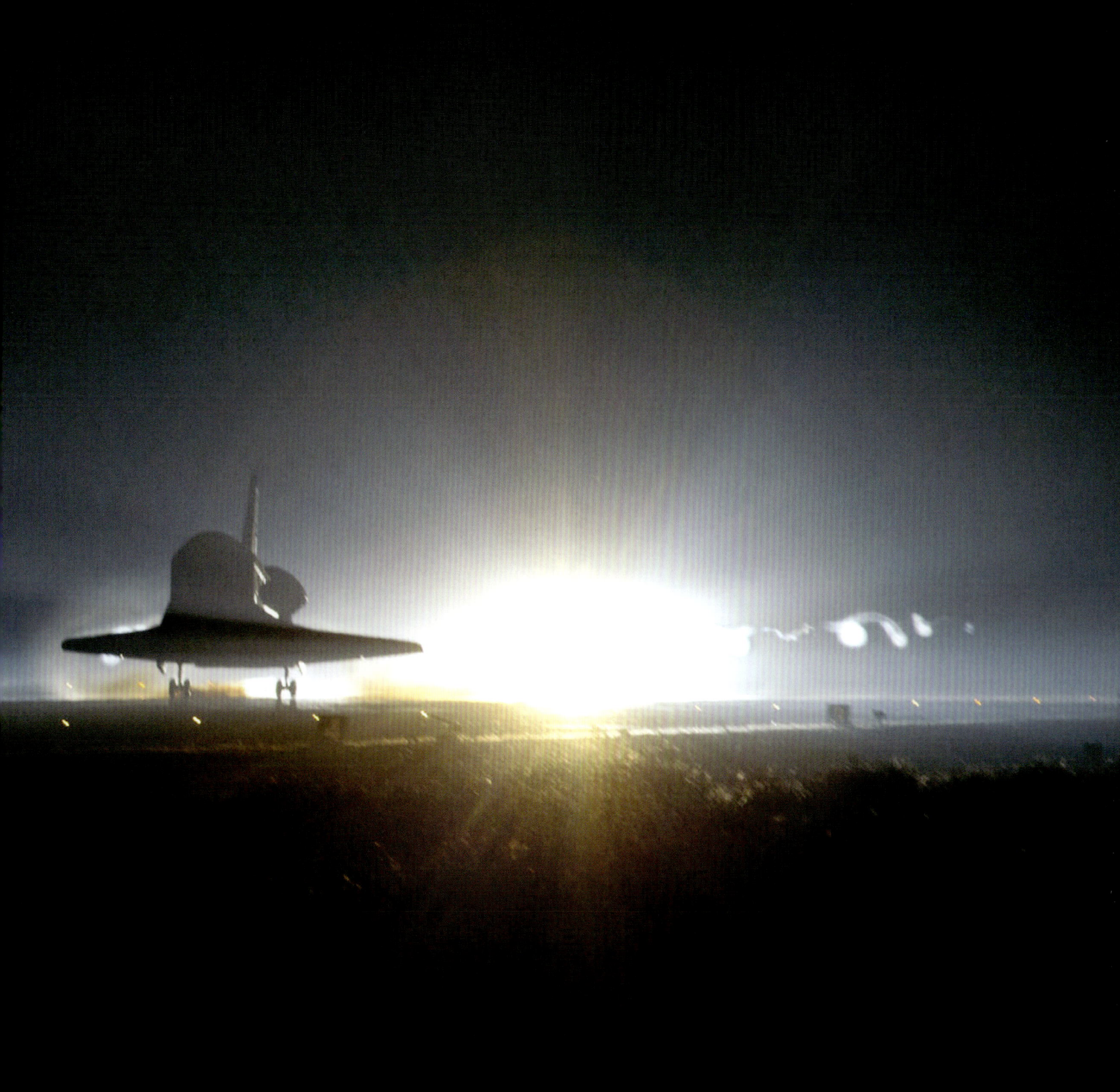

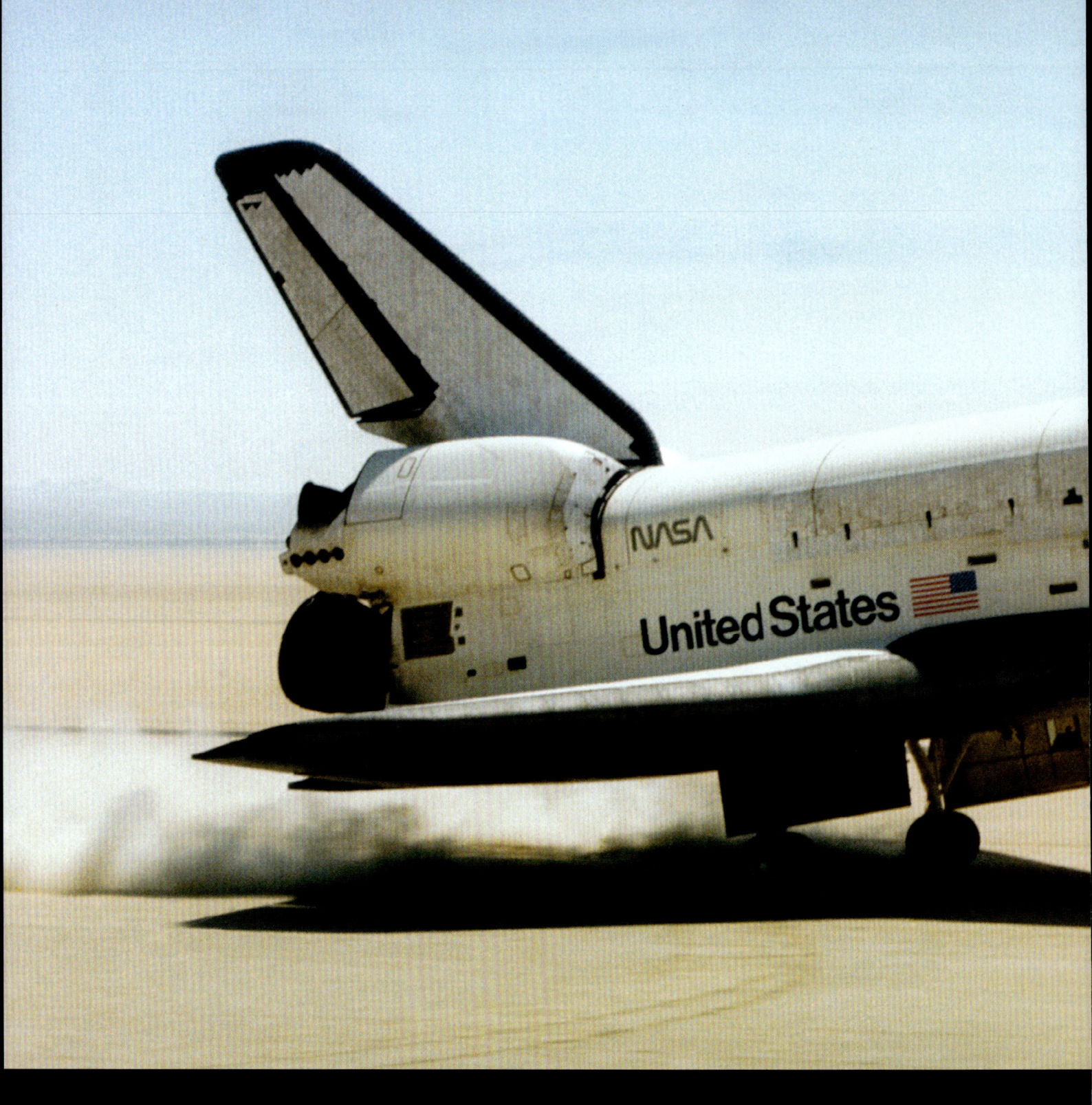

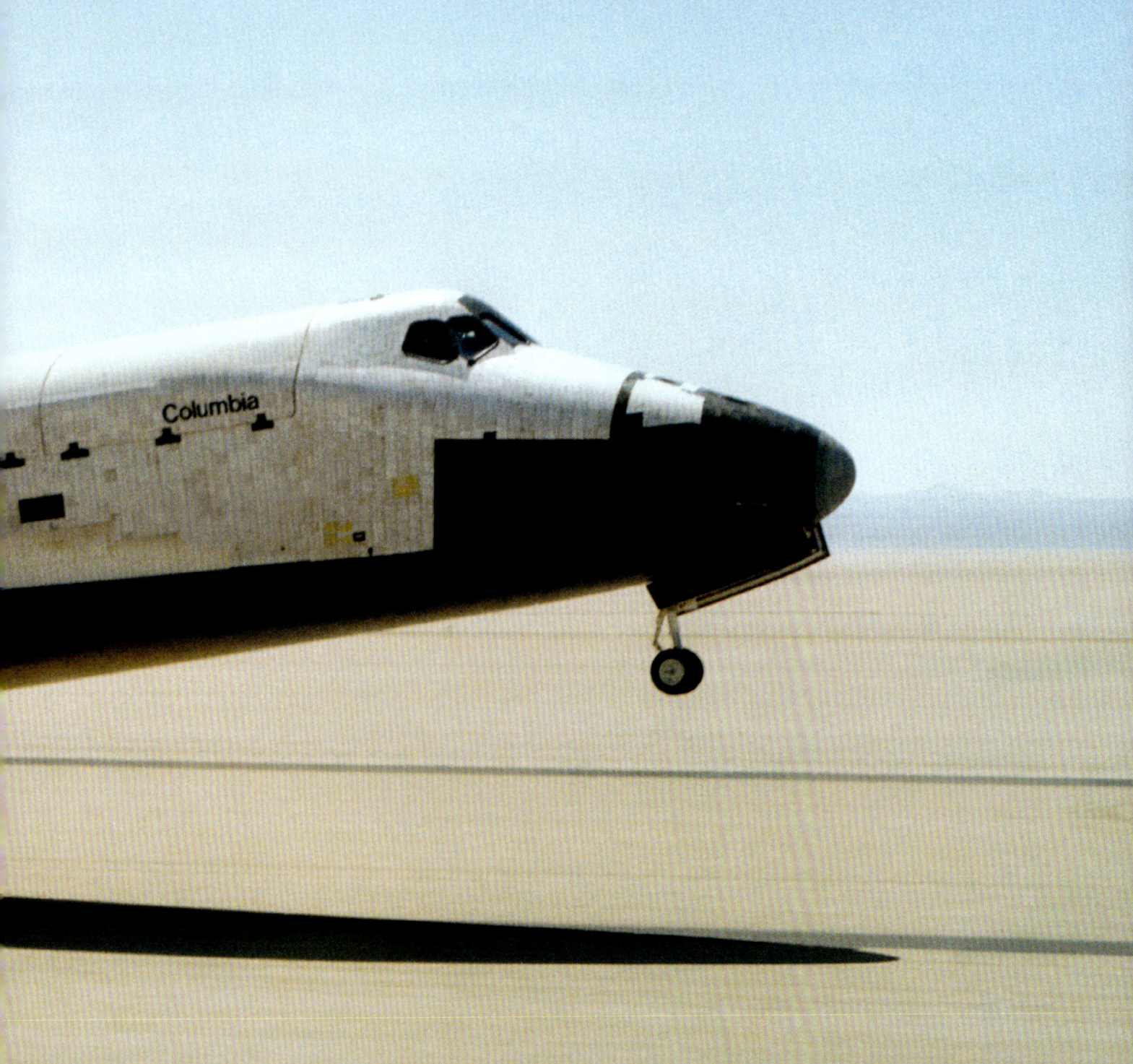

12–14 April 1981

(Previous spread) Columbia makes a triumphant touchdown at Rogers Dry Lakebed, Edwards Air Force Base, to complete the first orbital space shuttle mission. The dream is alive!

STS-126
14–30 November 2008

(Right) As Endeavour touches down on runway 04-L at Edwards Air Force Base, the main drag chute is deployed to help slow the space shuttle to a halt, bringing an end to a mission lasting 15 days, 20 hours, 29 minutes and 37 seconds. The crew travelled 6.6 million miles and completed 251 revolutions of Earth.

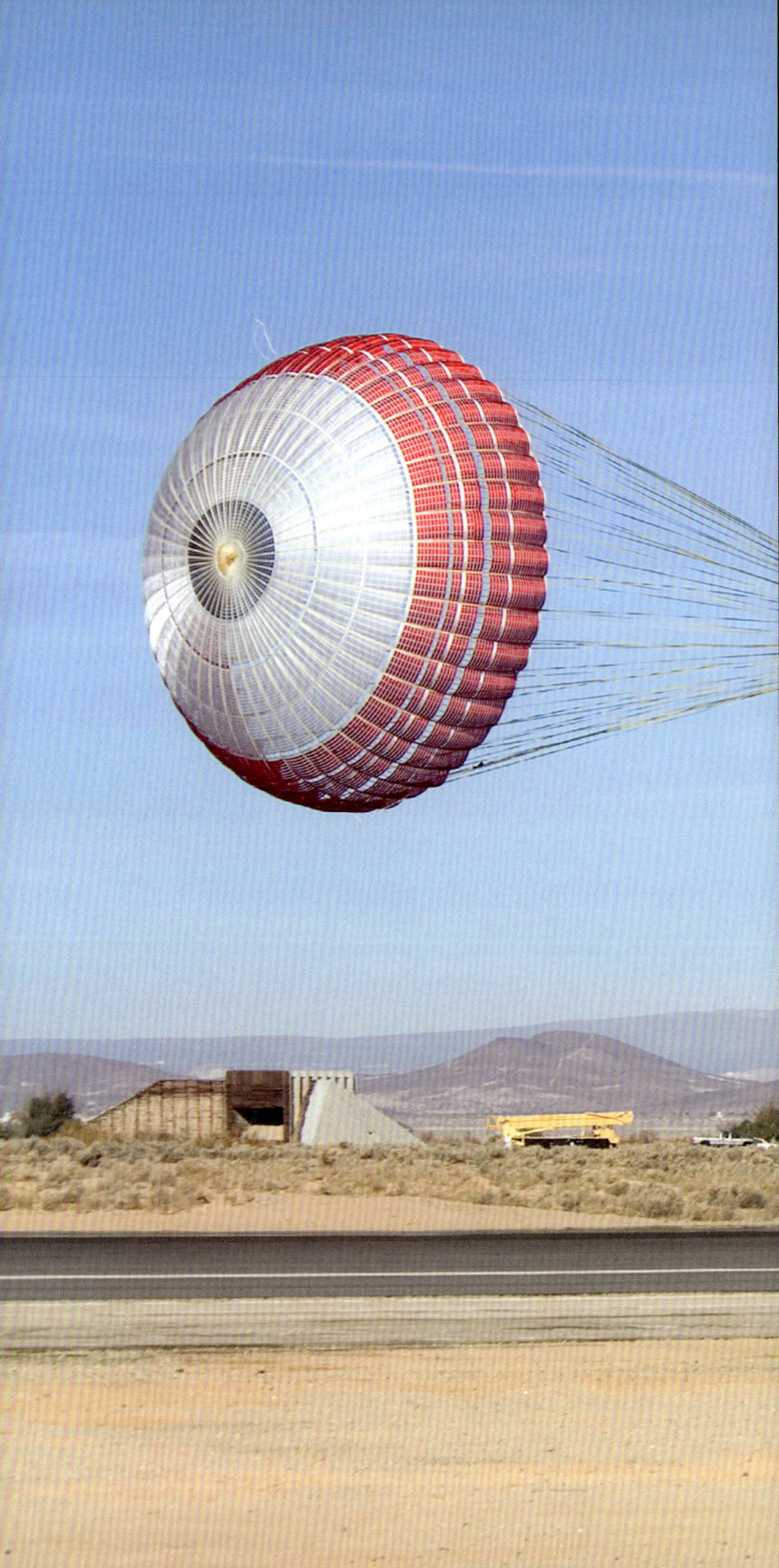

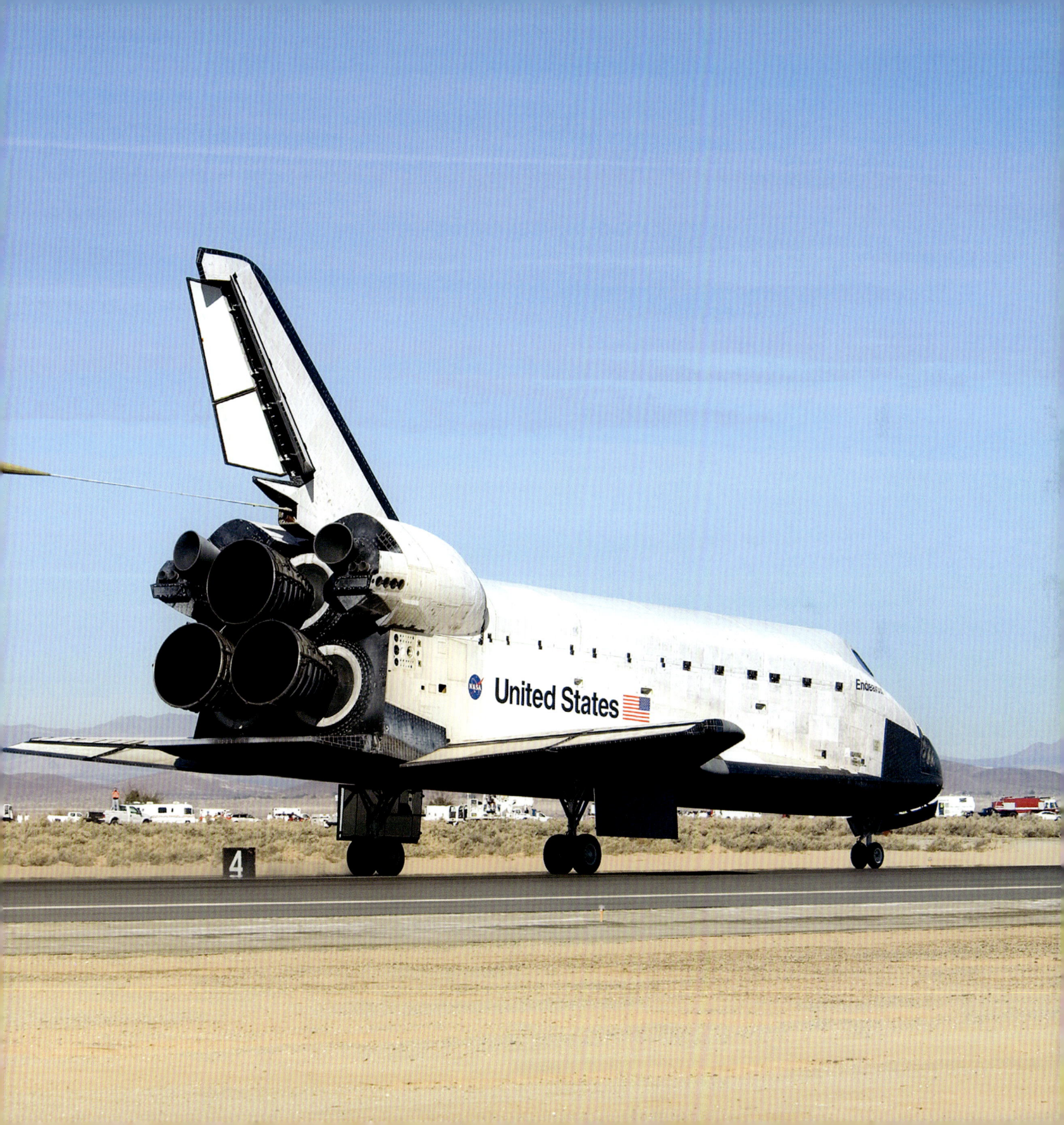

Endeavour arrives home safely at Kennedy Space Center during the early morning hours. The dazzling runway lights provide an effective face-on silhouette of the triumphant orbiter. The shuttle and her crew delivered the Alpha Magnetic Spectrometer and various spare parts to the International Space Station. The mission, which lasted nearly 16 days, marked the 25th and final space flight of space shuttle Endeavour.

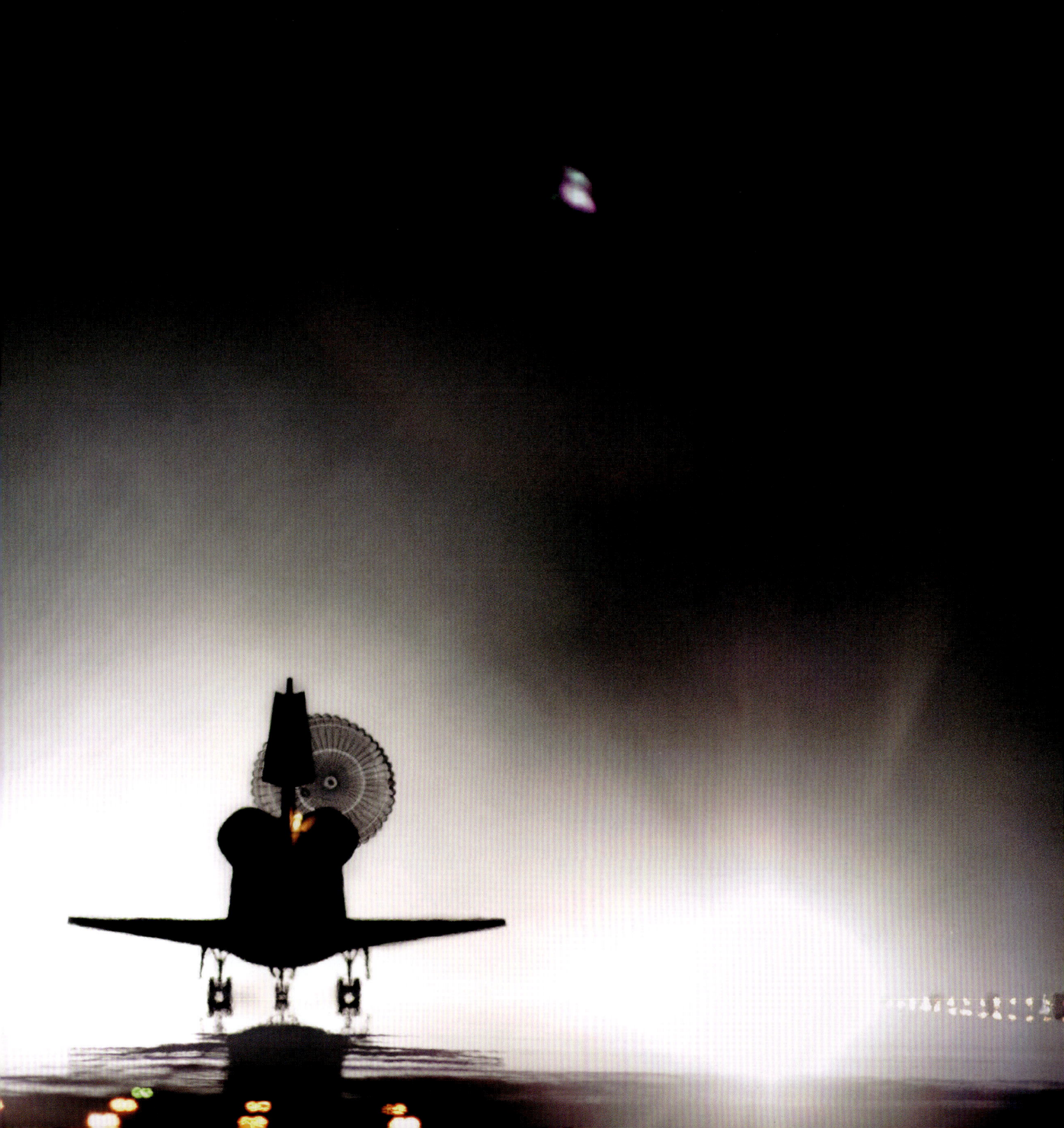

12–14 April 1981

Columbia rests on Runway 23 at Rogers Dry Lakebed, Edwards Air Force Base, after a successful maiden voyage. The two-man crew, consisting of John W. Young and Robert L. Crippen, travelled over one million miles and completed 37 Earth orbits. This mission was the first time a brand new spacecraft had been piloted on its maiden space flight.

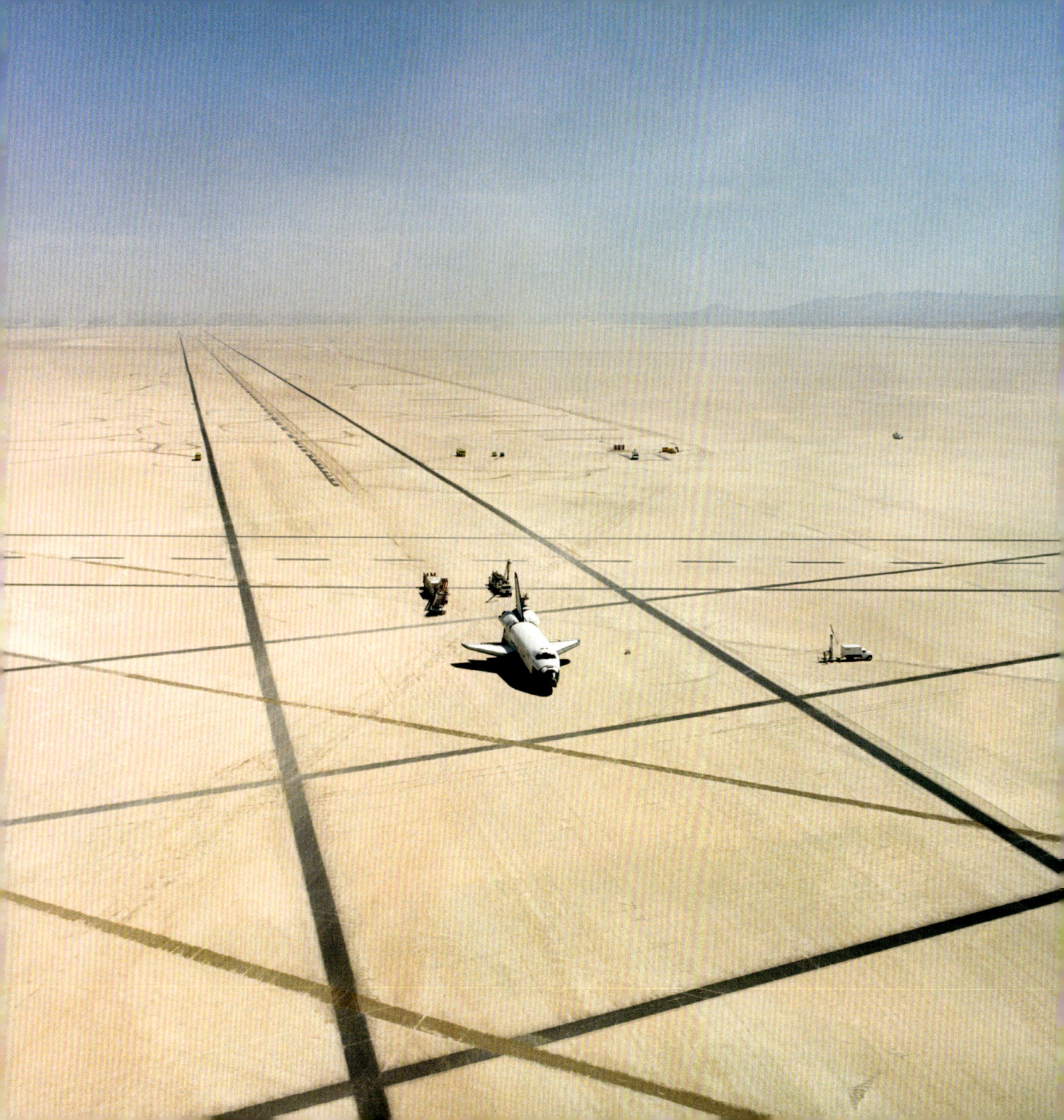

STS-128
28 August to 11 September 2009

(Right) As the sun sets over Edwards Air Force Base, Discovery's crew disembark from the resting spacecraft to conclude a 14-day mission to the International Space Station. The mission marked the 25th anniversary flight of Discovery since its maiden space voyage in 1984.

STS-117
8–22 June 2007

(Following spread) After completing a 14-day mission in orbit, Atlantis is towed to the Dryden Flight Research Center's Mate-Demate device, located at Edwards Air Force Base in the California desert. The shuttle will undergo post-mission processing before making its journey back to Kennedy Space Center, atop the Boeing 747 shuttle carrier aircraft.

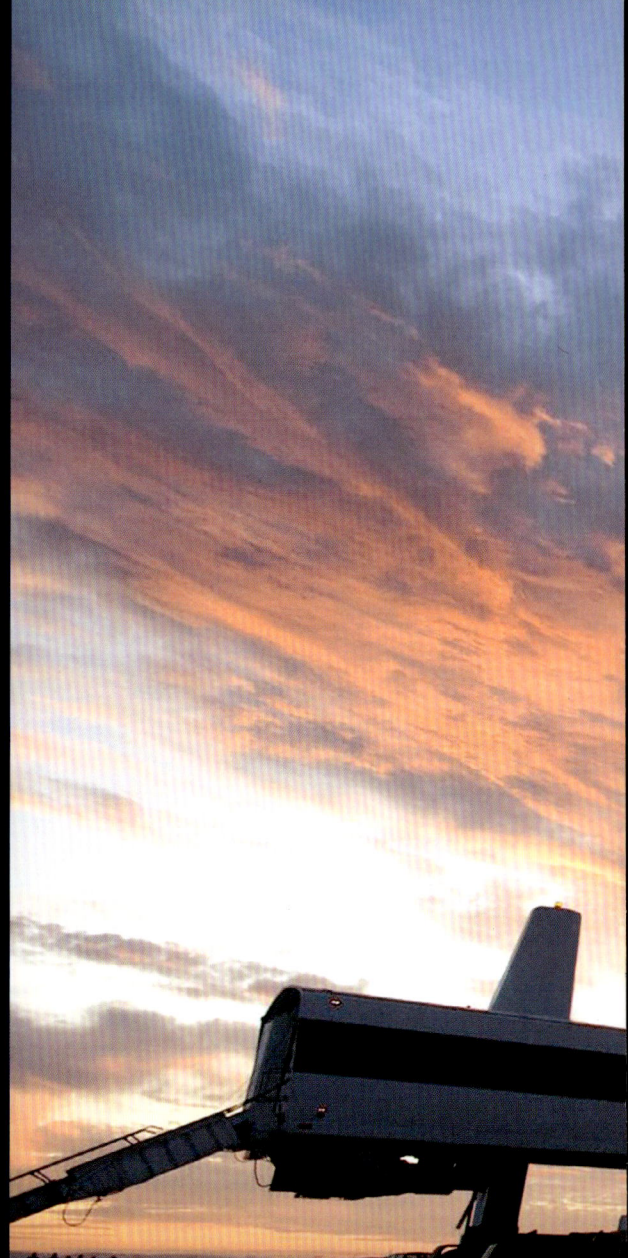

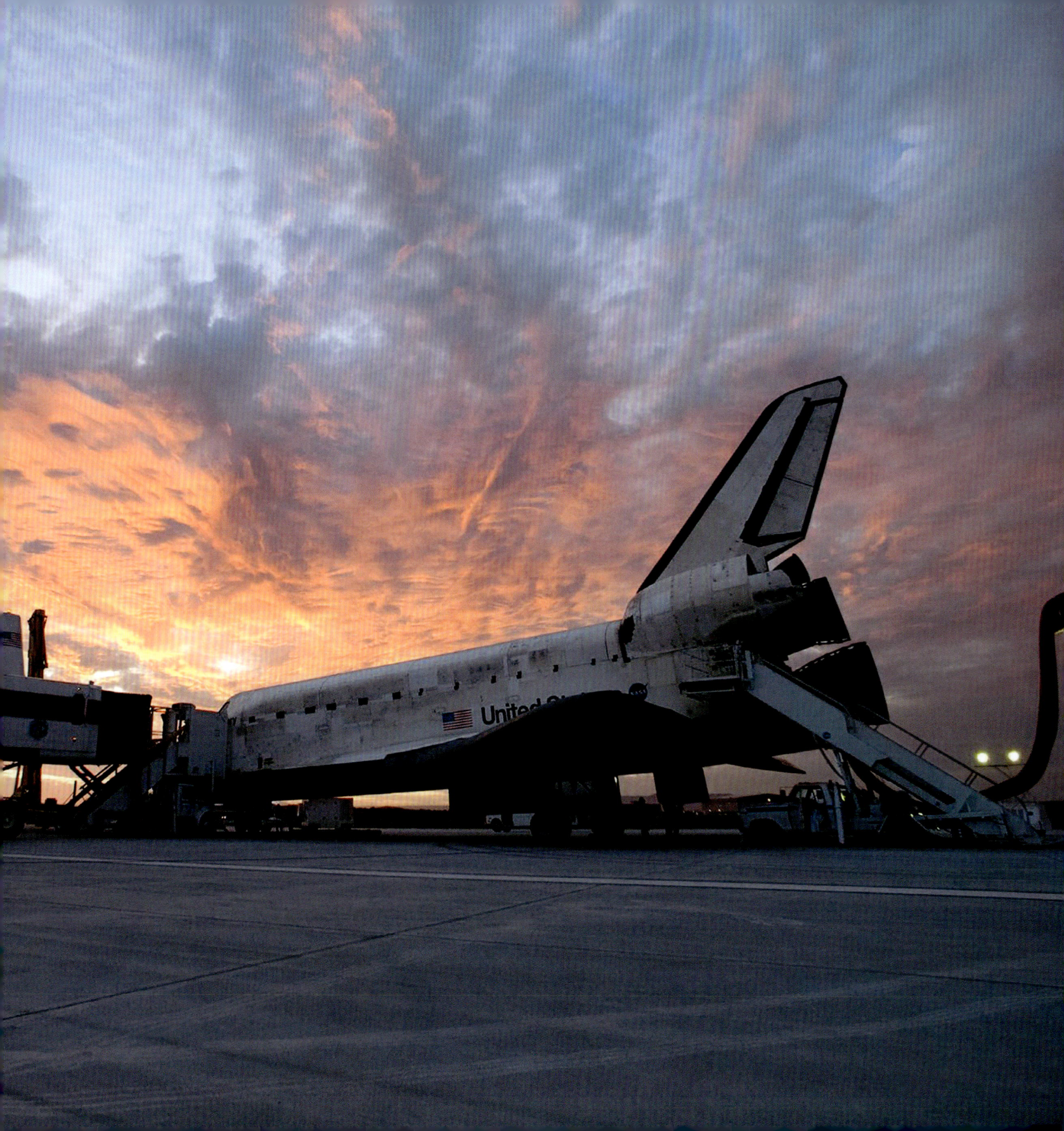

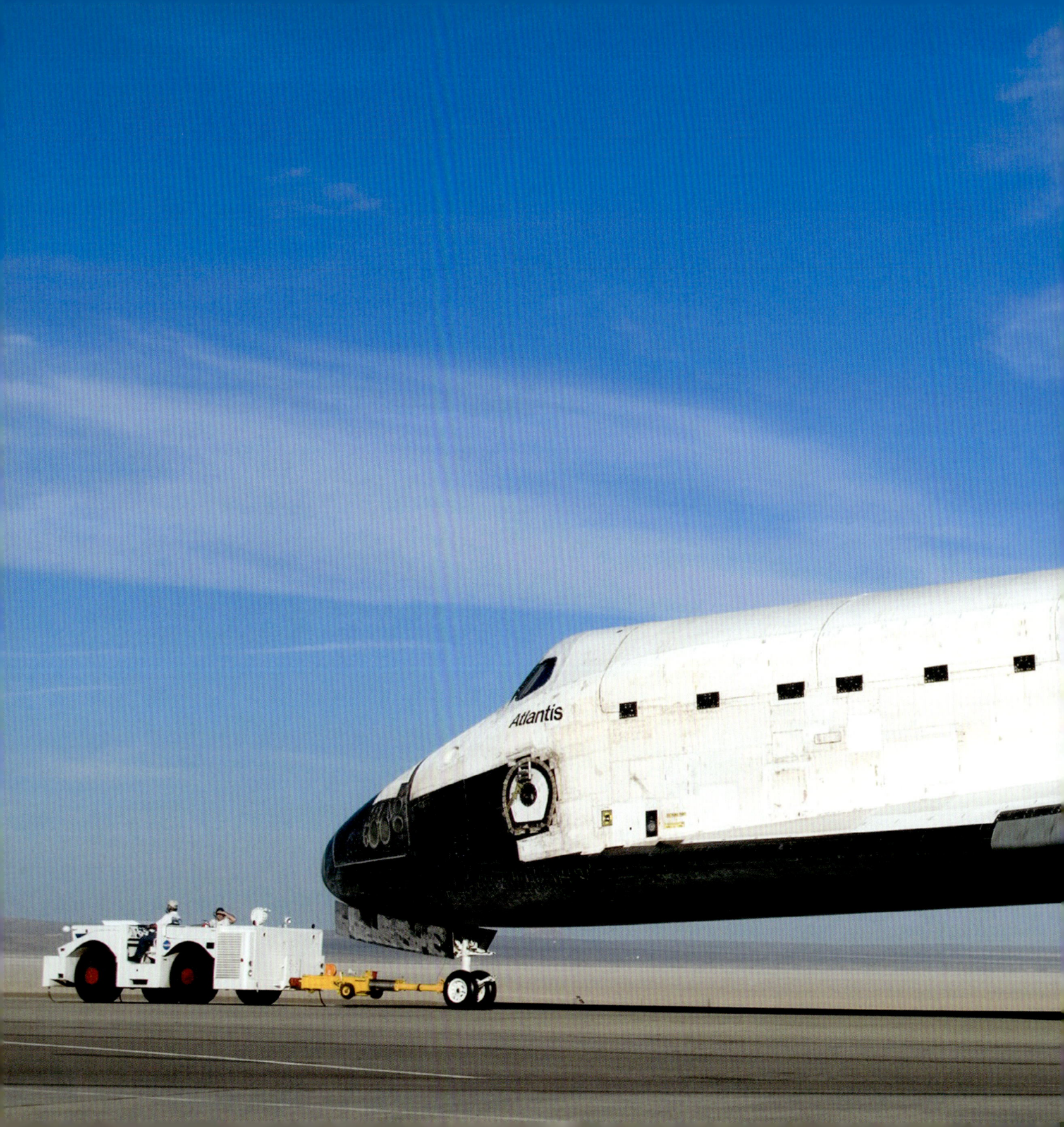

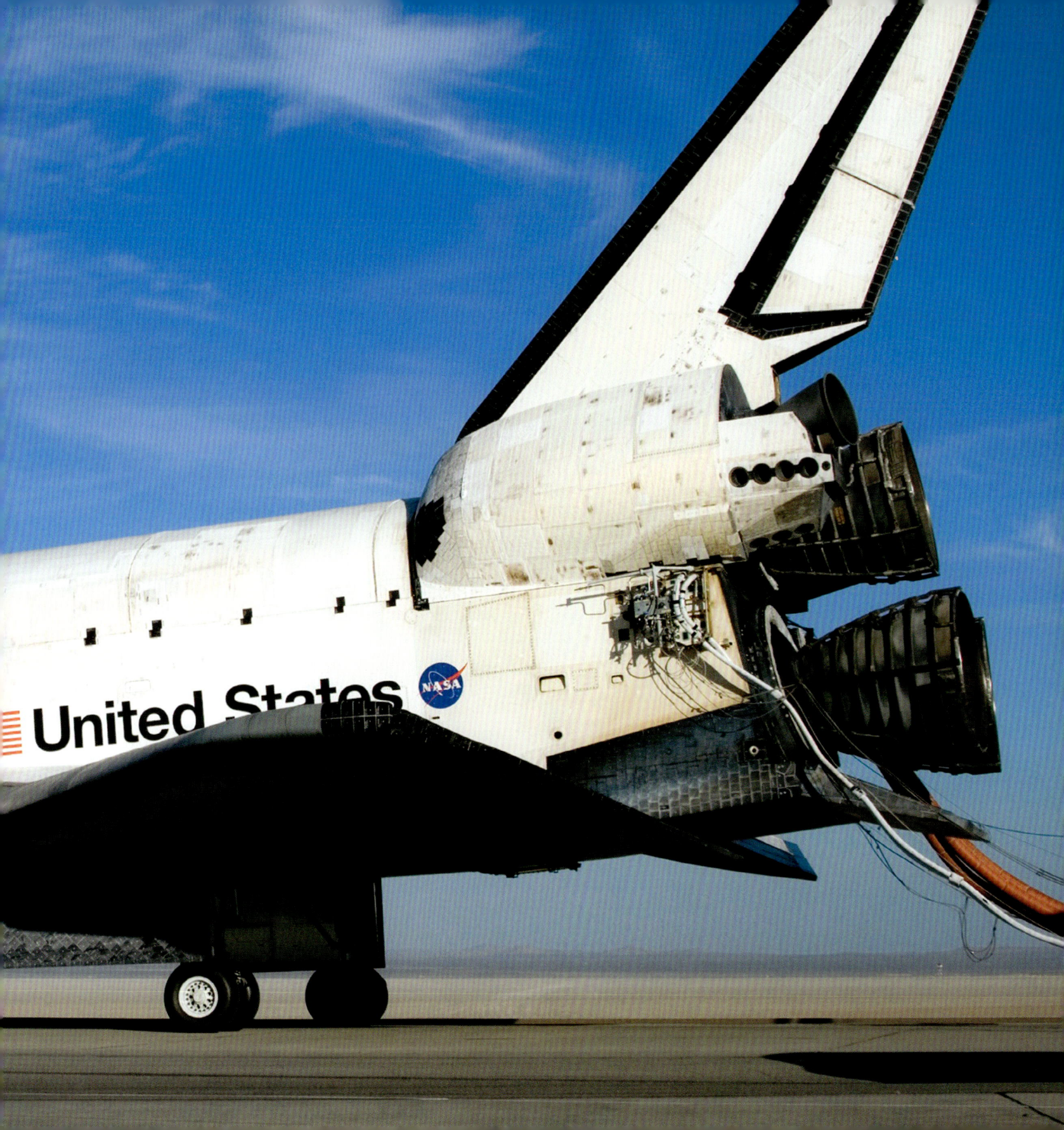

30 September to 11 October 1994

While Endeavour rests on Runway 22 at Edwards Air Force Base following an 11-day mission, Columbia makes a glorious fly-by atop the shuttle carrier aircraft. Columbia is being transferred from Kennedy Space Center to Air Force Plant 42 in Palmdale, California, where it will undergo a refit.

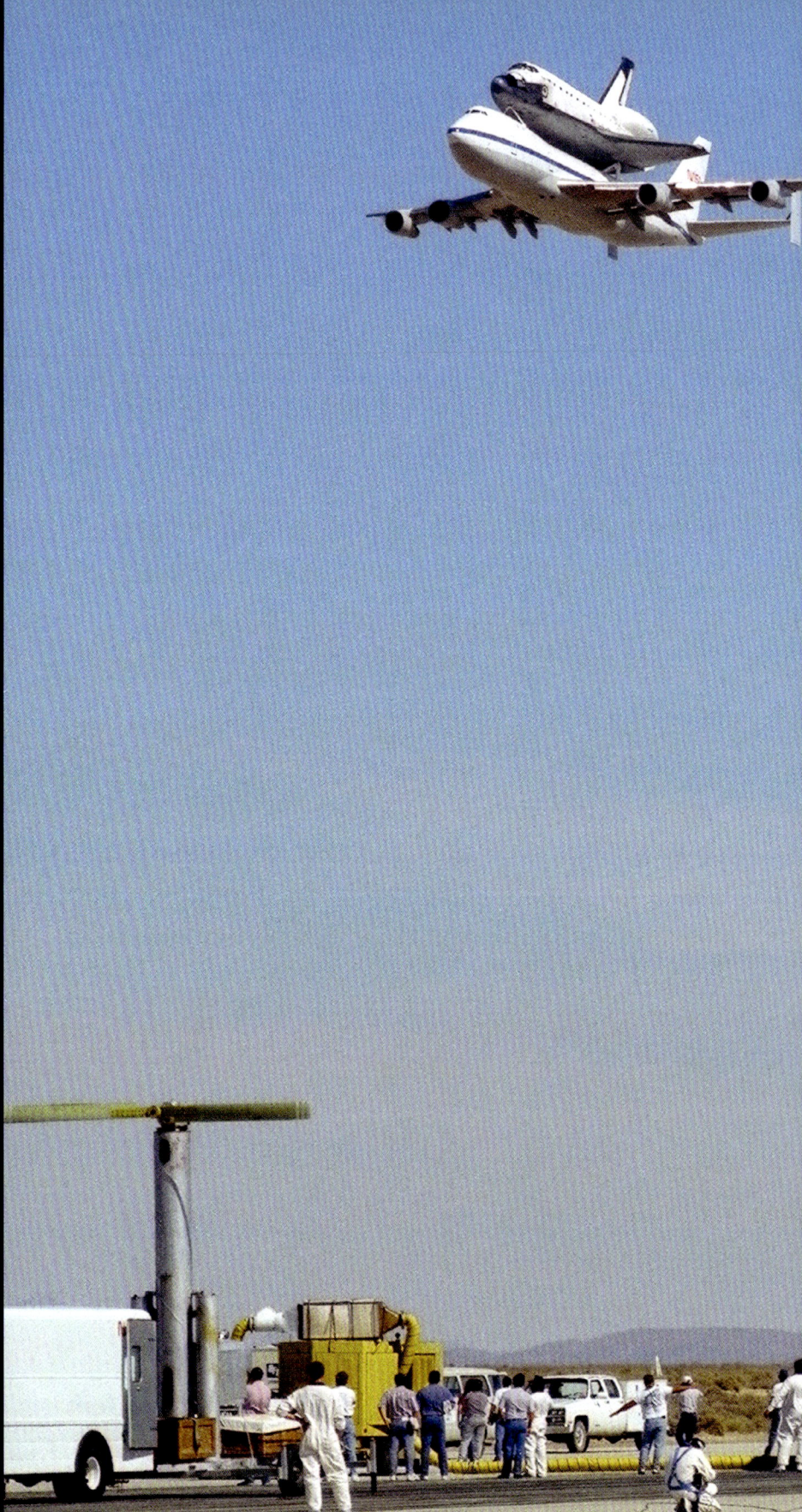

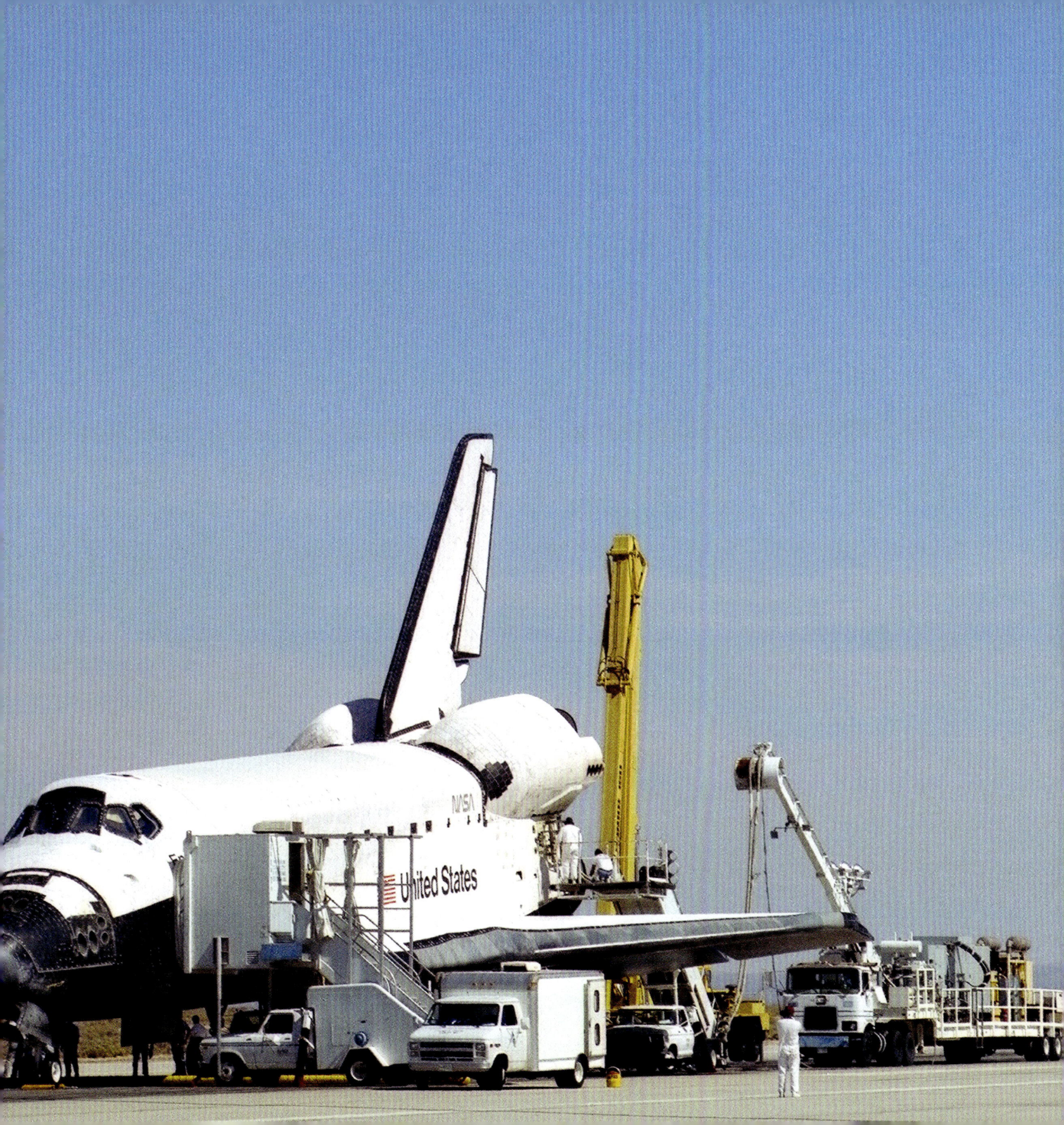

24 November to 1 December 1991

(Right and Following spread) Evening twilight at the Dryden Flight Research Center, Edwards Air Force Base. After a seven-day US Department of Defense (DoD) mission, Atlantis is mounted atop the NASA Boeing 747 shuttle carrier aircraft following post-flight servicing. The shuttle will be transported back to Kennedy Space Center in preparation for a brave new crew and another bold adventure into the hostile void of space.

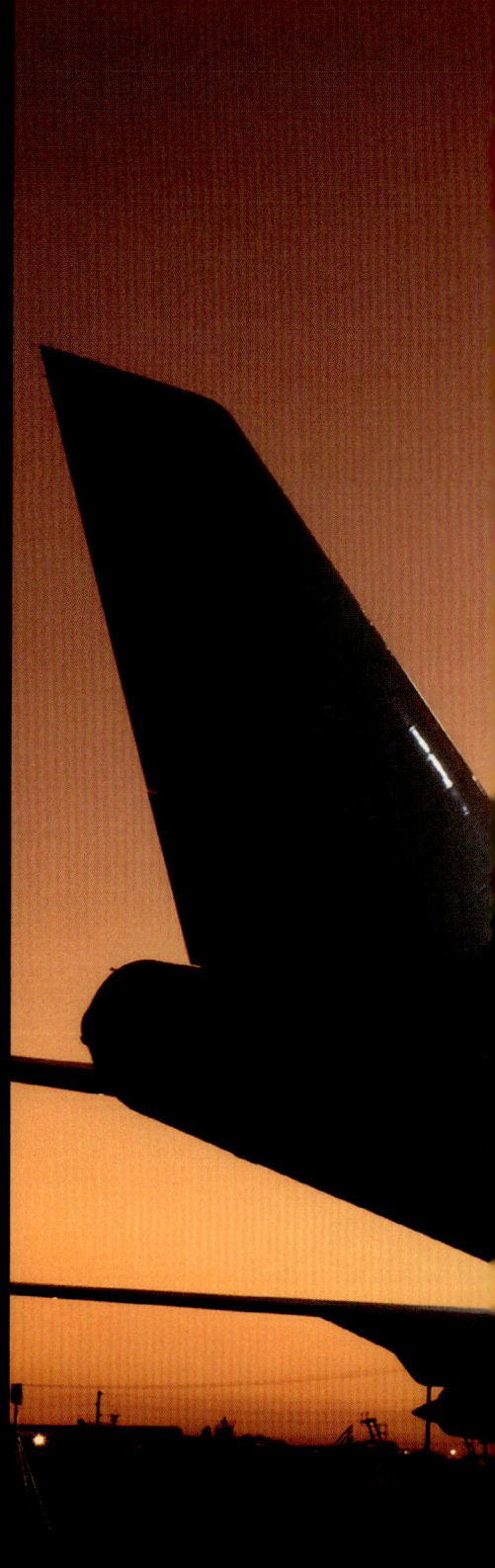

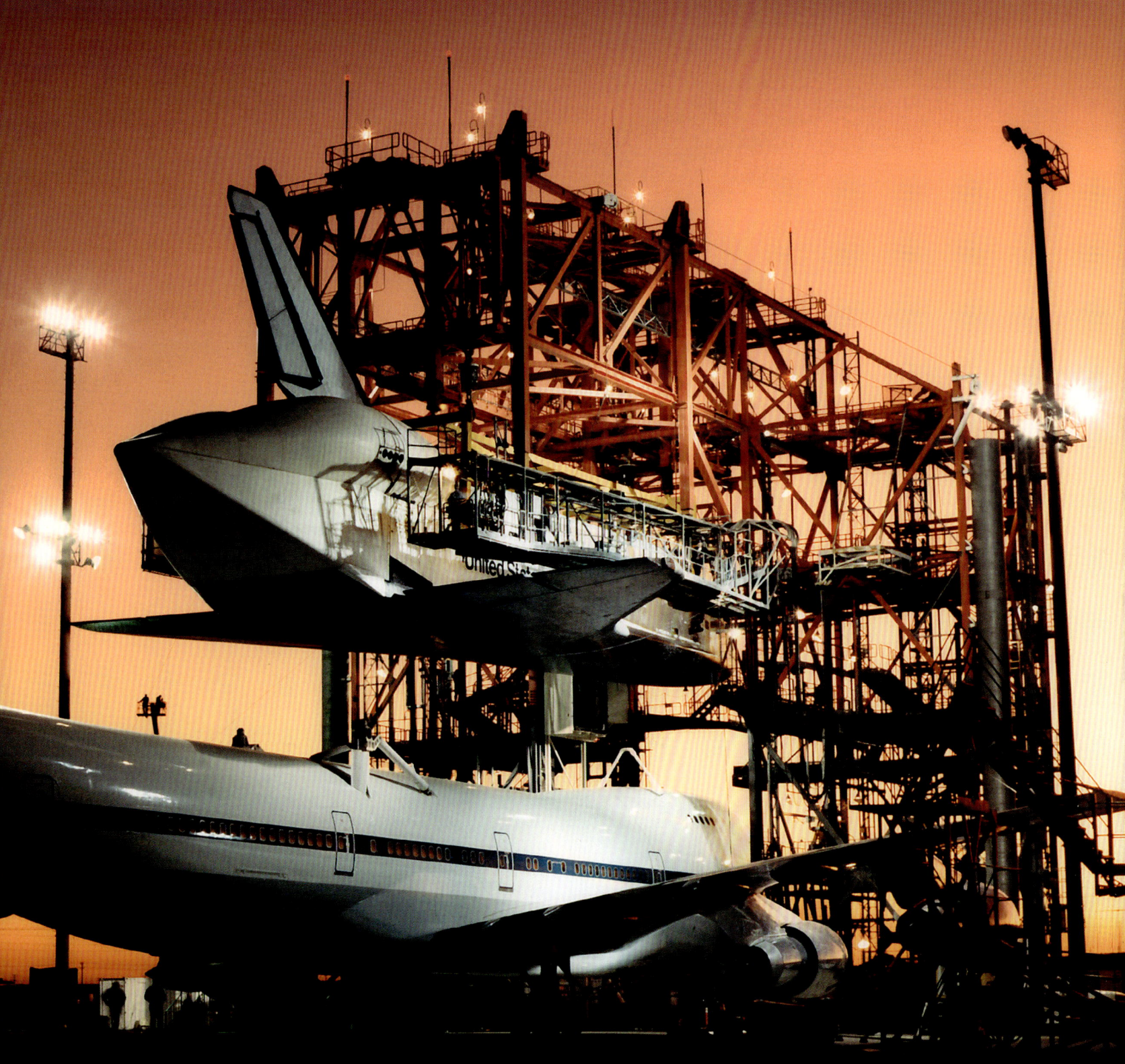

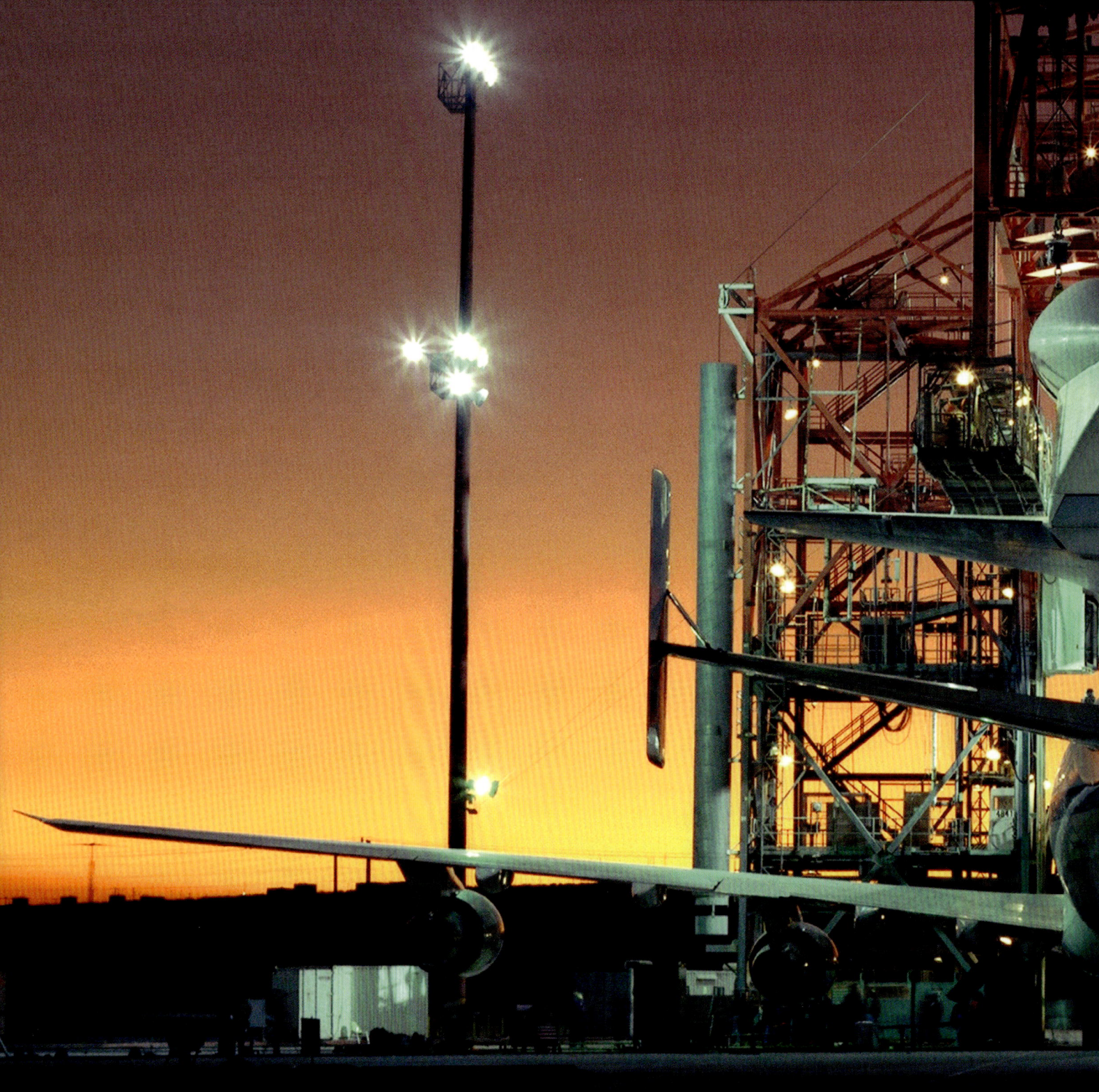

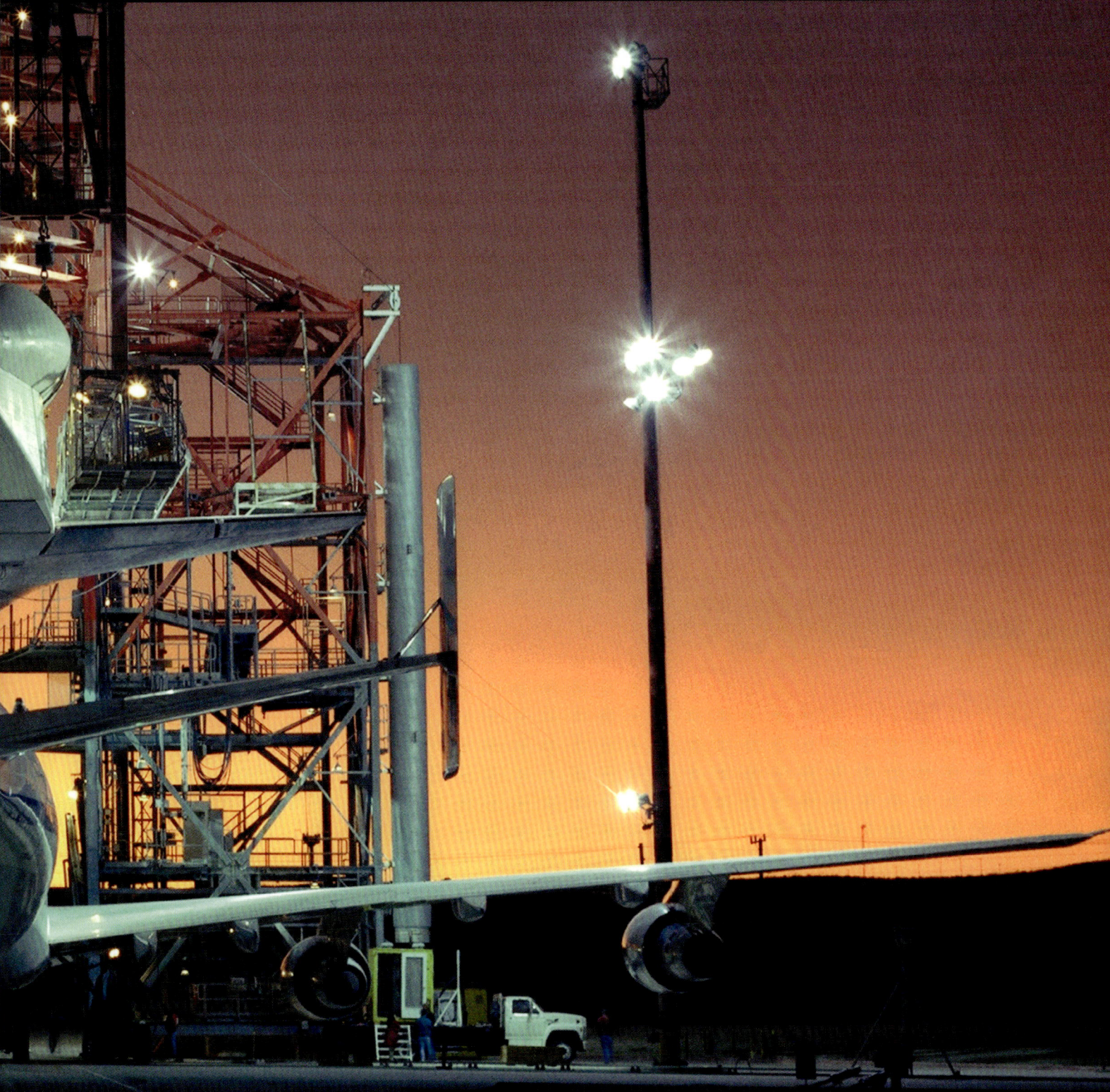

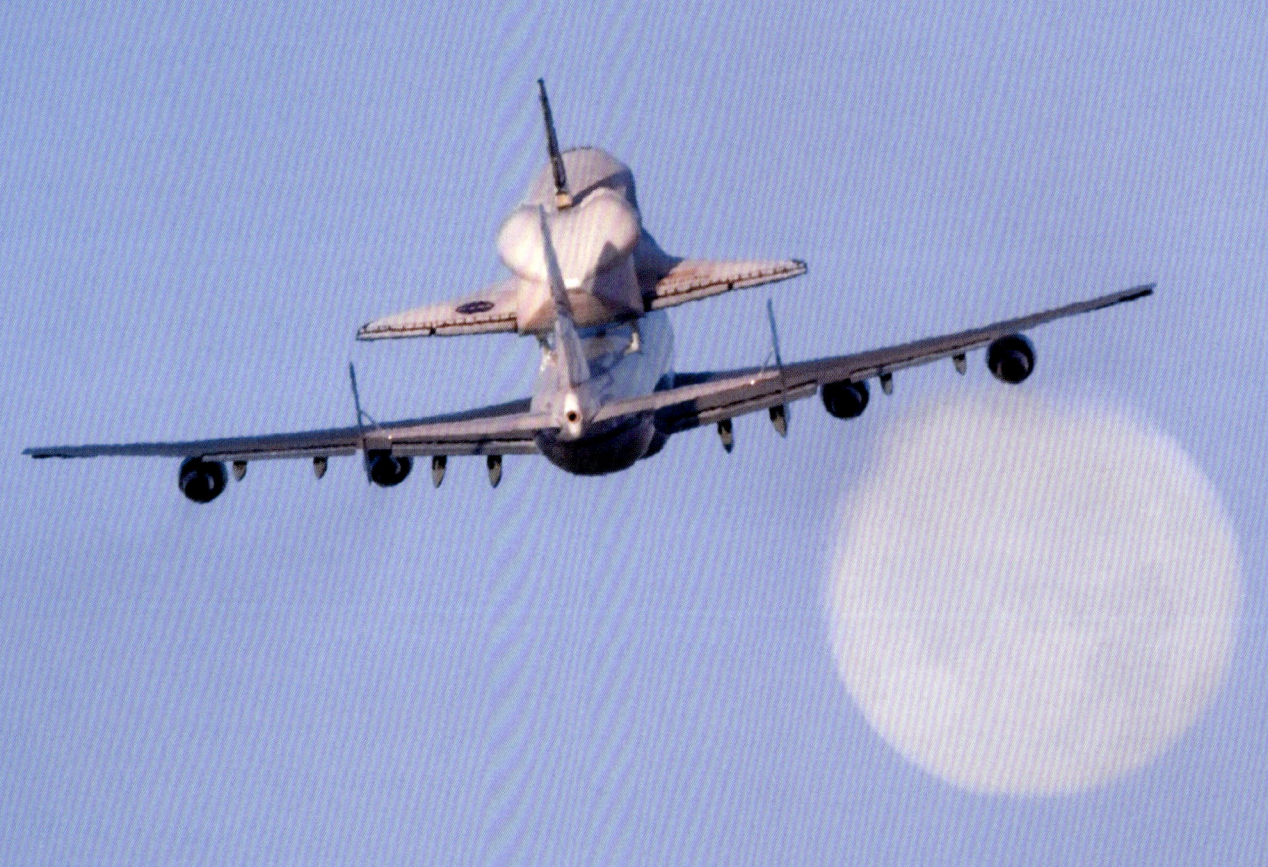

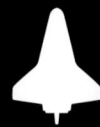

NASA SPACE TRANSPORTATION SYSTEM

1981–2011

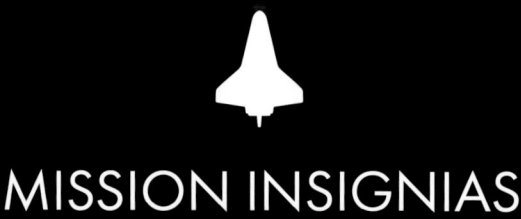

MISSION INSIGNIAS

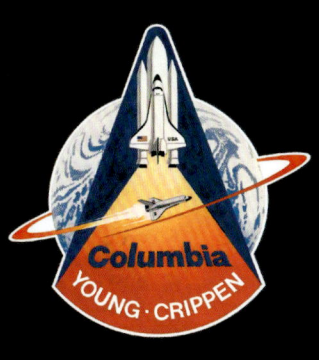

STS-1

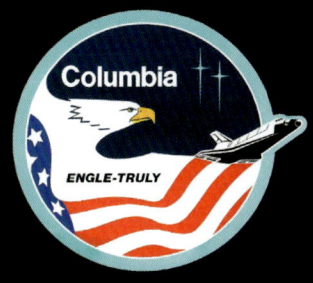

STS-2

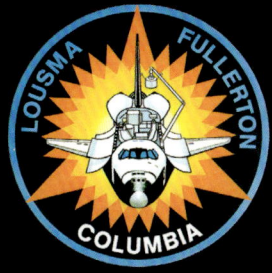

STS-3

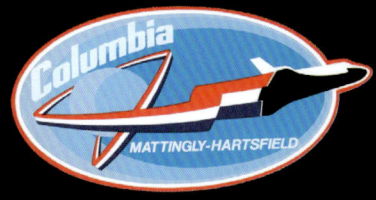

STS-4

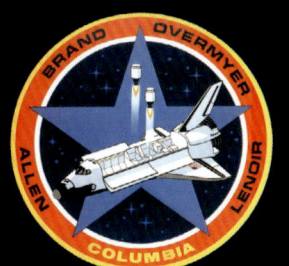

STS-5

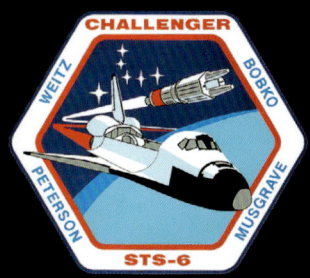

STS-6

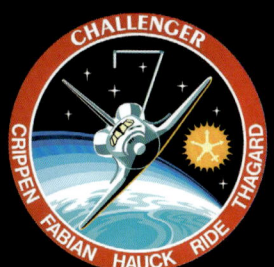

STS-7

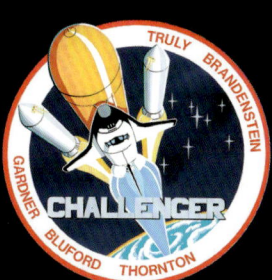

STS-8

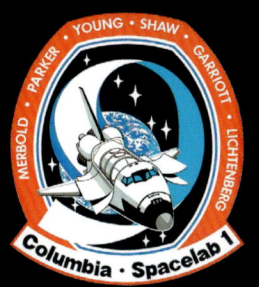

STS-9

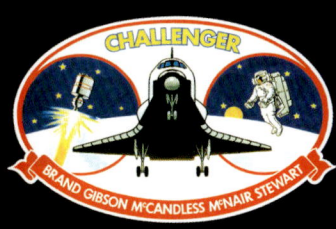

STS-41-B

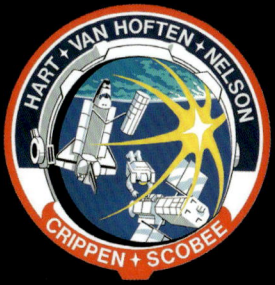

STS-41-C

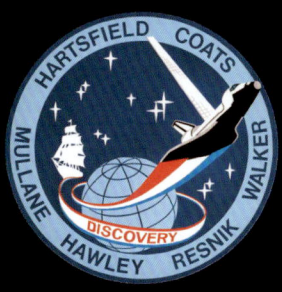

STS-41-D

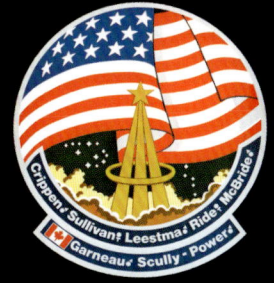

STS-41-G

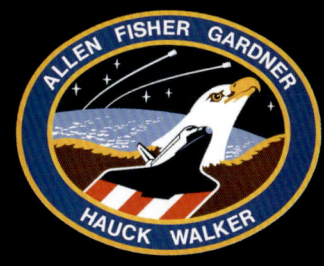

STS-51-A

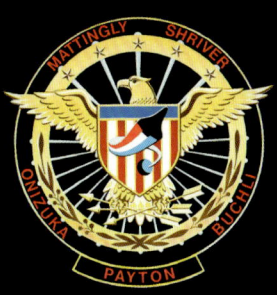

STS-51-C

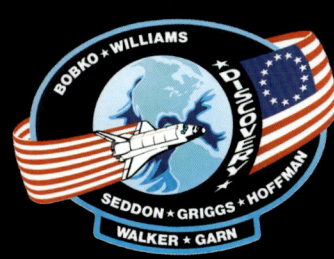

STS-51-D

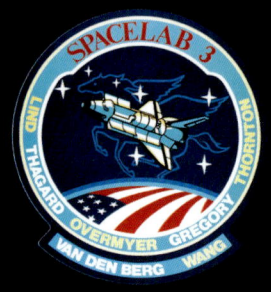

STS-51-B

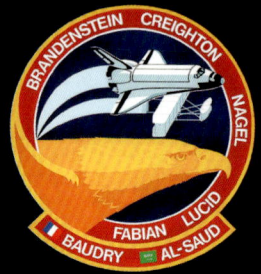

STS-51-G

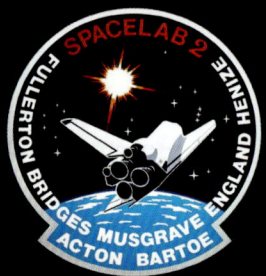

STS-51-F

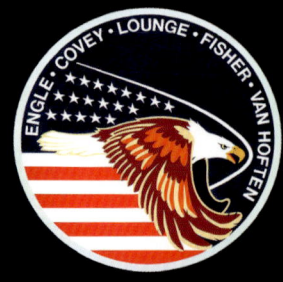

STS-51-I

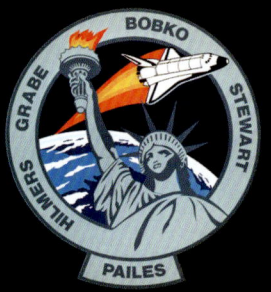

STS-51-J

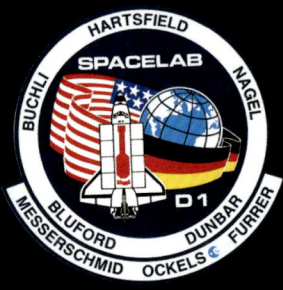

STS-61-A

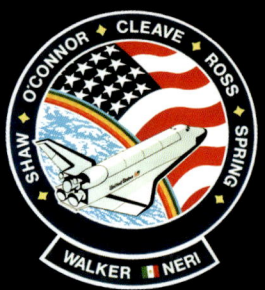

STS-61-B

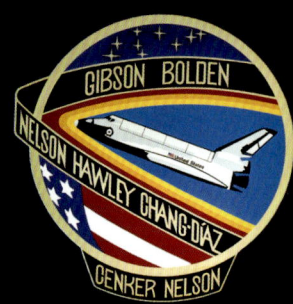

STS-61-C

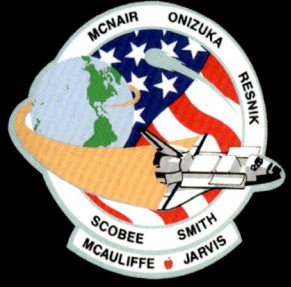

STS-51-L

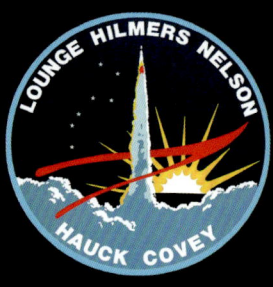

STS-26

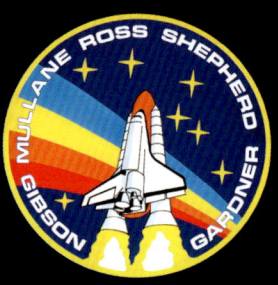

STS-27

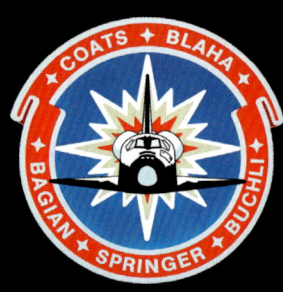

STS-29

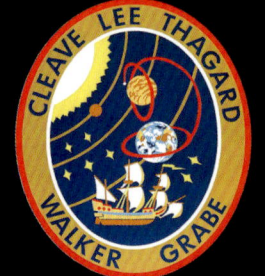

STS-30

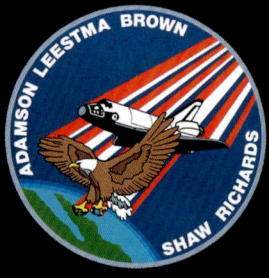

STS-28

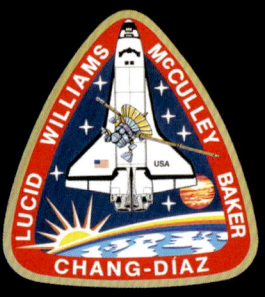

STS-34

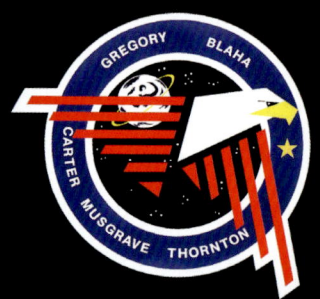

STS-33

214

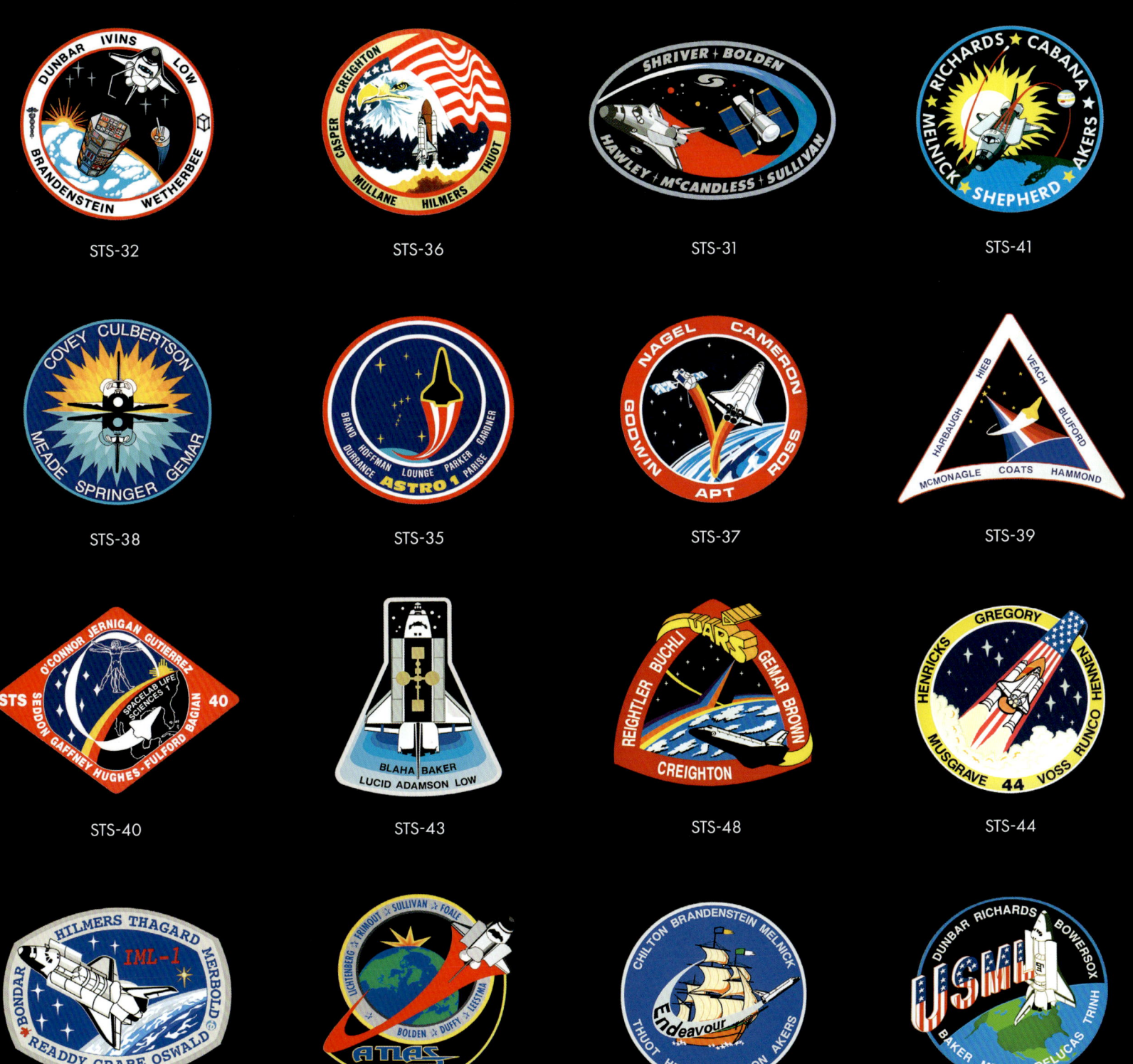

STS-32

STS-36

STS-31

STS-41

STS-38

STS-35

STS-37

STS-39

STS-40

STS-43

STS-48

STS-44

STS-42

STS-45

STS-49

STS-50

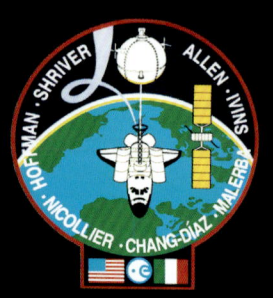

STS-46

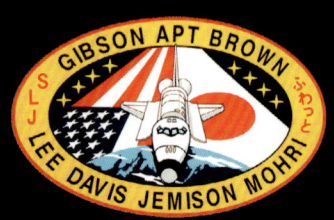

STS-47

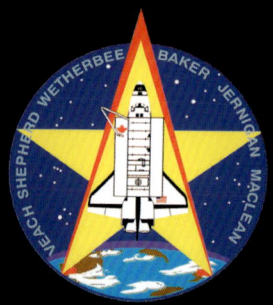

STS-52

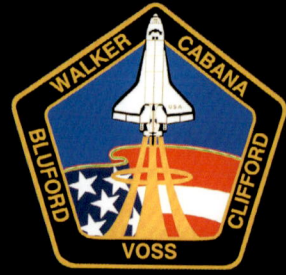

STS-53

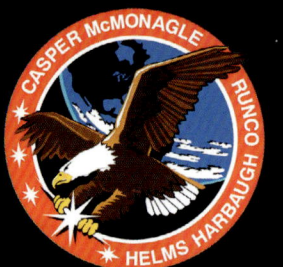

STS-54

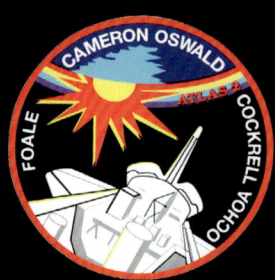

STS-56

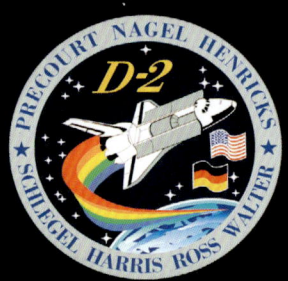

STS-55

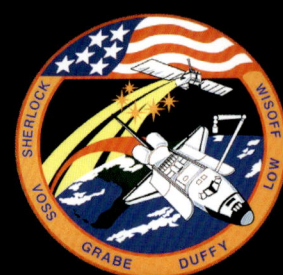

STS-57

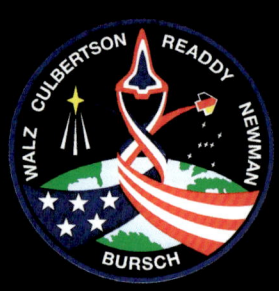

STS-51

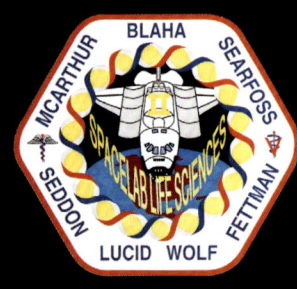

STS-58

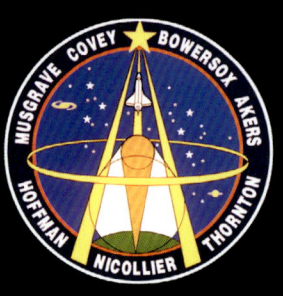

STS-61

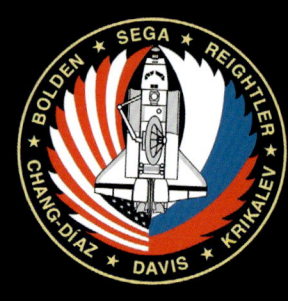

STS-60

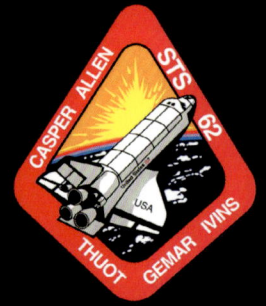

STS-62

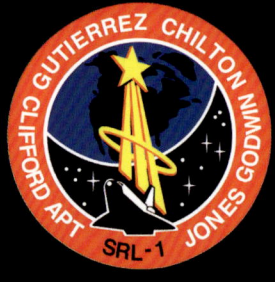

STS-59

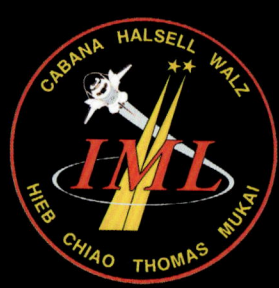

STS-65

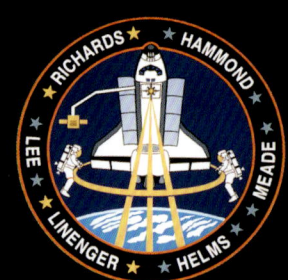

STS-64

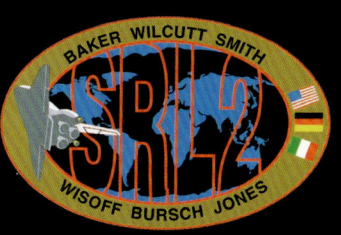

STS-68

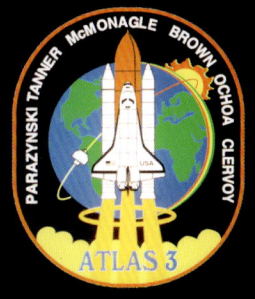

STS-66

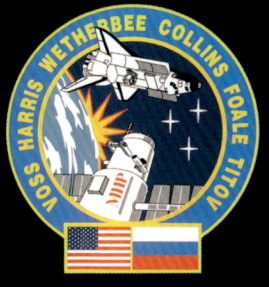

STS-63

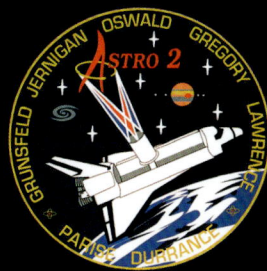

STS-67

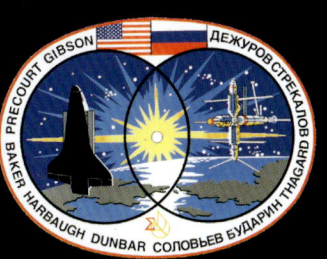

STS-71

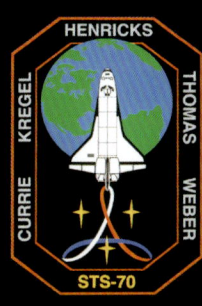

STS-70

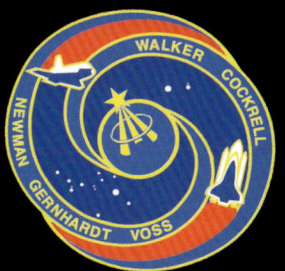

STS-69

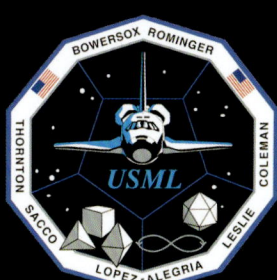

STS-73

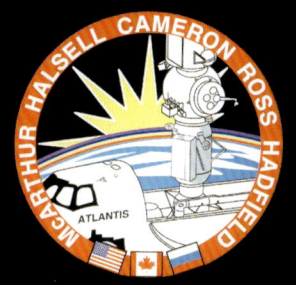

STS-74

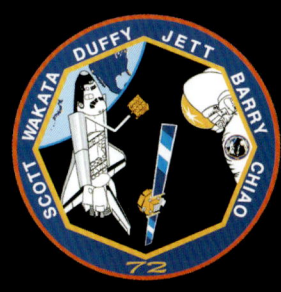

STS-72

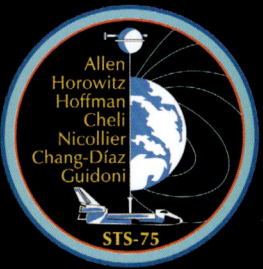

STS-75

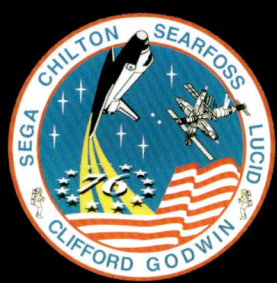

STS-76

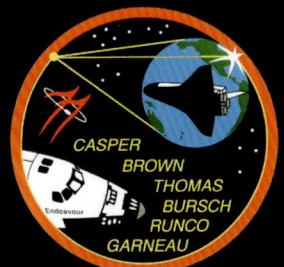

STS-77

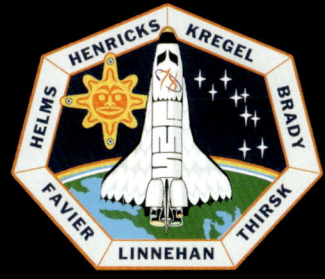

STS-78

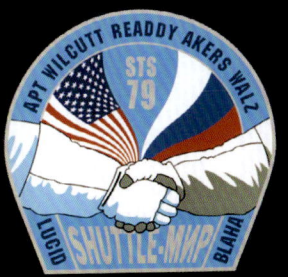

STS-79

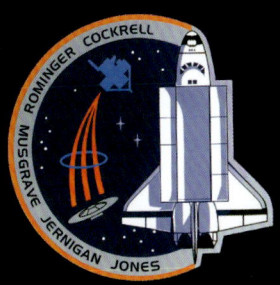

STS-80

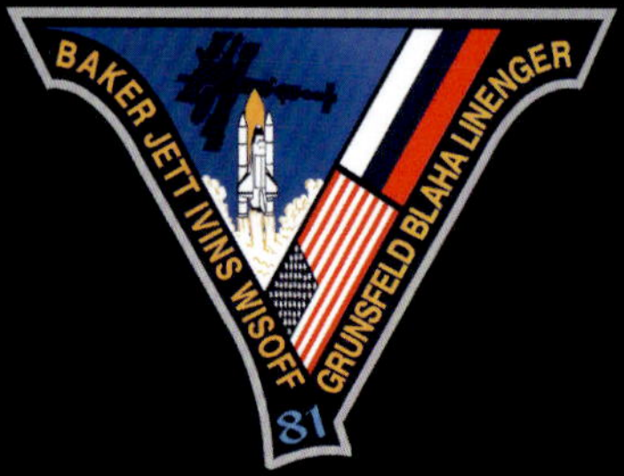

STS-81

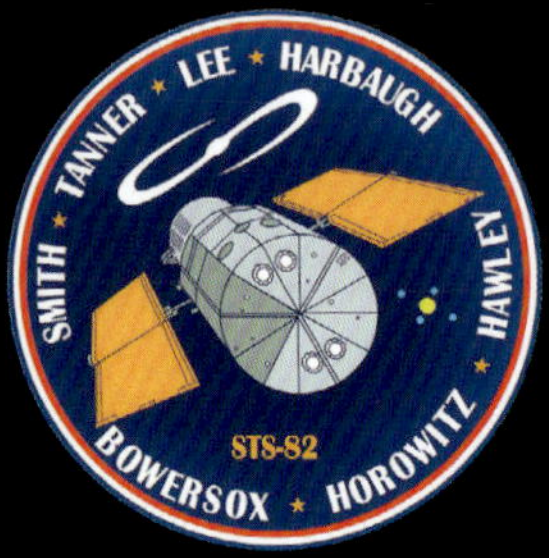

STS-82

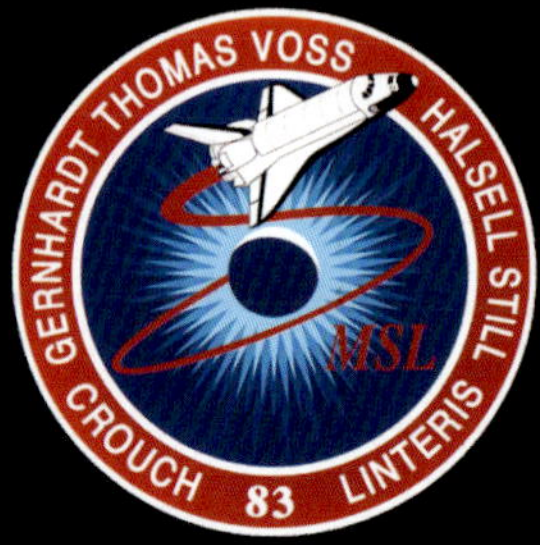

STS-83

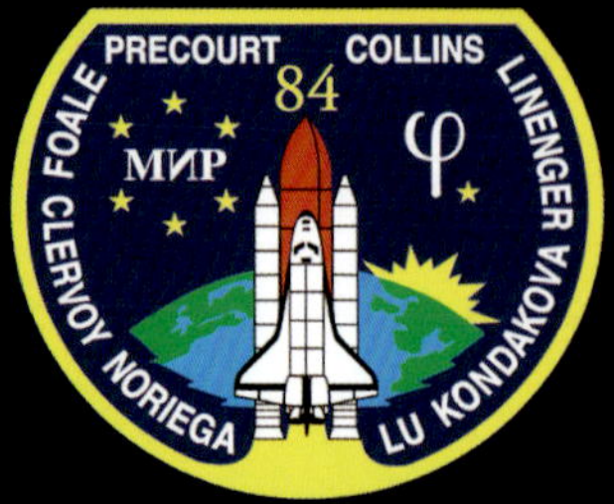

STS-84

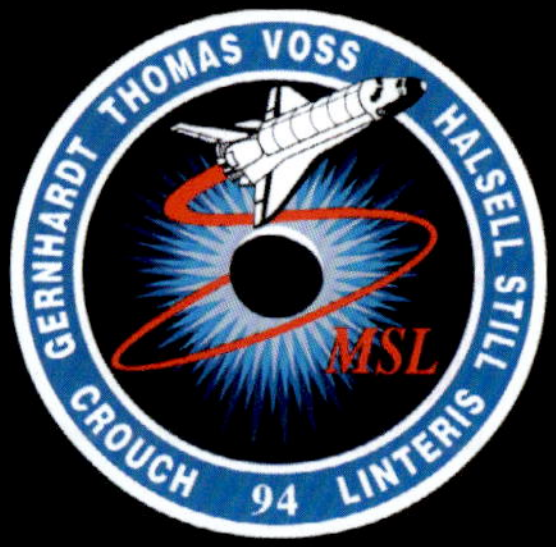

STS-94

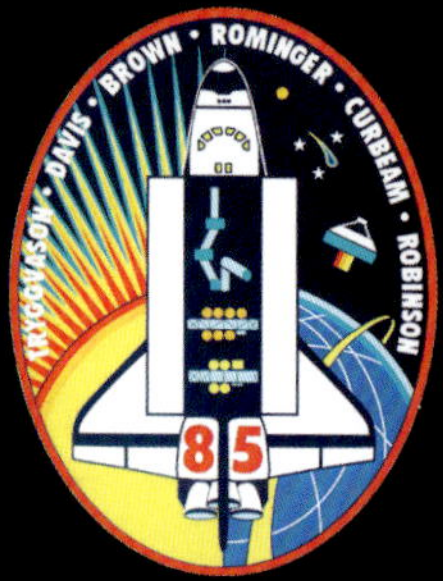

STS-85

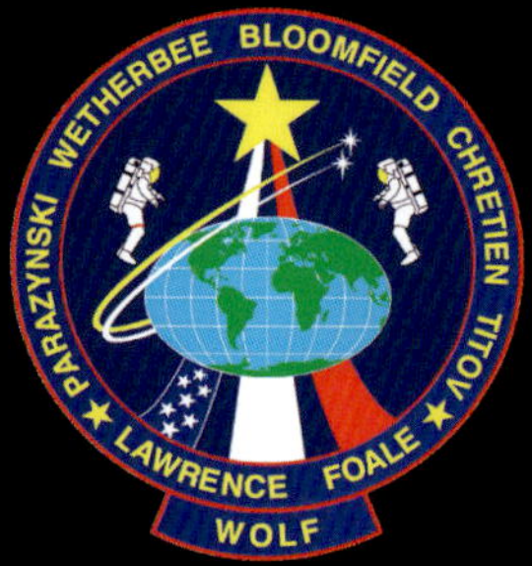

STS-86

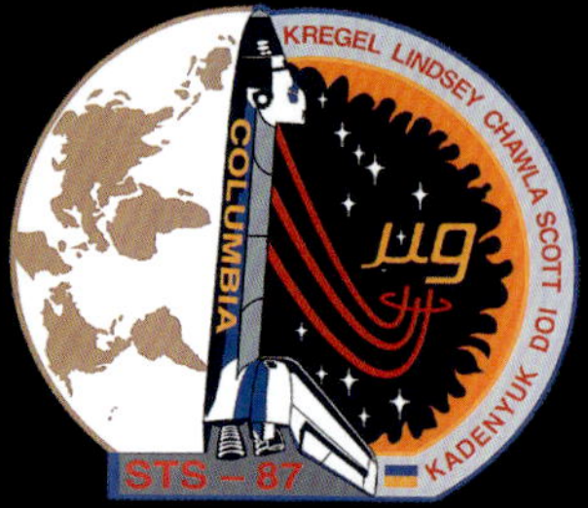

STS-87

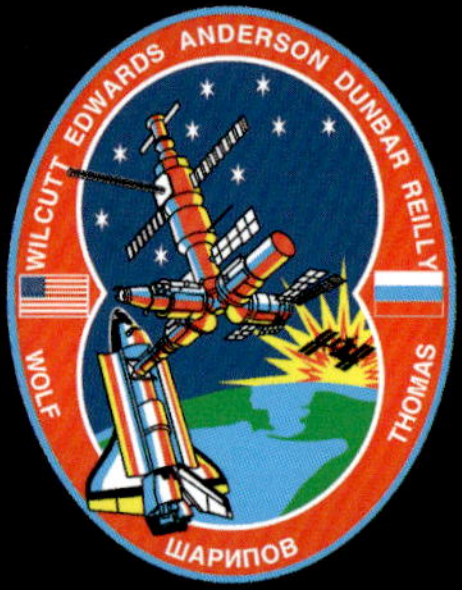

STS-89

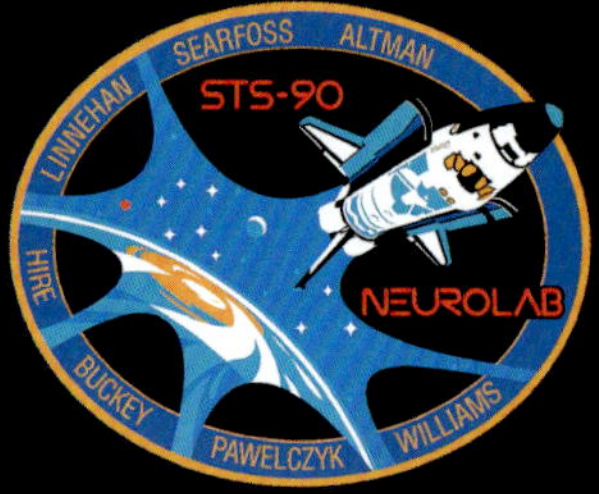

STS-90

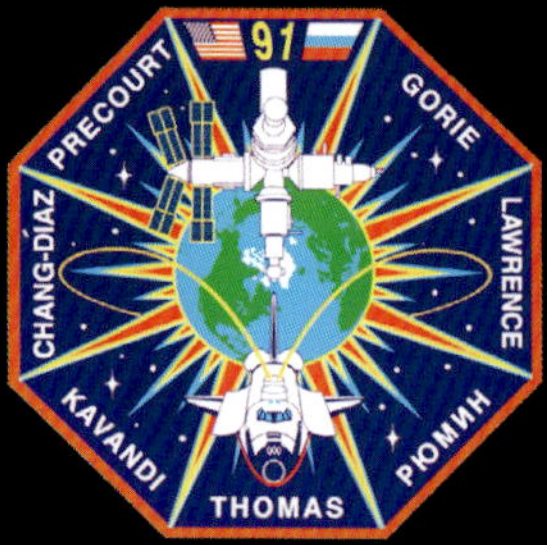

STS-91

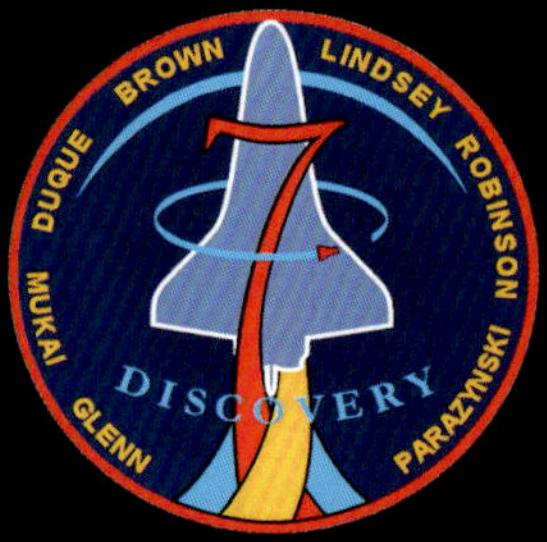

STS-95

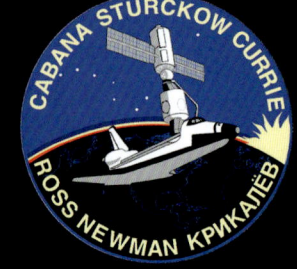

STS-88

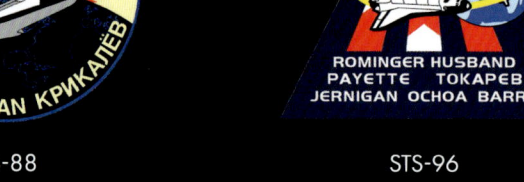

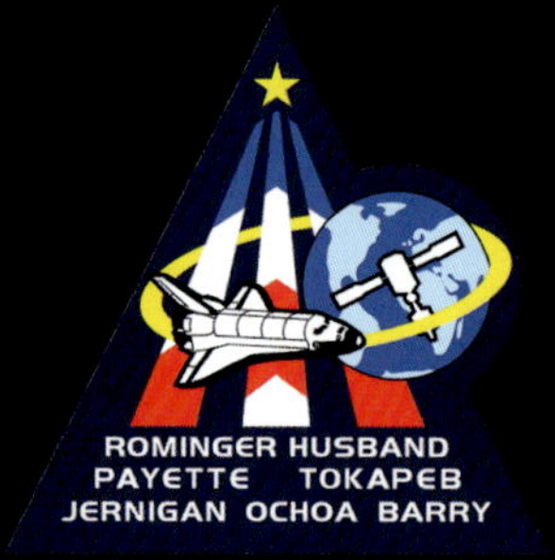

STS-96

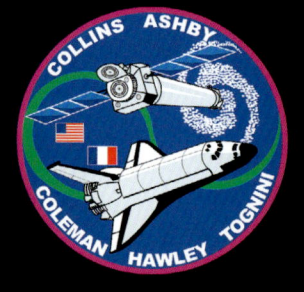

STS-93

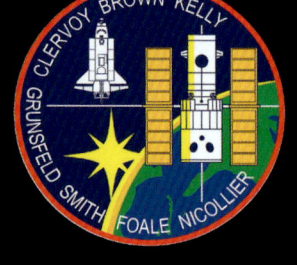

STS-103

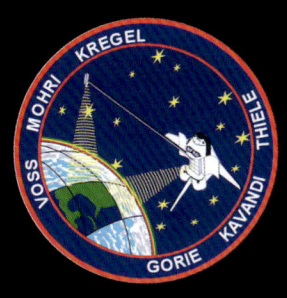

STS-99

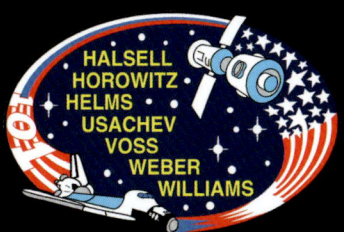

STS-101

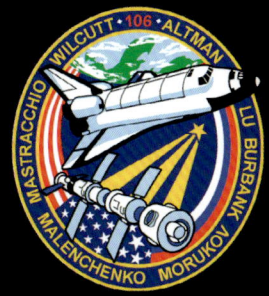

STS-106

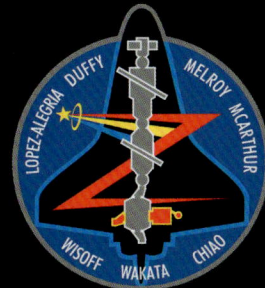

STS-92

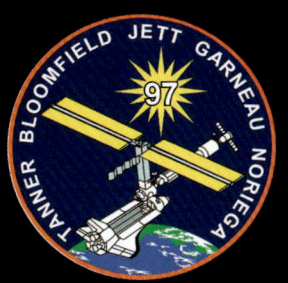

STS-97

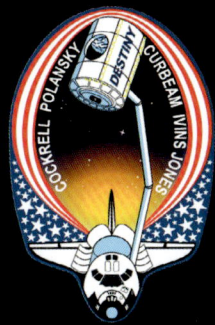

STS-98

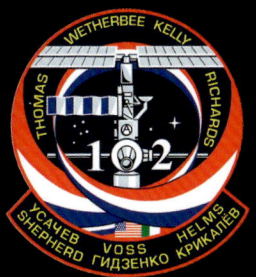

STS-102

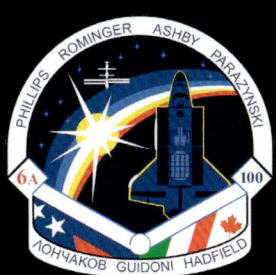

STS-100

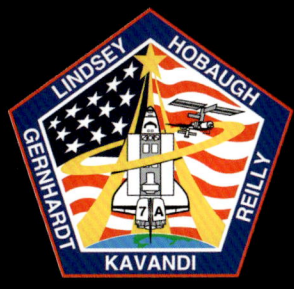

STS-104

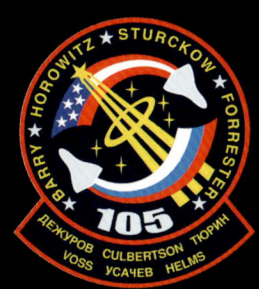

STS-105

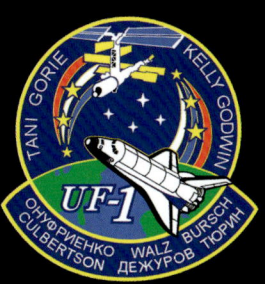

STS-108

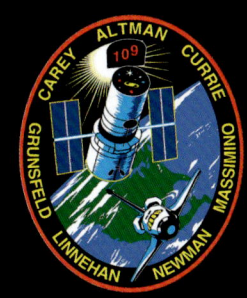

STS-109

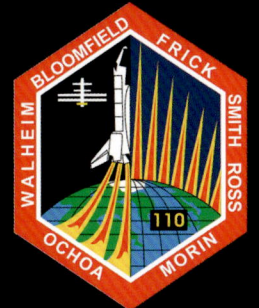

STS-110

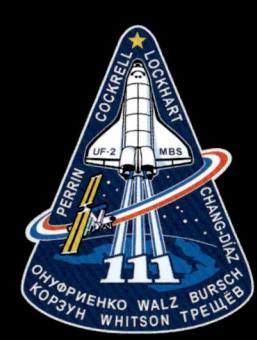

STS-111

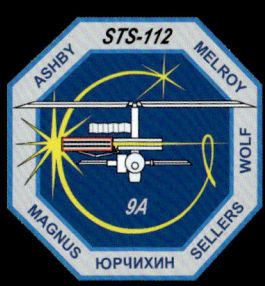

STS-112

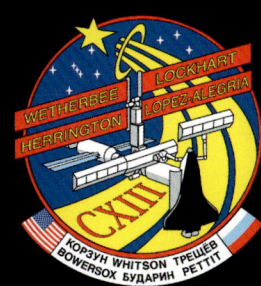

STS-113

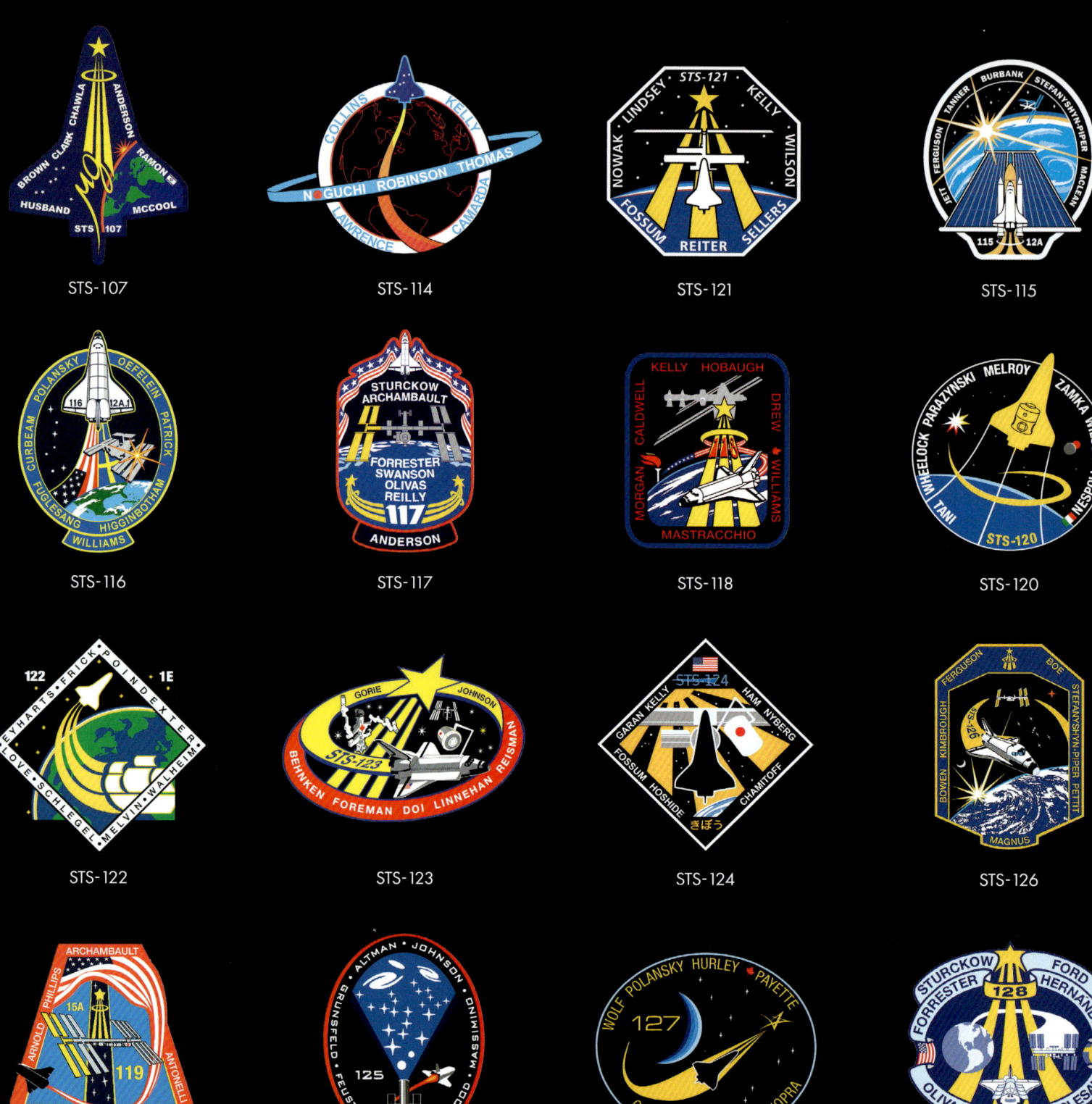

STS-107

STS-114

STS-121

STS-115

STS-116

STS-117

STS-118

STS-120

STS-122

STS-123

STS-124

STS-126

STS-119

STS-125

STS-127

STS-128

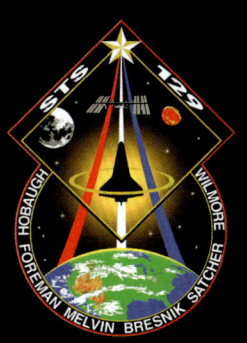

STS-129

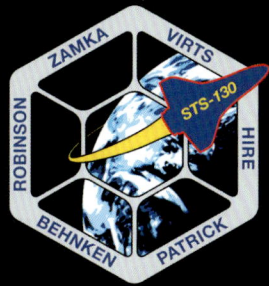

STS-130

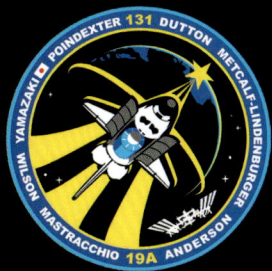

STS-131

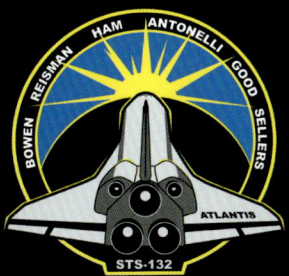

STS-132

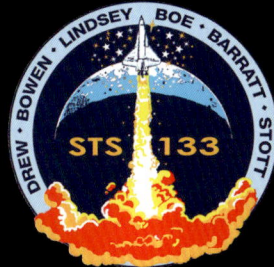

STS-133

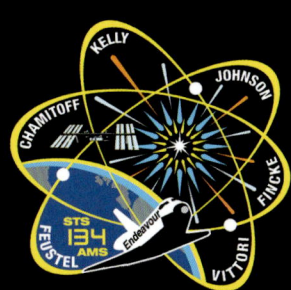

STS-134

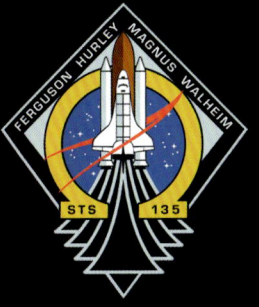

STS-135

These pages are dedicated to all the people who contributed
to designing the Mission Insignias.

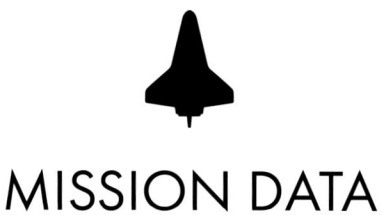

MISSION DATA

STS-1

Orbiter:	Columbia
Launch Date:	12 April 1981
Launch Time:	7:00:03 am EST
Launch Pad:	39A KSC Florida
Return Date:	14 April 1981
Mission Duration:	2 days, 6hr, 20min, 53sec
Orbits:	37
Miles Travelled:	1.074 million
Commander:	John W. Young
Pilot:	Robert L. Crippen

STS-2

Orbiter:	Columbia
Launch Date:	12 November 1981
Launch Time:	10:09:59 am EST
Launch Pad:	39A KSC Florida
Return Date:	14 November 1981
Mission Duration:	2 days, 6hr, 13min, 12sec
Orbits:	37
Miles Travelled:	1.075 million
Commander:	Joe H. Engle
Pilot:	Richard H. Truly

STS-3

Orbiter:	Columbia
Launch Date:	22 March 1982
Launch Time:	11:00:00 am EST
Launch Pad:	39A KSC Florida
Return Date:	30 March 1982
Mission Duration:	8 days, 4min, 46sec
Orbits:	130
Miles Travelled:	3.335 million
Commander:	Jack R. Lousma
Pilot:	C. Gordon Fullerton

STS-4

Orbiter:	Columbia
Launch Date:	27 June 1982
Launch Time:	11:00:00 am EDT
Launch Pad:	39A KSC Florida
Return Date:	4 July 1982
Mission Duration:	7 days, 1hr, 9min, 31sec
Orbits:	113
Miles Travelled:	2.9 million
Commander:	Thomas K. Mattingly II
Pilot:	Henry W. Hartsfield Jr

Partial Department of Defense Payload

STS-5

Orbiter:	Columbia
Launch Date:	11 November 1982
Launch Time:	7:19:00 am EST
Launch Pad:	39A KSC Florida
Return Date:	16 November 1982
Mission Duration:	5 days, 2hr, 14min, 26sec
Orbits:	82
Miles Travelled:	2.1 million
Commander:	Vance D. Brand
Pilot:	Robert F. Overmyer
Mission Specialists:	Joseph P. Allen
	William B. Lenoir

STS-6

Orbiter:	Challenger
Launch Date:	4 April 1983
Launch Time:	1:30:00 pm EST
Launch Pad:	39A KSC Florida
Return Date:	9 April 1983
Mission Duration:	5 days, 23min, 42sec
Orbits:	81
Miles Travelled:	2.1 million
Commander:	Paul J. Weitz
Pilot:	Karol J. Bobko
Mission Specialists:	Donald H. Peterson
	F. Story Musgrave

STS-7

Orbiter:	Challenger
Launch Date:	18 June 1983
Launch Time:	7:33:00 am EDT
Launch Pad:	39A KSC Florida
Return Date:	24 June 1983
Mission Duration:	6 days, 2hr, 23min, 59sec
Orbits:	98
Miles Travelled:	2.5 million
Commander:	Robert L. Crippen
Pilot:	Frederick H. Hauck
Mission Specialists:	John M. Fabian
	Sally K. Ride
	Norman E. Thagard

STS-8

Orbiter:	Challenger
Launch Date:	30 August 1983
Launch Time:	2:32:00 am EDT
Launch Pad:	39A KSC Florida
Return Date:	5 September 1983
Mission Duration:	6 days, 1hr, 8min, 43sec
Orbits:	98
Miles Travelled:	2.5 million
Commander:	Richard H. Truly
Pilot:	Daniel C. Brandenstein
Mission Specialists:	Dale A. Gardner
	Guion S. Bluford Jr
	William E. Thornton

STS-9

Orbiter:	Columbia
Launch Date:	28 November 1983
Launch Time:	11:00:00 am EST
Launch Pad:	39A KSC Florida
Return Date:	8 December 1983
Mission Duration:	10 days, 7hr, 47min, 24sec
Orbits:	167
Miles Travelled:	4.3 million
Commander:	John W. Young
Pilot:	Brewster H. Shaw
Mission Specialists:	Owen K. Garriott
	Robert A. Parker
Payload Specialists:	Ulf D. Merbold, ESA
	Byron K. Lichtenberg

STS-41-B

Orbiter:	Challenger
Launch Date:	3 February 1984
Launch Time:	8:00:00 am EST
Launch Pad:	39A KSC Florida
Return Date:	11 February 1984
Mission Duration:	7 days, 23hr, 15min, 55sec
Orbits:	128
Miles Travelled:	3.3 million
Commander:	Vance D. Brand
Pilot:	Robert L. Gibson
Mission Specialists:	Bruce McCandless II
	Ronald E. McNair
	Robert L. Stewart

STS-41-C

Orbiter:	Challenger
Launch Date:	6 April 1984
Launch Time:	8:58:00 am EST
Launch Pad:	39A KSC Florida
Return Date:	13 April 1984
Mission Duration:	6 days, 23hr, 40min, 7sec
Orbits:	108
Miles Travelled:	2.9 million
Commander:	Robert L. Crippen
Pilot:	Francis R. Scobee
Mission Specialists:	George D. Nelson
	James D. A. van Hoften
	Terry J. Hart

STS-41-D

Orbiter:	Discovery
Launch Date:	30 August 1984
Launch Time:	8:41:50 am EDT
Launch Pad:	39A KSC Florida
Return Date:	5 September 1984
Mission Duration:	6 days, 56min, 4sec
Orbits:	97
Miles Travelled:	2.5 million
Commander:	Henry W. Hartsfield Jr
Pilot:	Michael L. Coats
Mission Specialists:	Richard M. Mullane
	Steven A. Hawley
	Judith A. Resnick
Payload Specialist:	Charles D. Walker

STS-41-G

Orbiter:	Challenger
Launch Date:	5 October 1984
Launch Time:	7:03:00 am EDT
Launch Pad:	39A KSC Florida
Return Date:	13 October 1984
Mission Duration:	8 days, 5hr, 23min, 33sec
Orbits:	133
Miles Travelled:	3.3 million
Commander:	Robert L. Crippen
Pilot:	Jon A. McBride
Mission Specialists:	Kathryn D. Sullivan
	Sally K. Ride
	David C. Leestma
Payload Specialists:	Marc Garneau, CSA
	Paul D. Scully-Power

STS-51-A

Orbiter:	Discovery
Launch Date:	8 November 1984
Launch Time:	7:15:00 am EST
Launch Pad:	39A KSC Florida
Return Date:	16 November 1984
Mission Duration:	7 days, 23hr, 44min, 56sec
Orbits:	127
Miles Travelled:	3.3 million
Commander:	Frederick H. Hauck
Pilot:	David M. Walker
Mission Specialists:	Anna L. Fisher
	Dale A. Gardner
	Joseph P. Allen

STS-51-C

Orbiter:	Discovery
Launch Date:	24 January 1985
Launch Time:	2:50:00 pm EST
Launch Pad:	39A KSC Florida
Return Date:	27 January 1985
Mission Duration:	3 days, 1hr, 33min, 23sec
Orbits:	49
Miles Travelled:	1.3 million
Commander:	Thomas K. Mattingly II
Pilot:	Loren J. Shriver
Mission Specialists:	Ellison S. Onizuka
	James F. Buchli
Payload Specialist:	Gary E. Payton, MSE

Classified Department of Defense Mission

STS-51-D

Orbiter:	Discovery
Launch Date:	12 April 1985
Launch Time:	8:59:05 am EST
Launch Pad:	39A KSC Florida
Return Date:	19 April 1985
Mission Duration:	6 days, 23hr, 55min, 23sec
Orbits:	110
Miles Travelled:	2.9 million
Commander:	Karol J. Bobko
Pilot:	Donald E. Williams
Mission Specialists:	M. Rhea Seddon
	Jeffrey A. Hoffman
	S. David Griggs
Payload Specialists:	Charles D. Walker
	E. Jake Garn

STS-51-B

Orbiter:	Challenger
Launch Date:	29 April 1985
Launch Time:	12:02:18 pm EDT
Launch Pad:	39A KSC Florida
Return Date:	6 May 1985
Mission Duration:	7 days, 8min, 46sec
Orbits:	111
Miles Travelled:	2.9 million
Commander:	Robert F. Overmyer
Pilot:	Frederick D. Gregory
Mission Specialists:	Don L. Lind
	Norman E. Thagard
	William E. Thornton
Payload Specialists:	Lodewijk van den Berg
	Taylor G. Wang

STS-51-G

Orbiter:	Discovery
Launch Date:	17 June 1985
Launch Time:	7:33:00 am EDT
Launch Pad:	39A KSC Florida
Return Date:	24 June 1985
Mission Duration:	7 days, 1hr, 38min, 52sec
Orbits:	112
Miles Travelled:	2.9 million
Commander:	Daniel C. Brandenstein
Pilot:	John O. Creighton
Mission Specialists:	Shannon W. Lucid
	John M. Fabian
	Steven R. Nagel
Payload Specialists:	Patrick Baudry, CNES
	Sultan Salman Al-Saud (Royal Saudi Air Force)

STS-51-F

Orbiter:	Challenger
Launch Date:	29 July 1985
Launch Time:	5:00:00 pm EDT
Launch Pad:	39A KSC Florida
Return Date:	6 August 1985
Mission Duration:	7 days, 22hr, 45min, 26sec
Orbits:	127
Miles Travelled:	3.3 million
Commander:	C. Gordon Fullerton
Pilot:	Roy D. Bridges Jr
Mission Specialists:	F. Story Musgrave
	Anthony W. England
	Karl G. Henize
Payload Specialists:	Loren W. Acton
	John-David F. Bartoe

STS-51-I

Orbiter:	Discovery
Launch Date:	27 August 1985
Launch Time:	6:58:01 am EDT
Launch Pad:	39A KSC Florida
Return Date:	3 September 1985
Mission Duration:	7 days, 2hr, 17min, 42sec
Orbits:	112
Miles Travelled:	2.9 million
Commander:	Joe H. Engle
Pilot:	Richard O. Covey
Mission Specialists:	James D. A. van Hoften
	John M. Lounge
	William F. Fisher

STS-51-J

Orbiter:	Atlantis
Launch Date:	3 October 1985
Launch Time:	11:15:30 am EDT
Launch Pad:	39A KSC Florida
Return Date:	7 October 1985
Mission Duration:	4 days, 1hr, 44min, 38sec
Orbits:	64
Miles Travelled:	1.7 million
Commander:	Karol J. Bobko
Pilot:	Ronald J. Grabe
Mission Specialists:	David C. Hilmers
	Robert L. Stewart
	William A. Pailes, MSE

Classified Department of Defense Mission

STS-61-A

Orbiter:	Challenger
Launch Date:	30 October 1985
Launch Time:	12:00:00 pm EST
Launch Pad:	39A KSC Florida
Return Date:	6 November 1985
Mission Duration:	7 days, 44min, 51sec
Orbits:	112
Miles Travelled:	2.9 million
Commander:	Henry W. Hartsfield Jr
Pilot:	Steven R. Nagel
Mission Specialists:	Bonnie J. Dunbar
	James F. Buchli
	Guion S. Bluford Jr
Payload Specialists:	Reinhard Furrer, DFVLR
	Ernst Messerschmid, DFVLR
	Wubbo Ockels, ESA

STS-61-B

Orbiter:	Atlantis
Launch Date:	26 November 1985
Launch Time:	7:29:00 pm EST
Launch Pad:	39A KSC Florida
Return Date:	3 December 1985
Mission Duration:	6 days, 21hr, 4min, 49sec
Orbits:	109
Miles Travelled:	2.8 million
Commander:	Brewster H. Shaw Jr
Pilot:	Bryan D. O'Conner
Mission Specialists:	Mary L. Cleave
	Jerry L. Ross
	Sherwood C. Spring
Payload Specialists:	Rodolfo Neri Vela
	Charles D. Walker

STS-61-C

Orbiter:	Columbia
Launch Date:	12 January 1986
Launch Time:	6:55:00 am EST
Launch Pad:	39A KSC Florida
Return Date:	18 January 1986
Mission Duration:	6 days, 2hr, 3min, 51sec
Orbits:	98
Miles Travelled:	2.5 million
Commander:	Robert L. Gibson
Pilot:	Charles F. Bolden Jr
Mission Specialists:	Franklin R. Chang-Diaz
	Steven A. Hawley
	George D. Nelson
Payload Specialists:	Robert J. Cenker
	Clarence W. Nelson

STS-51-L

Orbiter:	Challenger
Launch Date:	28 January 1986
Launch Time:	11:38:00 am EST
Launch Pad:	39B KSC Florida
Return Date:	Destroyed on lift-off
Mission Duration:	1min, 13sec
Orbits:	0
Miles Travelled:	18
Commander:	Francis R. Scobee
Pilot:	Michael J. Smith
Mission Specialists:	Ellison S. Onizuka
	Judith A. Resnick
	Ronald E. McNair
Payload Specialists:	S. Christa McAuliffe
	Gregory B. Jarvis

STS-26

Orbiter:	Discovery
Launch Date:	29 September 1988
Launch Time:	11:37:00 am EDT
Launch Pad:	39B KSC Florida
Return Date:	3 October 1988
Mission Duration:	4 days, 1hr, 11sec
Orbits:	64
Miles Travelled:	1.7 million
Commander:	Frederick H. Hauck
Pilot:	Richard O. Covey
Mission Specialists:	John M. Lounge
	George D. Nelson
	David C. Hilmers

STS-27

Orbiter:	Atlantis
Launch Date:	2 December 1988
Launch Time:	9:30:34 am EST
Launch Pad:	39B KSC Florida
Return Date:	6 December 1988
Mission Duration:	4 days, 9hr, 5min, 37sec
Orbits:	68
Miles Travelled:	1.8 million
Commander:	Robert L. Gibson
Pilot:	Guy S. Gardner
Mission Specialists:	Richard M. Mullane
	Jerry L. Ross
	William M. Shepherd

Classified Department of Defense Mission

STS-29

Orbiter:	Discovery
Launch Date:	13 March 1989
Launch Time:	9:57:00 am EST
Launch Pad:	39B KSC Florida
Return Date:	18 March 1989
Mission Duration:	4 days, 23hr, 38min, 50sec
Orbits:	80
Miles Travelled:	2.1 million
Commander:	Michael L. Coats
Pilot:	John E. Blaha
Mission Specialists:	James P. Bagian
	James F. Buchli
	Robert C. Springer

STS-30

Orbiter:	Atlantis
Launch Date:	4 May 1989
Launch Time:	2:46:59 pm EDT
Launch Pad:	39B KSC Florida
Return Date:	8 May 1989
Mission Duration:	4 days, 56min, 28sec
Orbits:	65
Miles Travelled:	1.7 million
Commander:	David M. Walker
Pilot:	Ronald J. Grabe
Mission Specialists:	Norman E. Thagard
	Mary L. Cleave
	Mark C. Lee

STS-28

Orbiter:	Columbia
Launch Date:	8 August 1989
Launch Time:	8:37:00 am EDT
Launch Pad:	39B KSC Florida
Return Date:	13 August 1989
Mission Duration:	5 days, 1hr, 8sec
Orbits:	81
Miles Travelled:	2.1 million
Commander:	Brewster H. Shaw Jr
Pilot:	Richard N. Richards
Mission Specialists:	James C. Adamson
	David C. Leestma
	Mark N. Brown

Classified Department of Defense Mission

STS-34

Orbiter:	Atlantis
Launch Date:	18 October 1989
Launch Time:	12:53:40 pm EDT
Launch Pad:	39B KSC Florida
Return Date:	23 October 1989
Mission Duration:	4 days, 23hr, 39min, 20sec
Orbits:	79
Miles Travelled:	2 million
Commander:	Donald E. Williams
Pilot:	Michael J. McCulley
Mission Specialists:	Franklin R. Chang-Diaz
	Shannon W. Lucid
	Ellen S. Baker

STS-33

Orbiter:	Discovery
Launch Date:	22 November 1989
Launch Time:	7:23:30 pm EST
Launch Pad:	39B KSC Florida
Return Date:	27 November 1989
Mission Duration:	5 days, 6min, 49sec
Orbits:	79
Miles Travelled:	2 million
Commander:	Frederick D. Gregory
Pilot:	John E. Blaha
Mission Specialists:	F. Story Musgrave
	Manley L. Carter Jr
	Kathryn C. Thornton

Classified Department of Defense Mission

STS-32

Orbiter:	Columbia
Launch Date:	9 January 1990
Launch Time:	7:35:00 am EST
Launch Pad:	39A KSC Florida
Return Date:	20 January 1990
Mission Duration:	10 days, 21hr, 36sec
Orbits:	172
Miles Travelled:	4.5 million
Commander:	Daniel C. Brandenstein
Pilot:	James D. Wetherbee
Mission Specialists:	Bonnie J. Dunbar
	G. David Low
	Marsha S. Ivins

STS-36

Orbiter:	Atlantis
Launch Date:	28 February 1990
Launch Time:	2:50:22 am EST
Launch Pad:	39A KSC Florida
Return Date:	4 March 1990
Mission Duration:	4 days, 10hr, 18min, 22sec
Orbits:	72
Miles Travelled:	1.9 million
Commander:	John O. Creighton
Pilot:	John H. Casper
Mission Specialists:	Peirre J. Thuot
	David C. Hilmers
	Richard M. Mullane

Classified Department of Defense Mission

STS-31

Orbiter:	Discovery
Launch Date:	24 April 1990
Launch Time:	8:33:51 am EDT
Launch Pad:	39B KSC Florida
Return Date:	29 April 1990
Mission Duration:	5 days, 1hr, 16min, 6sec
Orbits:	80
Miles Travelled:	2.1 million
Commander:	Loren J. Shriver
Pilot:	Charles F. Bolden Jr
Mission Specialists:	Steven A. Hawley
	Bruce McCandless II
	Kathryn D. Sullivan

STS-41

Orbiter:	Discovery
Launch Date:	6 October 1990
Launch Time:	7:47:15 am EDT
Launch Pad:	39B KSC Florida
Return Date:	10 October 1990
Mission Duration:	4 days, 2hr, 10min, 4sec
Orbits:	66
Miles Travelled:	1.7 million
Commander:	Richard N. Richards
Pilot:	Robert D. Cabana
Mission Specialists:	William M. Shepherd
	Bruce E. Melnick
	Thomas D. Akers

STS-38

Orbiter:	Atlantis
Launch Date:	15 November 1990
Launch Time:	6:48:13 pm EST
Launch Pad:	39A KSC Florida
Return Date:	20 November 1990
Mission Duration:	4 days, 21hr, 54min, 31sec
Orbits:	79
Miles Travelled:	2 million
Commander:	Richard O. Covey
Pilot:	Frank L. Culberston Jr
Mission Specialists:	Robert C. Springer
	Carl J. Meade
	Charles D. Gernar

Classified Department of Defense Mission

STS-35

Orbiter:	Columbia
Launch Date:	2 December 1990
Launch Time:	1:49:01 am EST
Launch Pad:	39B KSC Florida
Return Date:	10 December 1990
Mission Duration:	8 days, 23hr, 8min, 8sec
Orbits:	144
Miles Travelled:	3.7 million
Commander:	Vance D. Brand
Pilot:	Guy S. Gardner
Mission Specialists:	Jeffrey A. Hoffman
	John M. Lounge
	Robert A. Parker
Payload Specialists:	Samuel T. Durrance
	Ronald A. Parise

STS-37

Orbiter:	Atlantis
Launch Date:	5 April 1991
Launch Time:	9:22:44 am EST
Launch Pad:	39B KSC Florida
Return Date:	11 April 1991
Mission Duration:	5 days, 23hr, 32min, 44sec
Orbits:	93
Miles Travelled:	2.5 million
Commander:	Steven R. Nagel
Pilot:	Kenneth D. Cameron
Mission Specialists:	Jerry L. Ross
	Jerome 'Jay' Apt
	Linda M. Godwin

STS-39

Orbiter:	Discovery
Launch Date:	28 April 1991
Launch Time:	7:33:14 am EDT
Launch Pad:	39A KSC Florida
Return Date:	6 May 1991
Mission Duration:	8 days, 7hr, 22min, 23sec
Orbits:	134
Miles Travelled:	3.5 million
Commander:	Michael L. Coats
Pilot:	L. Blaine Hammond Jr
Mission Specialists:	Guion S. Bluford Jr
	Gregory J. Harbaugh
	Richard J. Hieb
	Donald R. McMonagle
	Charles L. Veach

Unclassified Department of Defense Mission

STS-40

Orbiter:	Columbia
Launch Date:	5 June 1991
Launch Time:	9:24:51 am EDT
Launch Pad:	39B KSC Florida
Return Date:	14 June 1991
Mission Duration:	9 days, 2hr, 14min, 20sec
Orbits:	146
Miles Travelled:	3.8 million
Commander:	Bryan D. O'Connor
Pilot:	Sidney M. Gutierrez
Mission Specialists:	James P. Bagian
	Tamara E. Jernigan
	M. Rhea Seddon
Payload Specialists:	F. Drew Gaffney
	Millie Hughes-Fulford

STS-43

Orbiter:	Atlantis
Launch Date:	2 August 1991
Launch Time:	11:01:59 am EDT
Launch Pad:	39A KSC Florida
Return Date:	11 August 1991
Mission Duration:	8 days, 21hr, 21min, 25sec
Orbits:	142
Miles Travelled:	3.7 million
Commander:	John E. Blaha
Pilot:	Michael A. Baker
Mission Specialists:	Shannon W. Lucid
	James C. Adamson
	G. David Low

STS-48

Orbiter:	Discovery
Launch Date:	12 September 1991
Launch Time:	7:11:04 pm EDT
Launch Pad:	39A KSC Florida
Return Date:	18 September 1991
Mission Duration:	5 days, 8hr, 27min, 38sec
Orbits:	81
Miles Travelled:	2.2 million
Commander:	John O. Creighton
Pilot:	Kenneth S. Reightler Jr
Mission Specialists:	James F. Buchli
	Charles D. Gemar
	Mark N. Brown

STS-44

Orbiter:	Atlantis
Launch Date:	24 November 1991
Launch Time:	6:44:00 pm EST
Launch Pad:	39A KSC Florida
Return Date:	1 December 1991
Mission Duration:	6 days, 22hr, 50min, 44sec
Orbits:	110
Miles Travelled:	2.9 million
Commander:	Frederick D. Gregory
Pilot:	Terence T. Henricks
Mission Specialists:	F. Story Musgrave
	Mario Runco Jr
	James S. Voss
Payload Specialist:	Thomas J. Hennen

Unclassified Department of Defense Mission

STS-42

Orbiter:	Discovery
Launch Date:	22 January 1992
Launch Time:	9:52:33 am EST
Launch Pad:	39A KSC Florida
Return Date:	30 January 1992
Mission Duration:	8 days, 1hr, 14min, 44sec
Orbits:	129
Miles Travelled:	2.9 million
Commander:	Ronald J. Grabe
Pilot:	Stephen S. Oswald
Mission Specialists:	Norman E. Thagard
	David C. Hilmers
	William F. Readdy
Payload Specialists:	Roberta L. Bondar, CSA
	Ulf D. Merbold, ESA

STS-45

Orbiter:	Atlantis
Launch Date:	24 March 1992
Launch Time:	8:13:00 am EST
Launch Pad:	39A KSC Florida
Return Date:	2 April 1992
Mission Duration:	8 days, 22hr, 9min, 28sec
Orbits:	143
Miles Travelled:	3.2 million
Commander:	Charles F. Bolden Jr
Pilot:	Brian Duffy
Mission Specialists:	Kathryn D. Sullivan
	David C. Leestma
	C. Michael Foale
Payload Specialists:	Byron K. Lichtenberg
	Dirk D. Frimout, ESA

STS-49

Orbiter:	Endeavour
Launch Date:	7 May 1992
Launch Time:	7:40:00 pm EDT
Launch Pad:	39B KSC Florida
Return Date:	16 May 1992
Mission Duration:	8 days, 21hr, 17min, 38sec
Orbits:	141
Miles Travelled:	3.7 million
Commander:	Daniel C. Brandenstein
Pilot:	Kevin P. Chilton
Mission Specialists:	Kathryn C. Thornton
	Richard J. Hieb
	Thomas D. Akers
	Bruce E. Melnick
	Pierre J. Thuot

STS-50

Orbiter:	Columbia
Launch Date:	25 June 1992
Launch Time:	12:12:23 pm EDT
Launch Pad:	39A KSC Florida
Return Date:	9 July 1992
Mission Duration:	13 days, 19hr, 30min, 4min
Orbits:	221
Miles Travelled:	5.8 million
Commander:	Richard N. Richards
Pilot:	Kenneth D. Bowersox
Mission Specialists:	Bonnie J. Dunbar
	Ellen S. Baker
	Carl J. Meade
Payload Specialists:	Lawrence J. DeLucas
	Eugene H. Trinh

STS-46

Orbiter:	Atlantis
Launch Date:	31 July 1992
Launch Time:	9:56:48 am EDT
Launch Pad:	39B KSC Florida
Return Date:	8 August 1992
Mission Duration:	7 days, 23hr, 15min, 3sec
Orbits:	127
Miles Travelled:	3.3 million
Commander:	Loren J. Shriver
Pilot:	Andrew M. Allen
Mission Specialists:	Jeffrey A. Hoffman
	Franklin R. Chang-Diaz
	Claude Nicollier, ESA
	Marsha S. Ivins
Payload Specialist:	Franco Malerba, ASI

STS-47

Orbiter:	Endeavour
Launch Date:	12 September 1992
Launch Time:	10:23:00 am EDT
Launch Pad:	39B KSC Florida
Return Date:	20 September 1992
Mission Duration:	7 days, 22hr, 30min, 23sec
Orbits:	126
Miles Travelled:	3.3 million
Commander:	Robert L. Gibson
Pilot:	Curtis L. Brown Jr
Mission Specialists:	Mark C. Lee
	N. Jan Davis
	Jerome 'Jay' Apt
	Mae C. Jemison
Payload Specialist:	Mamoru Mohri, NASDA

STS-52

Orbiter:	Columbia
Launch Date:	22 October 1992
Launch Time:	1:09:39 pm EDT
Launch Pad:	39B KSC Florida
Return Date:	1 November 1992
Mission Duration:	9 days, 20hr, 56min, 13sec
Orbits:	159
Miles Travelled:	4.1 million
Commander:	James D. Wetherbee
Pilot:	Michael A. Baker
Mission Specialists:	Charles L. Veach
	William M. Shepherd
	Tamara E. Jernigan
Payload Specialist:	Steven G. Maclean, CSA

STS-53

Orbiter:	Discovery
Launch Date:	2 December 1992
Launch Time:	8:24:00 am EST
Launch Pad:	39A KSC Florida
Return Date:	9 December 1992
Mission Duration:	7 days, 7hr, 19min, 47sec
Orbits:	116
Miles Travelled:	3 million
Commander:	David M. Walker
Pilot:	Robert D. Cabana
Mission Specialists:	Guion S. Bluford Jr
	James S. Voss
	Michael R. Clifford

Classified Department of Defense Mission

STS-54

Orbiter:	Endeavour
Launch Date:	13 January 1993
Launch Time:	8:59:30 am EST
Launch Pad:	39B KSC Florida
Return Date:	19 January 1993
Mission Duration:	5 days, 23hr, 38min, 19sec
Orbits:	96
Miles Travelled:	2.5 million
Commander:	John H. Casper
Pilot:	Donald R. McMonagle
Mission Specialists:	Mario Runco Jr
	Gregory J. Harbaugh
	Susan J. Helms

STS-56

Orbiter:	Discovery
Launch Date:	8 April 1993
Launch Time:	1:29:00 am EDT
Launch Pad:	39B KSC Florida
Return Date:	17 April 1993
Mission Duration:	9 days, 6hr, 8min, 24sec
Orbits:	148
Miles Travelled:	3.9 million
Commander:	Kenneth D. Cameron
Pilot:	Stephen S. Oswald
Mission Specialists:	C. Michael Foale
	Kenneth D. Cockrell
	Ellen L. Ochoa

STS-55

Orbiter:	Columbia
Launch Date:	26 April 1993
Launch Time:	10:50:00 am EDT
Launch Pad:	39A KSC Florida
Return Date:	6 May 1993
Mission Duration:	9 days, 23hr, 39min, 59sec
Orbits:	160
Miles Travelled:	4.2 million
Commander:	Steven R. Nagel
Pilot:	Terence T. Henricks
Mission Specialists:	Jerry L. Ross
	Charles J. Precourt
	Bernard A. Harris Jr
Payload Specialists:	Ulrich Walter, DLR
	Hans Schlegel, DLR

STS-57

Orbiter:	Endeavour
Launch Date:	21 June 1993
Launch Time:	9:07:00 am EDT
Launch Pad:	39B KSC Florida
Return Date:	1 July 1993
Mission Duration:	9 days, 23hr, 44min, 54sec
Orbits:	155
Miles Travelled:	4.1 million
Commander:	Ronald J. Grabe
Pilot:	Brian Duffy
Mission Specialists:	G. David Low
	Nancy J. Sherlock (Currie)
	Peter J. K. Wisoff
	Janice E. Voss

STS-51

Orbiter:	Discovery
Launch Date:	12 September 1993
Launch Time:	7:45:00 am EDT
Launch Pad:	39B KSC Florida
Return Date:	22 September 1993
Mission Duration:	9 days, 20hr, 11min, 11sec
Orbits:	157
Miles Travelled:	4.1 million
Commander:	Frank L. Culbertson Jr
Pilot:	William F. Readdy
Mission Specialists:	James H. Newman
	Daniel W. Bursch
	Carl E. Walz

STS-58

Orbiter:	Columbia
Launch Date:	18 October 1993
Launch Time:	10:53:00 am EDT
Launch Pad:	39B KSC Florida
Return Date:	1 November 1993
Mission Duration:	14 days, 12min, 42sec
Orbits:	225
Miles Travelled:	5.8 million
Commander:	John E. Blaha
Pilot:	Richard A. Searfoss
Mission Specialists:	M. Rhea Seddon
	William S. McArthur
	David A. Wolf
	Shannon W. Lucid
Payload Specialist:	Martin J. Fettman

STS-61

Orbiter:	Endeavour
Launch Date:	2 December 1993
Launch Time:	4:27:00 am EST
Launch Pad:	39B KSC Florida
Return Date:	13 December 1993
Mission Duration:	10 days, 19hr, 58min, 33sec
Orbits:	163
Miles Travelled:	4.4 million
Commander:	Richard O. Covey
Pilot:	Kenneth D. Bowersox
Mission Specialists:	F. Story Musgrave
	Kathryn C. Thornton
	Claude Nicollier, ESA
	Jeffrey A. Hoffman
	Thomas D. Akers

STS-60

Orbiter:	Discovery
Launch Date:	3 February 1994
Launch Time:	7:10:00 am EST
Launch Pad:	39A KSC Florida
Return Date:	11 February 1994
Mission Duration:	8 days, 7hr, 9min, 22sec
Orbits:	130
Miles Travelled:	3.4 million
Commander:	Charles F. Bolden Jr
Pilot:	Kenneth S. Reightler Jr
Mission Specialists:	N. Jan Davis
	Ronald M. Sega
	Franklin R. Chang-Diaz
	Sergei K. Krikalev, RKA

STS-62

Orbiter:	Columbia
Launch Date:	4 March 1994
Launch Time:	8:53:00 am EST
Launch Pad:	39B KSC Florida
Return Date:	18 March 1994
Mission Duration:	13 days, 23hr, 16min, 41sec
Orbits:	224
Miles Travelled:	5.8 million
Commander:	John H. Casper
Pilot:	Andrew M. Allen
Mission Specialists:	Pierre J. Thuot
	Charles D. Gemar
	Marsha S. Ivins

STS-59

Orbiter:	Endeavour
Launch Date:	9 April 1994
Launch Time:	7:05:00 am EDT
Launch Pad:	39A KSC Florida
Return Date:	20 April 1994
Mission Duration:	11 days, 5hr, 49min, 30sec
Orbits:	183
Miles Travelled:	4.7 million
Commander:	Sidney M. Gutierrez
Pilot:	Kevin P. Chilton
Mission Specialists:	Linda M. Godwin
	Jerome 'Jay' Apt
	Michael R. Clifford
	Thomas D. Jones

STS-65

Orbiter:	Columbia
Launch Date:	8 July 1994
Launch Time:	12:43:01 pm EDT
Launch Pad:	39A KSC Florida
Return Date:	23 July 1994
Mission Duration:	14 days, 17hr, 55min
Orbits:	235
Miles Travelled:	6.1 million
Commander:	Robert D. Cabana
Pilot:	James D. Halsell
Mission Specialists:	Richard J. Hieb
	Carl E. Walz
	Leroy Chiao
	Donald A. Thomas
Payload Specialist:	Chiaki Mukai, NASDA

STS-64

Orbiter:	Discovery
Launch Date:	9 September 1994
Launch Time:	6:22:55 pm EDT
Launch Pad:	39B KSC Florida
Return Date:	20 September 1994
Mission Duration:	10 days, 22hr, 49min, 57sec
Orbits:	176
Miles Travelled:	4.5 million
Commander:	Richard N. Richards
Pilot:	L. Blaine Hammond Jr
Mission Specialists:	Jerry M. Linenger
	Susan J. Helms
	Carl J. Meade
	Mark C. Lee

STS-68

Orbiter:	Endeavour
Launch Date:	30 September 1994
Launch Time:	7:16:00 am EDT
Launch Pad:	39A KSC Florida
Return Date:	11 October 1994
Mission Duration:	11 days, 5hr, 46min, 8sec
Orbits:	182
Miles Travelled:	4.7 million
Commander:	Michael A. Baker
Pilot:	Terrence W. Wilcutt
Mission Specialists:	Thomas D. Jones
	Steven L. Smith
	Daniel W. Bursch
	Peter J. K. Wisoff

STS-66

Orbiter:	Atlantis
Launch Date:	3 November 1994
Launch Time:	11:49:53 am EST
Launch Pad:	39B KSC Florida
Return Date:	14 November 1994
Mission Duration:	10 days, 22hr, 34min, 2sec
Orbits:	174
Miles Travelled:	4.5 million
Commander:	Donald R. McMonagle
Pilot:	Curtis L. Brown Jr
Mission Specialists:	Ellen L. Ochoa
	Joseph R. Tanner
	Jean-Francois Clervoy, CNES
	Scott E. Parazynski

STS-63

Orbiter:	Discovery
Launch Date:	3 February 1995
Launch Time:	12:22:04 am EST
Launch Pad:	39B KSC Florida
Return Date:	11 February 1995
Mission Duration:	8 days, 6hr, 28min, 15sec
Orbits:	129
Miles Travelled:	3 million
Commander:	James D. Wetherbee
Pilot:	Eileen M. Collins
Mission Specialists:	C. Michael Foale
	Bernard A. Harris Jr
	Janice E. Voss
	Vladimir G. Titov, RKA

STS-67

Orbiter:	Endeavour
Launch Date:	2 March 1995
Launch Time:	1:38:13 am EST
Launch Pad:	39A KSC Florida
Return Date:	18 March 1995
Mission Duration:	16 days, 15hr, 8min, 48sec
Orbits:	262
Miles Travelled:	6.9 million
Commander:	Stephen S. Oswald
Pilot:	William G. Gregory
Mission Specialists:	John M. Grunsfeld
	Wendy B. Lawrence
	Tamara E. Jernigan
Payload Specialists:	Samuel T. Durrance
	Ronald A. Parise

STS-71

Orbiter:	Atlantis
Launch Date:	27 June 1995
Launch Time:	3:32:19 pm EDT
Launch Pad:	39A KSC Florida
Return Date:	7 July 1995
Mission Duration:	9 days, 19hr, 22min, 17sec
Orbits:	153
Miles Travelled:	4.1 million
Commander:	Robert L. Gibson
Pilot:	Charles J. Precourt
Mission Specialists:	Gregory J. Harbaugh
	Ellen S. Baker
	Bonnie J. Dunbar
	Anatoly Solovyev, RKA >
	Nikolai Budarin, RKA >
	Norman E. Thagard <
	Vladimir Dezhurov, RKA <
	Gennady Strekalov, RKA <

STS-70

Orbiter:	Discovery
Launch Date:	13 July 1995
Launch Time:	9:41:55 am EDT
Launch Pad:	39B KSC Florida
Return Date:	22 July 1995
Mission Duration:	8 days, 22hr, 20min, 5sec
Orbits:	143
Miles Travelled:	3.7 million
Commander:	Terence T. Henricks
Pilot:	Kevin R. Kregel
Mission Specialists:	Nancy J. Currie
	Donald A. Thomas
	Mary Ellen Weber

STS-69

Orbiter:	Endeavour
Launch Date:	7 September 1995
Launch Time:	11:09:00 am EDT
Launch Pad:	39A KSC Florida
Return Date:	18 September 1995
Mission Duration:	10 days, 20hr, 28min, 56sec
Orbits:	171
Miles Travelled:	4.5 million
Commander:	David M. Walker
Pilot:	Kenneth D. Cockrell
Mission Specialists:	James S. Voss
	James H. Newman
	Michael L. Gernhardt

STS-73

Orbiter:	Columbia
Launch Date:	20 October 1995
Launch Time:	9:53:00 am EDT
Launch Pad:	39B KSC Florida
Return Date:	5 November 1995
Mission Duration:	15 days, 21hr, 52min, 28sec
Orbits:	256
Miles Travelled:	6.6 million
Commander:	Kenneth D. Bowersox
Pilot:	Kent V. Rominger
Mission Specialists:	Kathryn C. Thornton
	Catherine G. Coleman
	Michael E. Lopez-Alegria
Payload Specialists:	Fred W. Leslie
	Albert Sacco Jr

STS-74

Orbiter:	Atlantis
Launch Date:	12 November 1995
Launch Time:	7:30:43 am EST
Launch Pad:	39A KSC Florida
Return Date:	20 November 1995
Mission Duration:	8 days, 4hr, 30min, 44sec
Orbits:	129
Miles Travelled:	3.4 million
Commander:	Kenneth D. Cameron
Pilot:	James D. Halsell
Mission Specialists:	Chris A. Hadfield, CSA
	Jerry L. Ross
	William S. McArthur Jr

STS-72

Orbiter:	Endeavour
Launch Date:	11 January 1996
Launch Time:	4:41:00 am EST
Launch Pad:	39B KSC Florida
Return Date:	20 January 1996
Mission Duration:	8 days, 22hr, 1min, 47sec
Orbits:	142
Miles Travelled:	3.7 million
Commander:	Brian Duffy
Pilot:	Brent W. Jett
Mission Specialists:	Leroy Chiao
	Winston E. Scott
	Koichi Wakata, NASDA
	Daniel T. Barry

STS-75

Orbiter:	Columbia
Launch Date:	22 February 1996
Launch Time:	3:18:00 pm EST
Launch Pad:	39B KSC Florida
Return Date:	9 March 1996
Mission Duration:	15 days, 17hr, 41min, 25sec
Orbits:	252
Miles Travelled:	6.5 million
Commander:	Andrew M. Allen
Pilot:	Scott J. Horowitz
Mission Specialists:	Jeffrey A. Hoffman
	Maurizio Cheli, ESA
	Claude Nicollier, ESA
	Franklin R. Chang-Diaz
Payload Specialists:	Umberto Guidoni, ASI

STS-76

Orbiter:	Atlantis
Launch Date:	22 March 1996
Launch Time:	3:13:04 am EST
Launch Pad:	39B KSC Florida
Return Date:	31 March 1996
Mission Duration:	9 days, 5hr, 15min, 53sec
Orbits:	145
Miles Travelled:	3.8 million
Commander:	Kevin P. Chilton
Pilot:	Richard A. Searfoss
Mission Specialists:	Ronald M. Sega
	Michael R. Clifford
	Linda M. Godwin
	Shannon W. Lucid >

STS-77

Orbiter:	Endeavour
Launch Date:	19 May 1996
Launch Time:	6:30:00 am EDT
Launch Pad:	39B KSC Florida
Return Date:	29 May 1996
Mission Duration:	10 days, 39min, 18sec
Orbits:	161
Miles Travelled:	4.1 million
Commander:	John H. Casper
Pilot:	Curtis L. Brown Jr
Mission Specialists:	Andrew S. W. Thomas
	Daniel W. Bursch
	Mario Runco Jr
	Marc Garneau, CSA

STS-78

Orbiter:	Columbia
Launch Date:	20 June 1996
Launch Time:	10:49:00 am EDT
Launch Pad:	39B KSC Florida
Return Date:	7 July 1996
Mission Duration:	16 days, 21hr, 47min, 45sec
Orbits:	272
Miles Travelled:	7 million
Commander:	Terence T. Henricks
Pilot:	Kevin R. Kregel
Mission Specialists:	Richard M. Linnehan
	Susan J. Helms
	Charles E. Brady Jr
Payload Specialists:	Jean-Jacques Favier, CNES
	Robert Brent Thirsk, CSA

STS-79

Orbiter:	Atlantis
Launch Date:	16 September 1996
Launch Time:	4:54:49 am EDT
Launch Pad:	39A KSC Florida
Return Date:	26 September 1996
Mission Duration:	10 days, 3hr, 18min, 26sec
Orbits:	160
Miles Travelled:	3.9 million
Commander:	William F. Readdy
Pilot:	Terrence W. Wilcutt
Mission Specialists:	Thomas D. Akers
	Jerome 'Jay' Apt
	Carl E. Walz
	John E. Blaha >
	Shannon W. Lucid <

STS-80

Orbiter:	Columbia
Launch Date:	19 November 1996
Launch Time:	2:55:47 pm EST
Launch Pad:	39B KSC Florida
Return Date:	7 December 1996
Mission Duration:	17 days, 15hr, 53min, 18sec
Orbits:	279
Miles Travelled:	7 million
Commander:	Kenneth D. Cockrell
Pilot:	Kent V. Rominger
Mission Specialists:	F. Story Musgrave
	Thomas David Jones
	Tamara E. Jernigan

STS-81

Orbiter:	Atlantis
Launch Date:	12 January 1997
Launch Time:	4:27:23 am EST
Launch Pad:	39B KSC Florida
Return Date:	22 January 1997
Mission Duration:	10 days, 4hr, 55min, 21sec
Orbits:	160
Miles Travelled:	4.1 million
Commander:	Michael A. Baker
Pilot:	Brent W. Jett Jr
Mission Specialists:	Peter J. K. Wisoff
	John M. Grunsfeld
	Marsha S. Ivins
	Jerry M. Linenger >
	John E. Blaha <

STS-82

Orbiter:	Discovery
Launch Date:	11 February 1997
Launch Time:	3:55:17 am EST
Launch Pad:	39A KSC Florida
Return Date:	21 February 1997
Mission Duration:	9 days, 23hr, 37min, 9sec
Orbits:	150
Miles Travelled:	4.1 million
Commander:	Kenneth D. Bowersox
Pilot:	Scott J. Horowitz
Mission Specialists:	Joseph R. Tanner
	Steven A. Hawley
	Gregory J. Harbaugh
	Mark C. Lee
	Steven L. Smith

STS-83

Orbiter:	Columbia
Launch Date:	4 April 1997
Launch Time:	2:20:32 pm EST
Launch Pad:	39A KSC Florida
Return Date:	8 April 1997
Mission Duration:	3 days, 23hr, 13min, 38sec
Orbits:	63
Miles Travelled:	1.5 million
Commander:	James D. Halsell
Pilot:	Susan L. Still (Kilrain)
Mission Specialists:	Michael L. Gernhardt
	Donald A. Thomas
Payload Commander:	Janice E. Voss
Payload Specialists:	Roger K. Crouch
	Gregory T. Linteris

STS-84

Orbiter:	Atlantis
Launch Date:	15 May 1997
Launch Time:	4:07:48 am EDT
Launch Pad:	39A KSC Florida
Return Date:	24 May 1997
Mission Duration:	9 days, 23hr, 19min, 56sec
Orbits:	144
Miles Travelled:	3.6 million
Commander:	Charles J. Precourt
Pilot:	Eileen M. Collins
Mission Specialists:	Jean-Francois Clervoy, ESA
	Carlos I. Noriega
	Edward T. Lu
	Yelena V. Kondakova, RKA
	C. Michael Foale >
	Jerry M. Linenger <

STS-94

Orbiter:	Columbia
Launch Date:	1 July 1997
Launch Time:	2:02:00 pm EDT
Launch Pad:	39A KSC Florida
Return Date:	17 July 1997
Mission Duration:	15 days, 16hr, 44min, 34sec
Orbits:	251
Miles Travelled:	6.2 million
Commander:	James D. Halsell
Pilot:	Susan L. Still (Kilrain)
Mission Specialists:	Michael L. Gernhardt
	Donald A. Thomas
Payload Commander:	Janice E. Voss
Payload Specialists:	Roger K. Crouch
	Gregory T. Linteris

STS-85

Orbiter:	Discovery
Launch Date:	7 August 1997
Launch Time:	10:41:00 am EDT
Launch Pad:	39A KSC Florida
Return Date:	19 August 1997
Mission Duration:	11 days, 19hr, 18min, 47sec
Orbits:	185
Miles Travelled:	4.7 million
Commander:	Curtis L. Brown Jr
Pilot:	Kent V. Rominger
Mission Specialists:	Stephen K. Robinson
	Robert L. Curbeam Jr
Payload Commander:	N. Jan Davis
Payload Specialist:	Bjarni Tryggvason, CSA

STS-86

Orbiter:	Atlantis
Launch Date:	25 September 1997
Launch Time:	10:34:19 pm EDT
Launch Pad:	39A KSC Florida
Return Date:	6 October 1997
Mission Duration:	10 days, 19hr, 22min, 12sec
Orbits:	170
Miles Travelled:	4.5 million
Commander:	James D. Wetherbee
Pilot:	Michael J. Bloomfield
Mission Specialists:	Vladimir G. Titov, RKA
	Scott E. Parazynski
	Jean-Loup Chrétien, CNES
	Wendy B. Lawrence
	David A. Wolf >
	C. Michael Foale <

STS-87

Orbiter:	Columbia
Launch Date:	19 November 1997
Launch Time:	2:46:00 pm EST
Launch Pad:	39B KSC Florida
Return Date:	5 December 1997
Mission Duration:	15 days, 16hr, 34min, 4sec
Orbits:	252
Miles Travelled:	6.5 million
Commander:	Kevin R. Kregel
Pilot:	Steven W. Lindsey
Mission Specialists:	Winston E. Scott
	Kalpana Chawla
	Takao Doi, NASDA
Payload Specialist:	Leonid Kadeniuk, NSAU

STS-89

Orbiter:	Endeavour
Launch Date:	22 January 1998
Launch Time:	9:48:15 pm EST
Launch Pad:	39A KSC Florida
Return Date:	31 January 1998
Mission Duration:	8 days, 19hr, 46min, 54sec
Orbits:	139
Miles Travelled:	3.6 million
Commander:	Terrence W. Wilcutt
Pilot:	Joe F. Edwards Jr
Mission Specialists:	James F. Reilly, II
	Michael P. Anderson
	Bonnie J. Dunbar
	Salizhan Sharipov, RKA
	Andrew S. W. Thomas >
	David A. Wolf <

STS-90

Orbiter:	Columbia
Launch Date:	17 April 1998
Launch Time:	2:19:00 pm EDT
Launch Pad:	39B KSC Florida
Return Date:	3 May 1998
Mission Duration:	15 days, 21hr, 50min, 58sec
Orbits:	256
Miles Travelled:	6.5 million
Commander:	Richard A. Searfoss
Pilot:	Scott D. Altman
Mission Specialists:	Richard M. Linnehan
	Kathryn P. Hire
	Dafydd Williams, CSA
Payload Specialists:	Jay C. Buckey
	James A. Pawelczyk

STS-91

Orbiter:	Discovery
Launch Date:	2 June 1998
Launch Time:	6:06:24 pm EDT
Launch Pad:	39A KSC Florida
Return Date:	12 June 1998
Mission Duration:	9 days, 19hr, 54min, 2sec
Orbits:	155
Miles Travelled:	3.8 million
Commander:	Charles J. Precourt
Pilot:	Dominic L. Pudwill Gorie
Mission Specialists:	Wendy B. Lawrence
	Franklin R. Chang-Diaz
	Janet L. Kavandi
	Valery Ryumin, RKA
	Andrew S. W. Thomas <

STS-95

Orbiter:	Discovery
Launch Date:	29 October 1998
Launch Time:	2:19:34 pm EST
Launch Pad:	39B KSC Florida
Return Date:	7 November 1998
Mission Duration:	8 days, 21hr, 44min, 56sec
Orbits:	134
Miles Travelled:	3.6 million
Commander:	Curtis L. Brown
Pilot:	Steven W. Lindsey
Mission Specialists:	Pedro Duque, ESA
	Scott E. Parazynski
Payload Commander:	Stephen K. Robinson
Payload Specialists:	Chiaki Mukai, NASDA
	John H. Glenn Jr

STS-88

Orbiter:	Endeavour
Launch Date:	4 December 1998
Launch Time:	3:35:34 am EST
Launch Pad:	39A KSC Florida
Return Date:	15 December 1998
Mission Duration:	11 days, 19hr, 18min, 47sec
Orbits:	185
Miles Travelled:	4.7 million
Commander:	Robert D. Cabana
Pilot:	Frederick W. Sturckow
Mission Specialists:	Jerry L. Ross
	Nancy J. Currie
	James H. Newman
	Sergei K. Krikalev, RKA

STS-96

Orbiter:	Discovery
Launch Date:	27 May 1999
Launch Time:	6:49:42 am EDT
Launch Pad:	39B KSC Florida
Return Date:	6 June 1999
Mission Duration:	9 days, 19hr, 13min, 57sec
Orbits:	154
Miles Travelled:	3.8 million
Commander:	Kent V. Rominger
Pilot:	Rick D. Husband
Mission Specialists:	Tamara E. Jernigan
	Ellen L. Ochoa
	Daniel T. Barry
	Julie Payette, CSA
	Valery I. Tokarev, RKA

STS-93

Orbiter:	Columbia
Launch Date:	23 July 1999
Launch Time:	12:31:00 am EDT
Launch Pad:	39B KSC Florida
Return Date:	27 July 1999
Mission Duration:	4 days, 22hr, 49min, 37sec
Orbits:	80
Miles Travelled:	1.8 million
Commander:	Eileen M. Collins
Pilot:	Jeffrey S. Ashby
Mission Specialists:	Steven A. Hawley
	Catherine G. Coleman
	Michel Tognini, CNES

STS-103

Orbiter:	Discovery
Launch Date:	19 December 1999
Launch Time:	7:50:00 pm EST
Launch Pad:	39B KSC Florida
Return Date:	27 December 1999
Mission Duration:	7 days, 23hr, 10min, 47sec
Orbits:	119
Miles Travelled:	3.2 million
Commander:	Curtis L. Brown Jr
Pilot:	Scott J. Kelly
Mission Specialists:	Steven L. Smith
	Jean-François Clervoy, ESA
	John M. Grunsfeld
	C. Michael Foale
	Claude Nicollier, ESA

STS-99

Orbiter:	Endeavour
Launch Date:	11 February 2000
Launch Time:	12:43:40 pm EST
Launch Pad:	39A KSC Florida
Return Date:	22 February 2000
Mission Duration:	11 days, 5hr, 38min, 41sec
Orbits:	181
Miles Travelled:	4.7 million
Commander:	Kevin R. Kregel
Pilot:	Dominic L. Pudwill Gorie
Mission Specialists:	Gerhard P. J. Thiele, ESA
	Janet L. Kavandi
	Janice E. Voss
	Mamoru Mohri, NASDA

STS-101

Orbiter:	Atlantis
Launch Date:	19 May 2000
Launch Time:	6:11:10 am EDT
Launch Pad:	39A KSC Florida
Return Date:	29 May 2000
Mission Duration:	9 days, 20hr, 9min, 9sec
Orbits:	155
Miles Travelled:	4.1 million
Commander:	James D. Halsell Jr
Pilot:	Scott J. Horowitz
Mission Specialists:	Mary E. Weber
	Jeffrey N. Williams
	James S. Voss
	Susan J. Helms
	Yury V. Usachev, RKA

STS-106

Orbiter:	Atlantis
Launch Date:	8 September 2000
Launch Time:	8:45:47 am EDT
Launch Pad:	39B KSC Florida
Return Date:	20 September 2000
Mission Duration:	11 days, 19hr, 12min, 15sec
Orbits:	185
Miles Travelled:	4.9 million
Commander:	Terrence W. Wilcutt
Pilot:	Scott D. Altman
Mission Specialists:	Edward T. Lu
	Richard A. Mastracchio
	Daniel C. Burbank
	Yuri Malenchenko, RKA
	Boris Morukov, RKA

STS-92

Orbiter:	Discovery
Launch Date:	11 October 2000
Launch Time:	7:17:00 pm EDT
Launch Pad:	39A KSC Florida
Return Date:	24 October 2000
Mission Duration:	12 days, 21hr, 40min, 25sec
Orbits:	202
Miles Travelled:	5.3 million
Commander:	Brian Duffy
Pilot:	Pamela A. Melroy
Mission Specialists:	Leroy Chiao
	William S. McArthur
	Peter J. K. Wisoff
	Michael E. Lopez-Alegria
	Koichi Wakata, NASDA

STS-97

Orbiter:	Endeavour
Launch Date:	30 November 2000
Launch Time:	10:06:01 pm EST
Launch Pad:	39B KSC Florida
Return Date:	11 December 2000
Mission Duration:	10 days, 19hr, 58min
Orbits:	171
Miles Travelled:	4.5 million
Commander:	Brent W. Jett
Pilot:	Michael J. Bloomfield
Mission Specialists:	Joseph R. Tanner
	Marc Garneau, CSA
	Carlos I. Noriega

STS-98

Orbiter:	Atlantis
Launch Date:	7 February 2001
Launch Time:	6:13:02 pm EST
Launch Pad:	39A KSC Florida
Return Date:	20 February 2001
Mission Duration:	12 days, 20hr, 20min, 4sec
Orbits:	203
Miles Travelled:	5.3 million
Commander:	Kenneth D. Cockrell
Pilot:	Mark L. Polansky
Mission Specialists:	Robert L. Curbeam Jr
	Marsha S. Ivins
	Thomas D. Jones

STS-102

Orbiter:	Discovery
Launch Date:	8 March 2001
Launch Time:	6:42:09 am EST
Launch Pad:	39B KSC Florida
Return Date:	21 March 2001
Mission Duration:	12 days, 19hr, 49min
Orbits:	202
Miles Travelled:	5.3 million
Commander:	James D. Wetherbee
Pilot:	James M. Kelly
Mission Specialists:	Andrew S. W. Thomas
	Paul W. Richards
	Yury V. Usachev, RKA †
	James S. Voss †
	Susan J. Helms †
	William M. Shepherd *
	Yuri P. Gidzenko, RKA *
	Sergei K. Krikalev, RKA *

STS-100

Orbiter:	Endeavour
Launch Date:	19 April 2001
Launch Time:	2:40:42 pm EDT
Launch Pad:	39A KSC Florida
Return Date:	1 May 2001
Mission Duration:	11 days, 12hr, 54min
Orbits:	186
Miles Travelled:	4.9 million
Commander:	Kent V. Rominger
Pilot:	Jeffrey S. Ashby
Mission Specialists:	Chris Hadfield, CSA
	John L. Phillips
	Scott E. Parazynski
	Umberto Guidoni, ESA
	Yuri Lonchakov, RKA

STS-104

Orbiter:	Atlantis
Launch Date:	12 July 2001
Launch Time:	5:03:59 am EDT
Launch Pad:	39B KSC Florida
Return Date:	24 July 2001
Mission Duration:	12 days, 18hr, 36min
Orbits:	200
Miles Travelled:	5.3 million
Commander:	Steven W. Lindsey
Pilot:	Charles O. Hobaugh
Mission Specialists:	Michael L. Gernhardt
	Janet L. Kavandi
	James F. Reilly

STS-105

Orbiter:	Discovery
Launch Date:	10 August 2001
Launch Time:	5:10:14 pm EDT
Launch Pad:	39A KSC Florida
Return Date:	22 August 2001
Mission Duration:	11 days, 19hr, 38min
Orbits:	186
Miles Travelled:	4.3 million
Commander:	Scott J. Horowitz
Pilot:	Frederick W. Sturckow
Mission Specialists:	Daniel T. Barry
	Patrick G. Forrester
	Frank L. Culbertson Jr †
	Mikhail V. Tyurin, RKA †
	Vladimir N. Dezhurov, RKA †
	Yury V. Usachev, RKA *
	James S. Voss *
	Susan J. Helms *

STS-108

Orbiter:	Endeavour
Launch Date:	5 December 2001
Launch Time:	5:19:28 pm EST
Launch Pad:	39B KSC Florida
Return Date:	17 December 2001
Mission Duration:	11 days, 19hr, 55min
Orbits:	186
Miles Travelled:	4.8 million
Commander:	Dominic L. Pudwill Gorie
Pilot:	Mark E. Kelly
Mission Specialists:	Linda M. Godwin
	Daniel M. Tani
	Yuri I. Onufrienko, RKA †
	Carl E. Walz †
	Daniel W. Bursch †
	Frank L. Culbertson Jr *
	Mikhail V. Tyurin, RKA *
	Vladimir N. Dezhurov, RKA *

STS-109

Orbiter:	Columbia
Launch Date:	1 March 2002
Launch Time:	6:22:00 am EST
Launch Pad:	39A KSC Florida
Return Date:	12 March 2002
Mission Duration:	10 days, 22hr, 11min, 9sec
Orbits:	165
Miles Travelled:	3.9 million
Commander:	Scott D. Altman
Pilot:	Duane G. Carey
Mission Specialists:	John M. Grunsfeld
	Nancy J. Currie
	Richard M. Linnehan
	James H. Newman
	Michael J. Massimino

STS-110

Orbiter:	Atlantis
Launch Date:	8 April 2002
Launch Time:	4:44:19 pm EDT
Launch Pad:	39B KSC Florida
Return Date:	19 April 2002
Mission Duration:	10 days, 19hr, 42min, 44sec
Orbits:	171
Miles Travelled:	4.5 million
Commander:	Michael J. Bloomfield
Pilot:	Stephen N. Frick
Mission Specialists:	Jerry L. Ross
	Steven L. Smith
	Ellen L. Ochoa
	Lee M. E. Morin
	Rex J. Walheim

STS-111

Orbiter:	Endeavour
Launch Date:	5 June 2002
Launch Time:	5:22:49 pm EDT
Launch Pad:	39A KSC Florida
Return Date:	19 June 2002
Mission Duration:	13 days, 20hr, 35min, 56sec
Orbits:	217
Miles Travelled:	5.8 million
Commander:	Kenneth D. Cockrell
Pilot:	Paul S. Lockhart
Mission Specialists:	Franklin Chang-Diaz
	Philippe Perrin, CNES
	Valery G. Korzun, RKA †
	Peggy A. Whitson †
	Sergei Y. Treshchev, RKA †
	Yuri I. Onufrienko, RKA *
	Carl E. Walz *
	Daniel W. Bursch *

STS-112

Orbiter:	Atlantis
Launch Date:	7 October 2002
Launch Time:	3:45:51 pm EDT
Launch Pad:	39B KSC Florida
Return Date:	18 October 2002
Mission Duration:	10 days, 19hr, 58min, 44sec
Orbits:	170
Miles Travelled:	4.5 million
Commander:	Jeffrey S. Ashby
Pilot:	Pamela A. Melroy
Mission Specialists:	David A. Wolf
	Sandra H. Magnus
	Piers J. Sellers
	Fyodor N. Yurchikhin, RKA

STS-113

Orbiter:	Endeavour
Launch Date:	23 November 2002
Launch Time:	7:49:47 pm EST
Launch Pad:	39A KSC Florida
Return Date:	7 December 2002
Mission Duration:	13 days, 18hr, 48min, 38sec
Orbits:	200+
Miles Travelled:	5.7 million
Commander:	James D. Wetherbee
Pilot:	Paul S. Lockhart
Mission Specialists:	Michael Lopez-Alegria
	John B. Herrington
	Kenneth D. Bowersox †
	Nikolai M. Budarin, RKA †
	Donald R. Pettit †
	Valery G. Korzun, RKA *
	Peggy A. Whitson *
	Sergei Y. Treshchev, RKA *

STS-107

Orbiter:	Columbia
Launch Date:	16 January 2003
Launch Time:	10:39:00 am EST
Launch Pad:	39A KSC Florida
Return Date:	Destroyed on re-entry
	1 February 2003
Mission Duration:	15 days, 22hr, 20min, 32sec
Orbits:	255
Miles Travelled:	6.6 million
Commander:	Rick D. Husband
Pilot:	William C. McCool
Mission Specialists:	David M. Brown
	Kalpana Chawla
	Laurel B. Clark
Payload Commander:	Michael P. Anderson
Payload Specialist:	Ilan Ramon

STS-114

Orbiter:	Discovery
Launch Date:	26 July 2005
Launch Time:	10:39:00 am EDT
Launch Pad:	39B KSC Florida
Return Date:	9 August 2005
Mission Duration:	13 days, 21hr, 32min, 48sec
Orbits:	219
Miles Travelled:	5.8 million
Commander:	Eileen M. Collins
Pilot:	James M. Kelly
Mission Specialists:	Soichi Noguchi, JAXA
	Stephen K. Robinson
	Andrew S. W. Thomas
	Wendy B. Lawrence
	Charlie J. Camarda

STS-121

Orbiter:	Discovery
Launch Date:	4 July 2006
Launch Time:	2:37:55 pm EDT
Launch Pad:	39B KSC Florida
Return Date:	17 July 2006
Mission Duration:	12 days, 18hr, 37min, 54sec
Orbits:	202
Miles Travelled:	5.3 million
Commander:	Steven W. Lindsey
Pilot:	Mark E. Kelly
Mission Specialists:	Michael E. Fossum
	Lisa M. Nowak
	Stephanie D. Wilson
	Piers J. Sellers
	Thomas Reiter, ESA †

STS-115

Orbiter:	Atlantis
Launch Date:	9 September 2006
Launch Time:	11:15:00 am EDT
Launch Pad:	39B KSC Florida
Return Date:	21 September 2006
Mission Duration:	11 days, 19hr, 6min
Orbits:	187
Miles Travelled:	4.9 million
Commander:	Brent W. Jett Jr
Pilot:	Chris J. Ferguson
Mission Specialists:	Steve MacLean, CSA
	Daniel C. Burbank
	Heidemarie Stefanyshyn-Piper
	Joseph R. Tanner

STS-116

Orbiter:	Discovery
Launch Date:	9 December 2006
Launch Time:	5:32:00 pm EDT
Launch Pad:	39B KSC Florida
Return Date:	22 December 2006
Mission Duration:	12 days, 20hr, 45min
Orbits:	202
Miles Travelled:	5.3 million
Commander:	Mark L. Polansky
Pilot:	William A. Oefelein
Mission Specialists:	Nicholas J. M. Patrick
	Robert L. Curbeam Jr
	Christer Fuglesang, ESA
	Joan Higginbotham
	Sunita Williams †
	Thomas Reiter, ESA *

STS-117

Orbiter:	Atlantis
Launch Date:	8 June 2007
Launch Time:	7:38:04 pm EDT
Launch Pad:	39A KSC Florida
Return Date:	22 June 2007
Mission Duration:	13 days, 20hr, 12min, 44sec
Orbits:	219
Miles Travelled:	5.8 million
Commander:	Frederick W. Sturckow
Pilot:	Lee J. Archambault
Mission Specialists:	Patrick G. Forrester
	Steven R. Swanson
	John D. Olivas
	James F. Reilly
	Clayton C. Anderson †
	Sunita L. Williams *

STS-118

Orbiter:	Endeavour
Launch Date:	8 August 2007
Launch Time:	6:36:42 pm EDT
Launch Pad:	39A KSC Florida
Return Date:	21 August 2007
Mission Duration:	12 days, 17hr, 55min, 34sec
Orbits:	201
Miles Travelled:	5.3 million
Commander:	Scott J. Kelly
Pilot:	Charles O. Hobaugh
Mission Specialists:	Tracy E. Caldwell
	Richard A. Mastracchio
	Dafydd R. Williams, CSA
	Barbara R. Morgan
	B. Alvin Drew

STS-120

Orbiter:	Discovery
Launch Date:	23 October 2007
Launch Time:	11:38:19 am EDT
Launch Pad:	39A KSC Florida
Return Date:	7 November 2007
Mission Duration:	15 days, 2hr, 23min
Orbits:	238
Miles Travelled:	6.3 million
Commander:	Pamela Melroy
Pilot:	George D. Zamka
Mission Specialists:	Douglas H. Wheelock
	Stephanie D. Wilson
	Scott E. Parazynski
	Paolo A. Nespoli, ESA
	Daniel M. Tani †
	Clayton C. Anderson *

STS-122

Orbiter:	Atlantis
Launch Date:	7 February 2008
Launch Time:	2:45:00 pm EST
Launch Pad:	39A KSC Florida
Return Date:	20 February 2008
Mission Duration:	12 days, 18hr, 21min, 50sec
Orbits:	202
Miles Travelled:	5.3 million
Commander:	Stephen N. Frick
Pilot:	Alan G. Poindexter
Mission Specialists:	Leland D. Melvin
	Rex J. Walheim
	Hans Schlegel, ESA
	Stanley G. Love
	Léopold Eyharts, ESA †
	Daniel M. Tani *

STS-123

Orbiter:	Endeavour
Launch Date:	11 March 2008
Launch Time:	2:28:14 am EST
Launch Pad:	39A KSC Florida
Return Date:	26 March 2008
Mission Duration:	16 days, 14hr, 12min, 27sec
Orbits:	250
Miles Travelled:	6.6 million
Commander:	Dominic L. Pudwill Gorie
Pilot:	Gregory H. Johnson
Mission Specialists:	Robert L. Behnken
	Michael J. Foreman
	Richard M. Linnehan
	Takao Doi, JAXA
	Garrett Reisman †
	Léopold Eyharts, ESA *

STS-124

Orbiter:	Discovery
Launch Date:	31 May 2008
Launch Time:	5:02:12 pm EDT
Launch Pad:	39A KSC Florida
Return Date:	14 June 2008
Mission Duration:	13 days, 18hr, 13min, 7sec
Orbits:	217
Miles Travelled:	5.7 million
Commander:	Mark E. Kelly
Pilot:	Kenneth T. Ham
Mission Specialists:	Karen L. Nyberg
	Ronald J. Garan Jr
	Michael E. Fossum
	Akihiko Hoshide, JAXA
	Gregory E. Chamitoff †
	Garrett Reisman *

STS-126

Orbiter:	Endeavour
Launch Date:	14 November 2008
Launch Time:	7:55:00 pm EST
Launch Pad:	39A KSC Florida
Return Date:	30 November 2008
Mission Duration:	15 days, 20hr, 29min, 37sec
Orbits:	251
Miles Travelled:	6.6 million
Commander:	Christopher J. Ferguson
Pilot:	Eric A. Boe
Mission Specialists:	Heidemarie Stefanyshyn-Piper
	Stephen G. Bowen
	Donald R. Pettit
	Robert S. Kimbrough
	Sandra H. Magnus †
	Gregory E. Chamitoff *

STS-119

Orbiter:	Discovery
Launch Date:	15 March 2009
Launch Time:	7:43:00 pm EDT
Launch Pad:	39A KSC Florida
Return Date:	28 March 2009
Mission Duration:	12 days, 19hr, 29min, 33sec
Orbits:	202
Miles Travelled:	5.3 million
Commander:	Lee J. Archambault
Pilot:	Dominic A. Antonelli
Mission Specialists:	Joseph M. Acaba
	Steven R. Swanson
	Richard R. Arnold
	John L. Phillips
	Koichi Wakata, JAXA †
	Sandra H. Magnus *

STS-125

Orbiter:	Atlantis
Launch Date:	11 May 2009
Launch Time:	2:01:56 pm EDT
Launch Pad:	39A KSC Florida
Return Date:	24 May 2009
Mission Duration:	12 days, 21hr, 37min, 9sec
Orbits:	197
Miles Travelled:	5.3 million
Commander:	Scott D. Altman
Pilot:	Gregory C. Johnson
Mission Specialists:	Michael T. Good
	K. Megan McArthur
	John M. Grunsfeld
	Michael J. Massimino
	Andrew J. Feustel

STS-127

Orbiter:	Endeavour
Launch Date:	15 July 2009
Launch Time:	6:03:10 pm EDT
Launch Pad:	39A KSC Florida
Return Date:	31 July 2009
Mission Duration:	15 days, 16hr, 44min, 58sec
Orbits:	248
Miles Travelled:	6.5 million
Commander:	Mark L. Polansky
Pilot:	Douglas G. Hurley
Mission Specialists:	Christopher J. Cassidy
	Julie Payette, CSA
	Thomas H. Marshburn
	David A. Wolf
	Timothy L. Kopra †
	Koichi Wakata, JAXA *

STS-128

Orbiter:	Discovery
Launch Date:	28 August 2009
Launch Time:	11:59:00 pm EDT
Launch Pad:	39A KSC Florida
Return Date:	11 September 2009
Mission Duration:	13 days, 20hr, 54min, 55sec
Orbits:	219
Miles Travelled:	5.8 million
Commander:	Frederick W. Sturckow
Pilot:	Kevin A. Ford
Mission Specialists:	Patrick G. Forrester
	José M. Hernández
	Christer Fuglesang, ESA
	John D. Olivas
	Nicole M. P. Stott †
	Timothy L. Kopra *

STS-129

Orbiter:	Atlantis
Launch Date:	16 November 2009
Launch Time:	2:28:09 pm EST
Launch Pad:	39A KSC Florida
Return Date:	27 November 2009
Mission Duration:	10 days, 19hr, 16min, 13sec
Orbits:	171
Miles Travelled:	4.5 million
Commander:	Charles O. Hobaugh
Pilot:	Barry E. Wilmore
Mission Specialists:	Leland D. Melvin
	Randolph J. Bresnik
	Michael J. Foreman
	Robert L. Satcher Jr
	Nicole M. P. Stott *

STS-130

Orbiter:	Endeavour
Launch Date:	8 February 2010
Launch Time:	4:14:00 am EST
Launch Pad:	39A KSC Florida
Return Date:	21 February 2010
Mission Duration:	13 days, 18hr, 6min, 24sec
Orbits:	217
Miles Travelled:	5.8 million
Commander:	George D. Zamka
Pilot:	Terry W. Virts
Mission Specialists:	Kathryn P. Hire
	Stephen K. Robinson
	Nicholas J. M. Patrick
	Robert L. Behnken

STS-131

Orbiter:	Discovery
Launch Date:	5 April 2010
Launch Time:	6:21:22 am EDT
Launch Pad:	39A KSC Florida
Return Date:	20 April 2010
Mission Duration:	15 days, 2hr, 47min, 10sec
Orbits:	238
Miles Travelled:	6.2 million
Commander:	Alan G. Poindexter
Pilot:	James Patrick Dutton Jr
Mission Specialists:	Richard A. Mastracchio
	Dorothy M. Metcalf-Lindenburger
	Stephanie D. Wilson
	Naoko Yamazaki, JAXA
	Clayton C. Anderson

STS-132

Orbiter:	Atlantis
Launch Date:	14 May 2010
Launch Time:	2:20:00 pm EDT
Launch Pad:	39A KSC Florida
Return Date:	26 May 2010
Mission Duration:	11 days, 18hr, 29min, 9sec
Orbits:	186
Miles Travelled:	4.9 million
Commander:	Kenneth T. Ham
Pilot:	Dominic A. Antonelli
Mission Specialists:	Garrett E. Reisman
	Michael T. Good
	Stephen G. Bowen
	Piers J. Sellers

STS-133

Orbiter:	Discovery
Launch Date:	24 February 2011
Launch Time:	4:53:24 pm EST
Launch Pad:	39A KSC Florida
Return Date:	9 March 2011
Mission Duration:	12 days, 19hr, 4min, 50sec
Orbits:	199
Miles Travelled:	5.3 million
Commander:	Steven W. Lindsey
Pilot:	Eric A. Boe
Mission Specialists:	Nicole M. P. Stott
	B. Alvin Drew
	Michael R. Barratt
	Stephen G. Bowen

STS-134

Orbiter:	Endeavour
Launch Date:	16 May 2011
Launch Time:	8:56:00 am EDT
Launch Pad:	39A KSC Florida
Return Date:	1 June 2011
Mission Duration:	15 days, 17hr, 38min, 51sec
Orbits:	245
Miles Travelled:	6.5 million
Commander:	Mark E. Kelly
Pilot:	Gregory H. Johnson
Mission Specialists:	E. Michael Fincke
	Roberto Vittori, ESA
	Andrew J. Feustel
	Gregory E. Chamitoff

STS-135

Orbiter:	Atlantis
Launch Date:	8 July 2011
Launch Time:	11:29:03 am EDT
Launch Pad:	39A KSC Florida
Return Date:	21 July 2011
Mission Duration:	12 days, 18hr, 28min, 50sec
Orbits:	196
Miles Travelled:	5.3 million
Commander:	Christopher J. Ferguson
Pilot:	Douglas G. Hurley
Mission Specialists:	Sandra H. Magnus
	Rex J. Walheim

ESA - European Space Agency
CSA - Canadian Space Agency
MSE - Manned Spaceflight Engineer (US Air Force)
DFVLR/DLR - German Spaceflight Center

CNES - National Center for Space Studies (France)
ASI - Italian Space Agency
RKA - Russian Federal Space Agency

NASDA - National Space Development Agency of Japan
JAXA - Japan Aerospace Exploration Agency
NSAU - National Space Agency of Ukraine

> Crew member delivered to Mir Space Station
< Crew member returned for Mir Space Station
† Crew member delivered to International Space Station
* Crew member returned from International Space Station

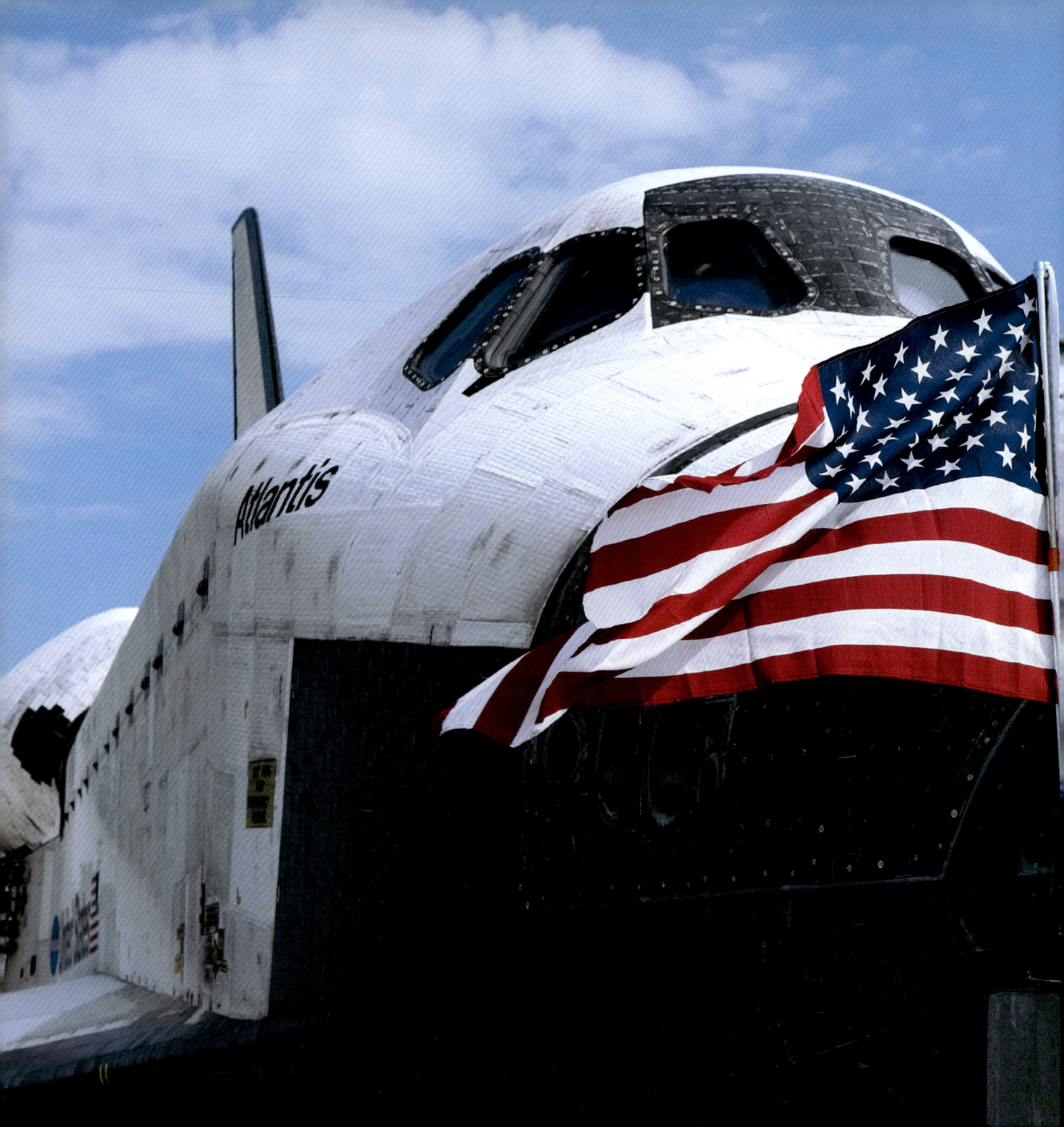

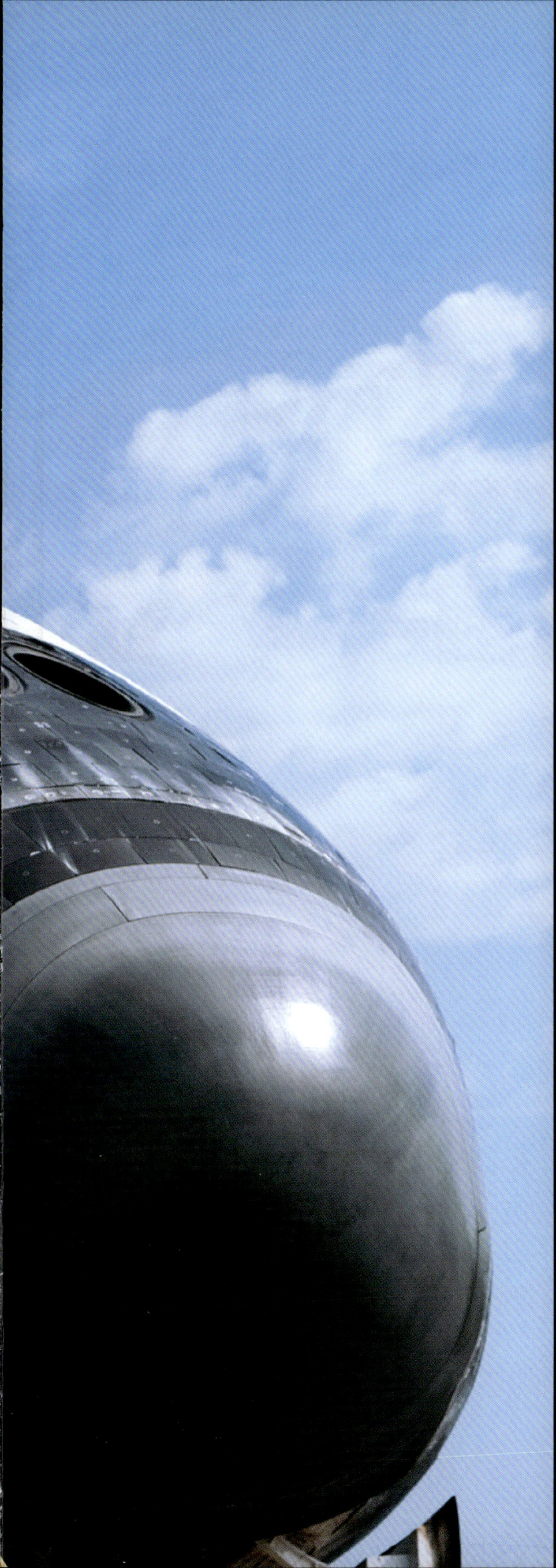

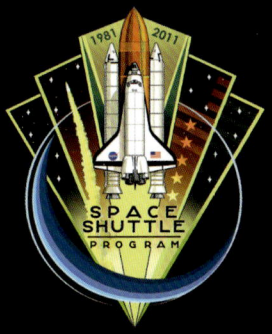

"Mission complete, Houston.
After serving the world for over 30 years, the space shuttle
has earned its place in history. It's come to a final stop."

Commander Christopher J. Ferguson
Atlantis STS-135
21 July 2011

ACKNOWLEDGMENTS

All photographs courtesy of National Aeronautics and Space Administration (NASA).
STS space shuttle mission insignias are courtesy of NASA.

Thanks to every single photographer and astronaut who has contributed to NASA's photographic history.
Special thanks to Gwen Pitman, Connie Moore and Bert Ulrich at NASA for their help with image sourcing.
Thanks also to Bruce McCandless II and Christopher J. Ferguson.

Front cover: STS-7, 18–24 June 1983; Back cover: STS-101, 19–29 May 2000; Page 2: Space shuttle main engine undergoing
testing at Stennis Space Center, Mississippi (1981); Page 210: Atlantis rides atop the shuttle carrier aircraft
on its way back to Kennedy Space Center (STS-117, 1 July 2007).

This edition published 2025 by
Ammonite Press
an imprint of Guild of Master Craftsman Publications Ltd
Castle Place, 166 High Street, Lewes, East Sussex, BN7 1XU, United Kingdom
www.ammonitepress.com

First Ammonite Press edition published 2019
First published by Hotegg Creative Design Ltd

Text and Design © Luke Wesley Price, 2025
Copyright in the Work © GMC Publications Ltd, 2025

ISBN: 978-1-78145-497-8

A catalogue record for this book is available from the British Library.

Publisher: Jason Hook
Design Manager: Robin Shields
Editor: Jamie Pumfrey
Image Research, Editing and Digital Image Processing/Retouching: Luke Wesley Price
Editing, Consulting and Digital Image Processing/Retouching: Sam Nathan Price

Colour reproduction by GMC Reprographics
Printed and bound in China

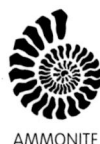

AMMONITE
PRESS

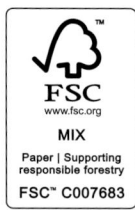

FSC
www.fsc.org
MIX
Paper | Supporting
responsible forestry
FSC™ C007683